I0014487

Sture Eriksson

Electrical Machine Development

Sture Eriksson

Electrical Machine Development

A study of four different machine types from a Swedish perspective

VDM Verlag Dr. Müller

Impressum/Imprint (nur für Deutschland/ only for Germany)
Bibliografische Information der Deutschen Nationalbibliothek: Die Deutsche Nationalbibliothek
verzeichnet diese Publikation in der Deutschen Nationalbibliografie; detaillierte bibliografische
Daten sind im Internet über http://dnb.d-nb.de abrufbar.
Alle in diesem Buch genannten Marken und Produktnamen unterliegen warenzeichen-, marken-
oder patentrechtlichem Schutz bzw. sind Warenzeichen oder eingetragene Warenzeichen der
jeweiligen Inhaber. Die Wiedergabe von Marken, Produktnamen, Gebrauchsnamen,
Handelsnamen, Warenbezeichnungen u.s.w. in diesem Werk berechtigt auch ohne besondere
Kennzeichnung nicht zu der Annahme, dass solche Namen im Sinne der Warenzeichen- und
Markenschutzgesetzgebung als frei zu betrachten wären und daher von jedermann benutzt
werden dürften.

Coverbild: www.purestockx.com

Verlag: VDM Verlag Dr. Müller Aktiengesellschaft & Co. KG
Dudweiler Landstr. 125 a, 66123 Saarbrücken, Deutschland
Telefon +49 681 9100-698, Telefax +49 681 9100-988, Email: info@vdm-verlag.de
Zugl.: Stockholm, Royal Institute of Technology, Dissertation, 2007

Herstellung in Deutschland:
Schaltungsdienst Lange o.H.G., Zehrensdorfer Str. 11, D-12277 Berlin
Books on Demand GmbH, Gutenbergring 53, D-22848 Norderstedt
Reha GmbH, Dudweiler Landstr. 99, D- 66123 Saarbrücken
ISBN: 978-3-639-09383-4

Imprint (only for USA, GB)
Bibliographic information published by the Deutsche Nationalbibliothek: The Deutsche
Nationalbibliothek lists this publication in the Deutsche Nationalbibliografie; detailed
bibliographic data are available in the Internet at http://dnb.d-nb.de.
Any brand names and product names mentioned in this book are subject to trademark, brand or
patent protection and are trademarks or registered trademarks of their respective holders. The use
of brand names, product names, common names, trade names, product descriptions etc. even
without
a particular marking in this works is in no way to be construed to mean that such names may be
regarded as unrestricted in respect of trademark and brand protection legislation and could thus
be used by anyone.

Cover image: www.purestockx.com

Publisher:
VDM Verlag Dr. Müller Aktiengesellschaft & Co. KG
Dudweiler Landstr. 125 a, 66123 Saarbrücken, Germany
Phone +49 681 9100-698, Fax +49 681 9100-988, Email: info@vdm-verlag.de

Copyright © 2008 VDM Verlag Dr. Müller Aktiengesellschaft & Co. KG and licensors
All rights reserved. Saarbrücken 2008

Produced in USA and UK by:
Lightning Source Inc., 1246 Heil Quaker Blvd., La Vergne, TN 37086, USA
Lightning Source UK Ltd., Chapter House, Pitfield, Kiln Farm, Milton Keynes, MK11 3LW, GB
BookSurge, 7290 B. Investment Drive, North Charleston, SC 29418, USA
ISBN: 978-3-639-09383-4

"Since then, I have designed a new dynamo-electric machine, which I belive will have several advantages before others."

Jonas Wenström, February 22, 1877

*to Jonas, Pontus, Simon, Markus, Hanna and Emil
in memory of their grandmother, my beloved wife Birgitta*

Abstract

Development and manufacturing of rotating electrical machines have been essential for the Swedish society and industry for more than a century. The dominating manufacturer has been Asea/ABB, even if a more diversified structure has emerged in recent years. The thesis deals primarily with Asea/ABB's development of four kinds of electrical machines, although reflected against a wider national and international background. The purpose of the research has been to study the development of these machines and the related industrial processes, focusing on factors with a major influence on the development.

The thesis contains introductory chapters presenting the research, the public importance of electrical machines, their initial history, as well as a technical introduction of the machines and the development process. The main chapters start with standard induction motors which are manufactured in large numbers and can be seen as a commodity with little product differentiation. The development focus has been on rational production and increased use for frequency controlled variable speed drives. Large directly water-cooled turbogenerators were developed for nuclear power plants, in the 1970's, and created initially many difficulties. Advanced technologies and strategic matters have been strongly interlinked in this development, which is described comprehensively in the thesis. Electrical machines for automotive drivelines have, in recent years, been subject to intensive international development and several new concepts have been introduced. The thesis analyzes the Swedish attempts that have been technically satisfactory but have not led to commercial products. ABB launched synchronous machines for very high voltages as a revolutionary product ten years ago, but without commercial success. This controversial development and the difficult business situation are subject for discussion in the fourth main chapter.

The study presents conclusions, concerning the development, individually for each machine type, but also comparisons based on divisions in large and small machines and in mature and new technologies. A common result is that the development, in retrospect, has been more successful from technical point of view than from commercial, independently whether the development has been market or technology driven. An important contribution of the thesis is that it presents the first comprehensive Swedish study of electrical machine development and which factors have been most influential. The thesis ends with a discussion of future prospects for the Swedish electrical machine industry and the possibilities and threats it is facing.

Keywords: electrical machines, development, history, induction motors, turbogenerators, hybrid vehicle motors, powerformer

Preface

At a lunch with some colleagues a couple of years ago, we happened to touch on the subject of my planned thesis. I started to tell a little about the background and my intentions, whereupon someone gave me the idea of calling it "A Romance with Electrical Machines". Even if I felt tempted, I was not brave enough to finally propose such a title.

Rotating electrical machines have been the common denominator throughout my professional life. I have been fascinated by their great importance to society, their versatility, but also by the somewhat elusive technical challenges they represent, apparent through the different technologies that they synthesize. Electro technology, mechanics and material technology are obvious examples of such areas but there are others. Industrial and commercial aspects have also contributed to make my work with these machines extra interesting.

For many years I had had the privilege to combine industrial and academic work, when my life was changed both privately and professionally. I had to make up new plans and the most important was when I decided, one day in April 2002, that I wanted to start a PhD project of my own. At my age, with almost 45 years experience from electrical machine development, but also several decades of dust on my schoolbooks in mathematics and other basic subjects, I realized that my approach would have to be very different compared with that of my younger colleagues. I had the possibility to write a thesis from a much different perspective – from the inside of the Swedish electrical machine industry.

My ambition has been to investigate and tell the development history for a few selected machine types focusing on which factors have been important from the technical perspective, how development was initiated and carried out, in which wider context the development process can be viewed, which problems have been faced and which results have been obtained. The relationships between the industry and the universities have also been part of my study. The intention is to learn from historical studies and, therefore, I have tried to draw some conclusions of importance for the future development. This is always a risk and I will perhaps be blamed for having made false predictions, but , at present, they are the best I can make.

While writing this thesis I have seen a few different categories of readers in front of me. The first is under- and postgraduate students with an interest in electrical machine related subjects. My thesis will hopefully give them a better understanding of the industrial development and the different kind of factors that influence this development. The next group of readers is scientists and other specialists looking for specific information. I would be glad if they succeed in finding useful material for their needs. I realize that a third category is old colleagues who are curious to read what I have written. I am anxious to receive their comments and I hope that they do not find any serious mistake. Finally I have had a fourth category of readers in my mind, people who are amateurs with respect to electrical machines. They could perhaps find parts of the thesis interesting and I would be happy if they understood the importance of the electrical machine, a product that serves daily everyone of us in so many different ways.

The thesis is comprehensive, even if it is focused on four machine types. The text shifts back and forth between technical descriptions, comments on the development process and resources, problems encountered, strategic decisions, international competition and cooperation, etc. There is, of course, a structure in my thesis but many readers may, nevertheless, find some parts a little confusing. It is made on purpose. The thesis should be a reflex of the reality; the process of developing electrical machines has many facets.

Overview of the thesis

Chapter 1 specifies the project objectives, refers to related research and discusses the selection of the machine types included in the study. It lists factors which have influence on the electrical machine development. It also includes information on the sources I have used and finally highlights some major contributions of the thesis.

Chapter 2 emphasizes the importance of electrical machine to the society at large. These machines, large and small, are used in all sectors of modern society and their expansion has been very rapid, especially in recent decades. Special attention is given to the situation in Sweden and the electrical machine related industries currently active in the country. This chapter does not provide direct contribution to the research, even if the listing of companies required direct contact and interviews with all of them. It is my hope that this chapter can reveal the importance of the electrical machines to many different readers.

Chapter 3 focuses on the 19th century history of electrical machines. It starts with the scientific discoveries leading to pioneering inventions and the initial steps of the electrical machine industry. The early activities in Sweden, including Jonas Wenström's inventions and his role in the invention of the 3-phase machines, are given special attention. This chapter has become more comprehensive than I first had intended, mainly because I became convinced during the study, that we can learn more from this early history than we usually expect.

Chapter 4 contains a technical overview of electrical machines and it is not in any respect considered as a contribution to the research project. It tries to explain the principles of operation for electrical machines, it describes basic concepts, components and materials, it emphasizes the importance of losses and cooling and, finally, it presents some common machine types. I have included this overview for convenience of those readers, who lack background in electrical machine technology. It can hopefully serve that purpose.

Chapter 5 deals with the development process which is the second part of the main title of my thesis. It discusses first the abstraction "development" and gives a brief summary of product development. The focus of the chapter is an introduction to electrical machine development and the engineering tools used.

Chapter 6 is the first of the main chapters and it is focused on standard induction motors, by far the most common electrical machine for industrial applications as well as many other drive systems. It has a long development history starting back in the late 19th century. The Swedish electro-technical company, Asea (later ABB), has successfully participated in the development and manufacturing of induction motors for more than 115 years. This chapter describes not only the product development, but also process development and business strategic considerations. Asea/ABB has been in focus for the study, but even other domestic and foreign manufacturers have been included. An increasing number of induction motors are used for variable speed drives, and such systems have, therefore, received much attention in my study. It can be argued that there is not much development of standard induction motors and that it would have been better to leave them outside this thesis. My argument for including them is their huge importance, and the fact that my research project should contain a mix of machines, representing various phases of development, product sizes and production volumes. It is not only the technical development, which determines whether the products become successful or not.

Chapter 7, which deals with the development of large turbogenerators in Sweden, is one of the major parts of the thesis. This development is closely connected to the installation of nuclear power in the Swedish power system during the 1970's. The domestic generator manufacturer, Asea, had long been a leading supplier of hydropower generators, but was faced with the challenge of developing much larger turbogenerators than it had experience with. Asea's development of large turbogenerators contains several interesting phases, starting with the decision to choose a concept quite different from what was common in the generator industry. This was followed by design, manufacturing and testing, a phase when problems started to appear. Some spectacular failures occurred in the power plants, and the investigations and remedies of these failures contributed to the state-of-art of turbogenerator technology. The technical problems were satisfactorily solved, but reduced market and new corporate structures have practically eliminated the commercial possibilities for the Swedish factory. The history of Asea's large turbogenerators holds many facets, and it has not been studied and presented earlier. It was, therefore, well motivated to include it in my research project. For a period, I had one of the key roles in Asea's generator development, but my work on this thesis is primarily based on written documentation and interviews with former colleagues and other people who had been part of this development.

Chapter 8 covers development of electrical machines for automotive drivelines, a subject which has become very important in recent years. Someone could question whether it was right to include this kind of machines in a thesis, which has a Swedish perspective on electrical machine development, when there is no regular production of such machines in the country. My motives for including them are their globally growing importance. A substantial share of current electrical machine R&D is focused on these machines and a number of interesting prototype projects have been carried out in Sweden, which has a large automotive industry requesting this kind of products for future hybrid vehicles. It has been necessary to give this chapter a more international profile than the others, but the Swedish attempts to develop electric "vehicle motors" are also an interesting history, both technically and commercially. Furthermore, there is another, more personal reason for this chapter and that is that electric drivelines for hybrid vehicles have been a major field of activity for me since the late 1980's, both at university and in industry. I have therefore had access to many persons with deep knowledge from the automotive industry, and I have tried to reflect some of this in the thesis. The automotive industry is special and in several respects different from the electro-technical industry.

Chapter 9 deals with Powerformer®, a generator for very high voltages, which was developed by ABB during the 1990's and has received a lot of public and professional attention. It was not from the very beginning my own idea to include this machine type. It was suggested by the Swedish Energy Agency, but I am glad that I did so. It has been exciting to try to find the essential factors in a very intense and, in several respects, unusual development process. A number of key persons, who have been engaged in various parts of the development, have given me facts and their, not always converging, personal views. I have tried to interpret and combine them with information obtained from published documentation, both technical journals and newspapers. Powerformer has proved its technical merits but so far without any commercial success. The opinions about this radically new machine are very different. Some see it as the greatest achievement in electrical machine technology in 100 years and as a technology that will conquer the entire power generation world while others consider it almost as the "emperor's new cloths". The truth is usually somewhere in between. Powerformer has, from my point of view, been a fascinating study case full of nuances, technical, industrial, strategic and emotional.

Chapter 10 contains synthesis and conclusions. As separate conclusions are included in the previous chapters 6 – 9, this final chapter focuses on comparisons of the selected machine types and contains answers on some research questions stated in chapter 1. From the beginning of the project, I have planned to include projections concerning the future development of electrical machines, especially in Sweden. At the time when I had to make such projections, I realized the difficulties. Nevertheless, based on my studies, the latter part of the chapter presents my personal views on the prerequisites, the possibilities and the threats for future Swedish electrical machine development. I am aware that they can be disputed.

The thesis contains an extensive number of references of different kinds. Many are in Swedish, which I understand might be a complication for some readers, but it is necessary for a study focusing on development that has been carried out in Sweden. There are also many references to internal Asea/ABB documents that require access to the company's archives, or for some parts to Alstom Power's archive, for further studies. The length of the reference list reflects a wish from my side to document the sources I have used and thereby avoid suspicions that my old memories have been the primary base of information.

The final part of the thesis comprises lists of figure sources and persons interviewed as well as some other information as can be seen from the list of content. Some parts are included for the convenience of readers who are less familiar with Sweden.

Acknowledgements

"Will you feel relief when you are done with your PhD project?" I have got this question from many friends the last year, and I am not quite sure of the answer. Of course, I am glad that I have been able to carry through what I had planned, but there are also a touch of sadness. I have enjoyed looking for information in archives - which more than once gave me thrilling surprises - but even the late hours at home, in front of the computer, have been pleasant. Contacts and meetings with many persons, who in different ways have helped me to reach my target, have been most stimulating. Separation from this project is a little worrying, but it opens hopefully new possibilities.

My life with electrical machines has brought me in touch with many people, inside and outside of Sweden: colleagues within Asea/ABB, customers, competitors, academicians, and others. They have all been part of the development that I have tried to describe in this thesis and I want to recognize their contributions to the electrical machine development.

Professor Chandur Sadarangani, head of KTH's department for electrical machines and power electronics, has been supervisor of my PhD studies. I am very grateful for his acceptance of my idea, for his positive interest in my unorthodox project, for the freedom I have had to choose my own ways, and for valuable discussions and guidance. I have had a reference group, which contributed with good advices, especially during the first part of the project. Members in this group have been, besides Chandur Sadarangani, Messrs. Gunnar Mellgren and Sven Sjöberg, earlier with ABB, and Mikael Fjällström, Swedish Energy Agency. I really appreciate their support. Arne Kaijser, professor in history of science and technology at KTH, participated initially as special advisor. He criticized my original approach for being too "technocratic" and recommended me to focus the project on development that I had direct experience from. I am very thankful for his wise advice, which I have tried to follow, at least to a large extent.

This thesis is partly based on knowledge I have received through extensive interviews with many experienced colleagues, but also on information obtained in more limited contacts with other key persons. I appreciate the time and the positive interest all of them have given me. Many interviews have been really enjoyable reunions. All these persons are mentioned by name in the "list of interviews" at page 468, and I want to thank every one of them.

20 years, to and from, at KTH's electrical machine department have been inspiring and even fun. Many colleagues have passed during the years, some have remained. My room-mates, first Techn. Lic. Mats Leksell and then Dr. Peter Thelin, have always supported me with useful KTH know-how. Professors Hans-Peter Nee and Stefan Östlund, as well as Dr. Juliette Soulard have been helpful answering my questions. In addition, I want to mention the great "4QT"-team, besides Peter Thelin, the doctors Sylvain Chatelet, Freddy Magnussen and Erik Nordlund. Thanks also to Ms. Eva Pettersson and Mr.

Peter Lönn for good assistance with the administration and the computer system.

Especially warm thanks to Techn Lic. Maddalena Cirani, my co-worker for a number of years, who have proof-read most of this thesis. I have appreciated working together with you, your commitment and the Italian intensity in your argumentation. A couple of other colleagues, who I want to recognize for good cooperation and encouraging interest in my work, are Dr. Göran Johansson, Volvo, and Professor Mats Alaküla, Lund University. Dr. Anders Malmquist belongs also to this qualified group of hybrid vehicle promoters.

Mr. Karl-Erik Sjöström, Alstom Power, has read the turbogenerator chapter and his long experience from the development of these machines has been of great value. He and Mr. Bo Malmros, ABB Motors, have always patiently tried to help me with various questions. The same can be said of Techn. Lic. Jan Boivie, Alstom Power. I am grateful to them. Ms. Aina Nilsson, Elanders Infologistics, deserves thanks for guidance and assistance when I have tried to find documents in Asea/ABB's large archive in Västerås. I am grateful to both ABB and Alstom Power for allowing me access to their archives.

The Swedish Energy Agency is acknowledged for financial support of most of the expenses for the project. ABB and Alstom Power have also generously contributed with certain support.

I appreciate that many of my friends, old and new, have been much encouraging and seriously interested, often asking me about my progress. I owe explicit thanks to Psychologist Ing-Marie Matstoms for good discussions and advices in connection with some of my interviews.

My family has been very supporting from the beginning of this project. It is my son Peter Eriksson, my daughter Katarina Tullstedt, my sister Birgitta Hoff and their families. To some extent, the work with this thesis has become a family project. My American "second niece" Kara Erickson has helped me with the language, at least most of it. One of my grandsons, Pontus Eriksson, who is student at KTH, has spent many hours converting my Word- and PowerPoint-files and other documents to the format required for this final version of the thesis. Another grandson, Simon Tullstedt, has been document photographer. Peter, my son, has been my helpdesk more than once. I have dedicated this thesis to my grandchildren, but the rest of my family is included. I love you all.

..... and finally, Anna-Carin – I am very happy that you have come into my life, and that I have been allowed to become a part of your life. I am delighted and very grateful for all inspiration and all your support during the final years of my project. Looking up from my laptop, I suddenly become aware that once again, the spring has arrived.

Västerås in March 2007

Sture Eriksson

Acknowledgements

Content

1 Research objectives and implementation

The preface has briefly explained the origin of this research project. It has also given an overview of the structure of the thesis including some short comments on the various chapters. This initial chapter presents the background and refers to other published works, which represent previous research of relevance for this study. Necessary limitations and selection of specific study cases are discussed and made. Some research questions are specified as well as a list of factors, which have influence on development of electrical machines. Access to information is absolutely fundamental for this type of project and important sources are therefore presented. Finally, the major contributions of the thesis are indicated.

1.1 Background and aim

Is it possible to perform an academic study on the development of rotating electrical machines and is it worth the effort? Does it make sense to do it from a Swedish perspective? The author's opinion is "yes" and the major reasons are as follows:

The electrical machine has played, and will continue to play, a very important role in modern society. It is a cornerstone for production of electricity and for performance of mechanical work. It is, by no means, a trivial or insignificant product. In addition, electrical machines are subject to continuous development in spite of the long time that has elapsed since its introduction.

Many types of electrical machines are technically complicated products. The customers are both competent and demanding. Cost effective designs and manufacturing processes are necessary due to fierce competition. The development of electrical machines requires deep knowledge and wide competence engaging a significant number of qualified engineers.

The development process looks very different for various types of electrical machines. It is influenced by a number of technical and commercial factors and interaction with other fields of technology is essential. The process often requires choices among several alternatives and the results are not given in advance.

Sweden is a highly electrified nation. The Swedish electro-technical industry has held a frontline position in electrical machine development for more than 100 years. It is important to understand how this has been achieved and, if possible, learn from both successes and failures.

The aim of this research project is to study the development of the rotating electrical machines and to investigate which factors have had most influence on this development. The research is done through the study of some representative cases, where the focus will often be more on development process related matters than specific technical solutions, though these too will be included. Development process and technical solutions are often, in many respects, inseparable. The goal is to analyze and document selected parts of this development and to draw some conclusions of importance for the future.

1.2 Previous research

The scope of this thesis is different from the traditional research projects that have been carried out within the electrical engineering departments at technical universities. A review of titles of the doctoral theses, published since 1929, at the relevant departments in Sweden, i.e. at the Royal Institute of Technology in Stockholm (KTH), Chalmers University of Technology in Gothenburg and the Faculty of Engineering at Lund University (LTH), confirms this statement. First were John Wennerberg (1886-1979) from Asea and Fredrik Dahlgren (1893-1971), later professor in electrical machines at KTH, who both recieved their doctor degrees in 1929. In total more than 65 theses can be related to electrical machines and drive systems containing such machines. They are all focused on very specific technical subjects and can briefly be divided into three categories. The first group deals with electrical machine theory as well as performance and design of various machine concepts. The second group deals with the drive systems, including power electronics and control systems. The third group is mainly related to study and development of computerized tools for simulation and optimization of electrical machines and drive systems. The diagrams in figure 1.1 show an approximate split in these categories as well as the distribution in time for the respective universities. Some typical examples on such theses are given in the reference list [1, 2, 3, 4].

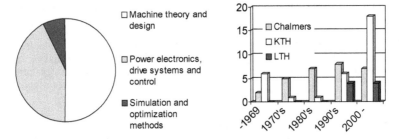

Figure 1.1 Electrical machine related doctoral theses in Sweden

An attempt to find any thesis that corresponds with this one, from electrical engineering departments at foreign universities has not yielded many results. One thesis, presented in 1993 at the Technical University in Dresden, Germany by Frank Dittmann entitled "Die Entwicklung der Technik elektrischer Antriebe in Deutschland von den Anfängen im 19. Jahrhundert bis zur Gegenwart" ("The development of the electrical drive technology in Germany from the beginning, back in the 19th century, until today") was found, but it has a different structure compared with this thesis. Dittmann's thesis has been reprinted by VDE in its series of books on the history of electro-technology [5]. The search for dissertations has been made through the web with the limitations that this implies.

The history of science and technology has become an important discipline at technical universities. This present work has obvious connections with this discipline, but without such an extensive socio-economical approach. This thesis will, to a large extent, be more focused on technical questions, which is one reason why it is carried out within the Electrical Engineering School. However, it is of interest to discover whether any historical research of importance for this study, has been made. The author has not been able to find any Swedish thesis or other scientific publication dealing specifically with the development of rotating electrical machines from an historical point of view, but there are a few books containing valuable information and useful theories. Mats Fridlund, KTH, has performed a study on the development cooperation between Sweden's leading electro-technical manufacturer, Asea, and the large power utility, Vattenfall. His study is documented in a doctoral thesis "Den gemensamma utvecklingen" ("The mutual development"), and this thesis is one of these books [6]. His discussion of definitions of "development" and "construction work" will be described in chapter 5. Of special interest are his examples on how Asea and Vattenfall cooperated on large development projects, but even more the partnership they maintained in spite of their supplier / customer relationship. Fridlund's work does not cover generators, but it will be apparent, from this thesis, that the development links between Asea and Vattenfall have been strong also in the case of power plant generators.

Another important work is Professor Jan Glete's large study "ASEA UNDER HUNDRA ÅR, 1883 - 1983, En studie i ett storföretags organisatoriska, tekniska och ekonomiska utveckling" ("ASEA OVER ONE HUNDRED YEARS, 1883 -1983, A study of a large company's organizational, technical and economic development") [7]. Glete's book contains many useful facts, but it also gives a deeper understanding of the context in which Asea's development was carried out. An important source of information, concerning Asea's earlier development, is also Martin Helén's (1879-1958) comprehensive history of the company, "ASEA:s HISTORIA, 1883 - 1948", published in three volumes [8]. This history is rich in details both concerning the products developed and the organization of the company and it also includes biographies of a number of key persons.

The history of electrical machines, from an international perspective, has been documented in several books and papers, but most of them cover a certain period, company or type of machine. The period before the previous turn of the century is very well documented in general books on electrical history [9], because the machines represented a big portion of the electrical industry in those years. The early development is also described in detail in many biographies. Two good examples, recently written, are "Edison, a life of invention" [10] and "Michael von Dolivo-Dobrowolsky und der Drehstrom" [11].

The 20th century development of small electrical machines is dealt with in a book with the title "Alles bewegt sich" ("Everything is moving") [12], included in the German comprehensive series of books on electro-technical history mentioned above. The author claims that it is the first overview of important development lines for small electrical motors. There also exist many papers, which summarize the development of various machine types. One example, of interest for this work, is a 33 page paper on "Recent trends in turbogenerators" published 1974 in IEE Proceedings [13]. Other papers of this type are included in the reference list and referred to in chapters 6, 7 and 8. Characteristic of the papers, which the author has studied, is their focus on the technical development of the products; very little attention is paid to the development process at large and questions related to this process.

There has of course been a lot of research in areas such as industrial organization, strategic planning, management of technology, design methodology, etc. and some results could have an impact on this study, but only on very specific items. The present thesis is not considered a contribution to those subjects and they can neither be viewed as previous research for this work.

1.3 Limitations

It is impossible to include, in a work like this, a comprehensive study of the global development of all types of rotating electrical machines. It is necessary to cut just "a piece of the cake", which can be digested both by the author and the readers. Therefore, the following limitations have been made:

Sweden is an "electrified country" in the sense that the electric power consumption per capita is high and has been so for a long time. As a consequence, the Swedish electro-technical industry has always belonged to the leading manufacturers in the world, also in the case of electrical machines. The production has been much larger than the size of the country would imply. It has therefore been considered as sufficient to <u>concentrate the study on Swedish development,</u> but reflected against an international background. An advantage is that the depth of the study will be larger due to the author's own experience and the access to documentation and key persons. A certain drawback is that the study may become somewhat less interesting from an international point of view.

There is practically no development and manufacture in Sweden of really small electric machines, such as motors for computers, wristwatches, cameras, toys and similar products. The development of such machines is therefore automatically excluded from the study and the <u>focus will hence be on machines from 0.1 kW and up</u>. It has been considered preferable to perform more comprehensive studies of a few machine types instead of brief overviews of all types, which have been manufactured in the country. A selection of these machine types will be made in the next section.

It is also necessary to make a limitation in time, which is more difficult. The results from an analysis of recent development can be considered to be more useful for prediction of future trends. This means that <u>more recent development will be prioritized</u>, but the early history cannot be completely omitted. The circumstances vary considerably between different types of machines, so it is therefore impossible to choose the same time frame for all cases to be studied. The time frames will be specified later for each selected case.

1.4   Selection of study cases

Many types of electrical machines have been developed in Sweden and several of them are still being developed. It is difficult to find any suitable objective method to select the cases to be studied in this project. It cannot be made randomly because the conditions are very different for the machine types, so statistical methods are likely to fail. The author has instead preferred to make a subjective selection and the motivation for each selection will be given. It is realized that the choices can be disputed, but it can, nevertheless, be claimed that the total result will be sufficiently representative of the development of important and technically demanding electrical machines.

The list, in table 1.1, includes machine types, which are developed in Sweden or at least have been developed in the country during the last 20 - 25 years. They can be divided into three main groups: "generators", "motors" and "special machines". This is not a stringent division, only a practical way of listing them based on the most common use of the different machine types.

Some of the machine types can be excluded due to their insignificance in Sweden, e.g. diesel and wind power generators, which both are important products for the Finnish electrical machine industry. Switched reluctance motors and high-speed machines can also be considered as too marginal for this study. There has been a large production of motors for domestic appliances, but important parts of this production has been shutdown and moved abroad. The Swedish manufacture of servomotors is very limited. None of these latter machines have been the subject to extensive development and they are, due to the reasons mentioned, deleted from the "short list" of possible machines to be studied.

Table 1.1 *Electrical machine types developed and/or manufactured in Sweden since 1980*

Generators	Motors
Hydropower generators	Universal motors
Steam and gas turbine-driven turbogenerators	Small DC motors
Steam and gas turbine-driven salient pole generators	DC motors
Diesel engine-driven generators	LV induction motors
Wind power generators	HV induction motors
	Synchronous motors
	Servo motors
Special machines	Switched reluctance motors
Synchronous condensers	High speed induction motors
Very high voltage machines	High speed PM motors
Motor/generators for automotive applications[*]	Traction motors
	Vehicle motors (automotive)[*]

[*]Only prototypes

The large machines can best be represented by hydropower generators or turbogenerators. The salient pole generators, driven by gas and steam turbines, are certainly important but have, in comparison, a limited history. ("Salient pole" is explained in chapter 4.) The author's first choice is to study large turbogenerators due to the following reasons: they are technically very demanding products requiring extensive development efforts; the introduction of nuclear power, and in connection with this very large turbogenerators, meant a very interesting quantum leap in both size and technical concept of such generators in Sweden. The development of hydropower generators has been well documented in the earlier mentioned works by Helén and Glete. Both authors have given the hydropower generators more than six times as much space as the turbogenerators. In addition, the hydropower generators have received extensive coverage through many reports, e.g. in Asea Journal, while the turbogenerators are much more of a "white field" in this respect. The development of these two types of generators has been partly interlaced, so a study of one of them will also include some information on the other.

In order to balance the large turbogenerators the author has decided on <u>series manufactured induction motors</u> as another study case. They have a long history, they are very common, they are standardized and to some extent mass-produced. Their development is very much linked to a shift of motor generations. The increased use of this kind of motors, in variable speed drives, has led the development engineers to be faced with new challenges. Around 90 percent of all electric motors, in the range above 1 kW, are induction motors and that is the main reason why DC and synchronous motors are excluded from the study. Furthermore, the importance of the DC motors is diminishing and the synchronous motors are very large and would not be a suitable antipode to the turbogenerators.

Two other less comprehensive cases will be included in this study. The first of them is <u>machines for electric vehicles</u>, mainly motors. So far only prototypes have been built in Sweden but the development of such machines is interesting because it means new applications, new customers, new requirements and new concepts for electrical machines as well as the development process. These products could have a major impact for the production of electrical machines in the future. Comparisons will be made with both industrial and traction motors. The last case is <u>machines for extra high voltages</u>, which were developed by ABB[*] in Sweden during the 1990's. The technical concept is still unique and the development approach was very special. The concept has received a lot of public attention and it is definitely of interest to include this example of recent, radical electrical machine development.

1.5   Research objectives

The study cases chosen represent two main groups. One consists of standard induction motors and electric vehicle machines, i.e. comparatively small, series produced machines. The second group includes turbogenerators and extra high-voltage machines, which means large machines with a considerable amount of order specific design and development. It is also possible to identify two other groups. Induction motors and turbogenerators constitute a group of well-established machines, while the electric vehicle motors and the extra high-voltage machines are new developments. Figure 1.2 shows a simple matrix with the four machine types included in the study.

The main structure of this research project has briefly been the following. The first, most time-consuming, phase has consisted of a search, chiefly from ABB sources, for information concerning the chosen products and the development of those products. The second phase included a search for relevant

---

[*]. ABB was established 1988 as a result of a merger between Asea and the
      Swiss electro-technical manufacturer Brown Boveri & Cie (BBC).

information from other manufacturers, to be used for comparison. The third task has been an attempt to answer a number of research questions through analysis of the information obtained, including review of similarities and differences between the groups specified above. A fourth and difficult task has been to synthesize the results and try to draw some conclusions of value for future development. Finally, the fifth part has had its focus on documentation. It is probably unnecessary to mention that the whole process has been very iterative and has included many parallel activities.

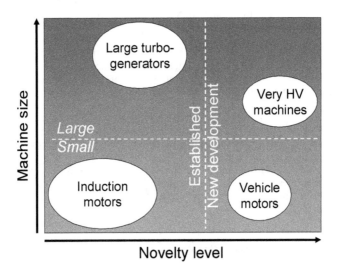

Figure 1.2 Size and technical maturity of the machine types chosen

The major questions, which the author will try to answer through this research project, are listed below. Each of them will of course later decompose into a number of sub-questions, but that will be apparent in later chapters.

- Which has the design of the selected machine types been and how has it been developed over time? Which properties have been most characteristic for each of these machines?

- How has the development been accomplished? Which kinds of problems have been faced, how have they been solved and what have results been?

- In which business context has the machine types been developed? How can the Swedish development be related to the international situation?

- Which scenarios are probable for the future develement of the studied machine types, especially with respect to the possibilities for Swedish manufacturers.

- Which factors have had the most influence on the development of different machine types? Is the development of smaller machines more market driven and the development of large machines more technology driven?

- How has knowledge and competence been acquired and maintained? Has the transfer of know-how, between development teams for large and small machines, been negligible?

- How has the development of the selected machine types changed as a function of time? Has the development of smaller machines become more innovative in later years compared to the development of large machines?

The answers can, to the largest extent, be found implicit in chapters 6 - 9 but are also summarized in chapter 10, especially the comparisons between the machine types.

1.6   Factors of influence

There are many factors that will influence the development of an electrical machine and they can of course also vary over time. Some of these factors are technical, others are commercial or strategical. Some can be characterized as company internal, others as external. It will be convenient, for the subsequent analyses, to have in mind some sort of structure of influencing factors, e.g. as proposed in table 1.2 in the following pages. However, the intention is not to use it as a checklist and the conclusions will be presented at a somewhat aggregated level.

The photos in figures 1.3 and 1.4 illustrate two type of factors which influence the machine design. The first shows a transport of a large turbogenerator stator. The design engineers must usually take transport limitations into considerations when they develop large machines. The second figure illustrates that many electrical machines, especially motors, are installed for operation under adverse environmental conditions. Such motors must be designed and built to withstand the environnmental stresses caused by exposure to corrosive agents, dirt, water, and other hazardous attacks.

Table 1.2   Factors which can influence electrical machine development

Strategic factors	Company internal	Product portfolio
		Facility limitations
		Technology acquisition
		Outsourcing
		Management preferences
	External	Public programs and funding
		Cooperation with other companies
Commercial and legal factors	Market needs and requirements	Local market rules: e.g. preference for local production
		Application specific requirements
		Customers: e. g. requirements from OEM customers
		Laws and decrees
		Standardization
		Competition: other manufacturers or other technology
	Comparative factors	Performance
		Maintenance
		Reliability
		Adaptability: integration with other machinery
	Costs	Material and components
		Labor
		Energy
		Investments
		Tariffs

Table 1.2   Factors which can influence electrical machine development

Technical factors	Performance factors	Rated and maximum power and torque: absolute and specific values
		Efficiency
		Speed: maximum speed and speed range
		Voltage
		Operation temperature
		Noise
		Controllability
		Environment: ambient conditions, protection forms, vibrations
		Electro-magnetic compatibility (EMC)
	Material development	Magnetic materials: electrical sheet and permanent magnets
		Conductors: conventional such as Cu and Al but even super conductors
		Insulation materials
		Structural materials: metals and composites
	Connected systems and components	Converters: topologies, semi-conducting devices and control electronics
		Cooling systems
		Bearings
		Sensors
	Media	Coolants
		Lubricants
	Design tools	Analytical calculations
		Numerical calculations and simulations
		Computer Aided Design (CAD)
	Design concepts	New machine topologies
		New design solutions
	Production technology	New operation methods: manual and automatic
		Production lines
	External conditions	Transport limitations
		Installation prerequisites

Figure 1.3 Transport of stator for an 825 MVA turbogenerator. The weight of the stator is around 350 tons. Often transport restrictions have big impact on the design of large machines.

Figure 1.4 Induction motor driving rotating magnetic separator. Many electrical machines must be built for operation in severe environments.

1.7 Sources

This research project has an historical perspective on the development of the electrical machine and the access to relevant information is therefore absolutely fundamental. Fortunately, there is no lack of documentation even though the information is not very coherent. It has been easy to obtain information regarding the actual machine designs, but often more difficult to find answers concerning the development process. The results are based on a combination of written as well as oral sources.

Some written information has been found in books and scientific papers in KTH's library, but the most important source has been Asea/ABB's archive in Västerås*. The latter contains internal protocols, reports, memoranda, correspondence, drawings, photos and published material such as journals, pamphlets, catalogues etc. Access to the archives of ABB Motors and Alstom Power, both in Västerås, has also been very helpful. Supplementary information from the Swedish Energy Agency's library has also been used. Professor Stig Ekelöf's (1904-1993) collection of old "Books and Papers in Electricity and Magnetism" at Chalmers' library is a goldmine with regard to early history [14]. Literature studies have also been made at Deutsches Museum's library in Munich. Various web pages have contributed with information. A number of books and other publications in the author's private library have been used frequently. A large number of references have been included in the thesis, but it must be admitted that there is a language problem. Many sources are of course in Swedish, but others are in English and quite a few are also in German.

Interviews with a number of key persons have contributed with a lot of valuable input, especially concerning motives for different decisions, implementation methods and critical review of results. Persons, who have been engaged in Asea/ABB's product development, have been interviewed first of all, but also some customer representatives have contributed in this way. The author has also had opportunities to meet and discuss with a few experts from other companies and from universities. All interviews have been documented in written notes. A list of all persons interviewed is included as an appendix.

Finally, another source of information is the author's own professional experience from four decades of electrical machine development, especially related to two of the studied machine types. This has of course been an advantage, but he has endeavored to base the entire study on written sources and the interviews and not on his own memories.

---

*. Now administered by Elanders Infologistics Väst AB in Västerås

1.8   Major contributions of thesis

Can a historic study contribute with new knowledge to a discipline like electrical machine technology? Of course, not in the same way as other projects, which are focused on electrical machine theory and design of new concepts. It can, nevertheless, improve the understanding of the industrial development of these products and the conditions and factors that have had a major influence on the development. This kind of knowledge could, in many cases, have at least the same importance as specific technical background information when new R&D efforts are considered. The major contributions of this thesis are listed below.

The thesis contains the first comprehensive presentation of electrical machine development in Sweden, especially with a focus on recent decades.

The thesis combines presentations of actual product designs with detailed information on the industrial contexts in which the development took place.

The study concludes that induction motors will remain dominant, primarily due to their superior properties at fixed frequency operation, i.e. grid connected.

The backgrounds for introduction of the unconventional direct water-cooling of large turbogenerator rotors, the design solutions, the difficult problems and the remedies have never before been subject to an academic study. Only a few papers addressing limited issues have been published previously.

The chapter on "Vehicle motors" contains comparisons between different motor concepts not available in already existing documents.

The chapter on Powerformer®, the generator for extra high voltages, presents the first objective report on this very special and even controversial development project. Earlier papers have been written by persons who had been engaged in the development or by journalists.

2 The importance of electrical machines to society

Rotating electrical machines are very important to society. This chapter intends to introduce and highlight their role in public and private life. It will also show that electrical machines are surprisingly fascinating, but in many respects very anonymous.

2.1 The use of electrical machines

2.1.1 Imagine ....

Imagine that God had forgotten to create the particular law of nature, which makes electrical machines possible. Could a modern, industrialized society then exist or would we all be back in the 19th century? We could still obtain electricity from batteries and solar cells, but much less and at considerably higher costs. Electricity would mainly be used for telecommunication and lighting. Steam and diesel engines could drive air compressors enabling us to use pneumatic tools in workshops, service centres, hospitals and even at home. Lifts and cranes could be run with hydraulic motors. We could still have refrigerators, cars and record players but hardly any practical vacuum cleaners, food processors or CD-players. Imagine a washing machine driven by a combustion engine or a hydraulic computer disk drive! It is possible to speculate for fun around many more examples.

However, mankind is innovative and would certainly have developed many substitutes, but there is no doubt that our lives would be very different without rotating electrical machines. The versatility of the electric motor surpasses all other types of machinery in helping us carry out mechanical work.

Practically all people, at least in the industrialized countries, are very dependent on the existence of rotating electrical machines. The need for electricity is generally well understood, but the great importance of a product, such as the rotating electrical machine, is usually not explicitly recognized. To most people, it therefore comes as a surprise when they start to realize to which extent they actually use electrical machines, predominantly motors, in their daily lives. The development, manufacture and use of these machines are worth more attention then they usually receive. The primary importance of rotating electrical machines is, of course, the work they can perform and the examples of this are almost uncountable.

2.1.2   The infrastructure sector

Electrical machines play important parts in our daily activities, at home, at work and elsewhere in society. Most of them appear as motors of different sizes used for all sorts of applications. Others are used as generators for producing the electric power we need. Almost 100 percent of all electric power is generated by large generators, which are driven by different types of turbines but also, to some extent, by reciprocating engines such as diesel engines. The generators represent the common denominator in hydropower plants, nuclear power plants, fossil fueled power plants, wind power plants and diesel generator sets. Electrical machines are, in this respect, a basic prerequisite for the modern electrified society. Figure 2.1 shows as an example a large steam turbine-driven generator in a Swedish nuclear power plant.

Power generation is not the only public use of the electrical machine in our societies. They are also used as motors for driving pumps in municipal water-supply plants and sewage plants, as traction motors for electrical trains and tramways and for many other infrastructure applications.

Figure 2.1 710 MVA, 3000 rpm turbogenerator and steam turbine at Barsebäck nuclear power plant installed in 1975

2.1.3   The industry and the service sector

Electrical motors perform most of the mechanical work in factories, offices, service centres, hospitals and schools. This is illustrated by the fact that approximately 65 percent of electric power consumption in Swedish industry is used for motor applications [15]. In industry, electrical motors are used for driving compressors, fans and pumps, machine tools, robots, transport equipment to name a few. Figure 2.2 shows electrical motors installed in an industrial robot. In offices, schools and service centres, electric motors are frequently used for lifts, air-conditioning equipment, office-machinery etc.

In hospitals, a lot of medical equipment is powered by electrical motors. Diesel engine-driven stand-by power generators, to be used in case of power supply failures, are especially common in hospitals but are becoming more and more widely used in buildings such as banks, shopping centres and hotels.

Figure 2.2 ABB robot, provided with six PM motors, in a foundry

2.1.4   The domestic sector

The domestic use of electrical motors has grown rapidly and is today much more extensive than most people believe. Private households, in the industrialized western countries, often own more than one hundred electrical motors without being aware of most of them. Through very simple interviews, with a number of persons with different background, a significant discrepancy was found between the numbers of motors they spontaneously estimated they had and the approximate number a brief review indicated they really had. The results are shown in the diagram in figure 2.3 below. These interview results cannot be seen as statistically correct, but that is not the point; this general conclusion can be easily verified by almost anyone.

For what purposes are all these electrical motors used? Water pumps, ventilation fans, domestic appliances and electrical tools are obvious examples. Others are computers, printers, CD-players, modern cameras etc; the list can go on and on. The car is another product which contains a large number of electrical machines, the generator, start motor and motors for the windshield wipers to name a few traditional, but there are also motors for many other functions. Table 2.1 below gives some examples of the number of motors in various products for private use.

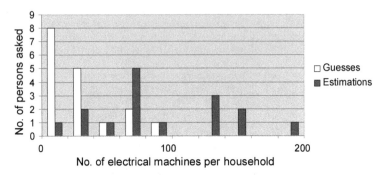

Figure 2.3 Estimated number of electrical motors per household obtained through mini-interviews

Table 2.1   Number of electrical motors in common domestic products

Product	No. of motors	Introduced in Sweden[*]
Vacuum cleaner	1	1920's
Refrigerator with compressor	1	1930's
Washing machine	2 - 3	1940's
Central heating system	> 1	1940's
Food processor	1	1940's
Drilling machine	1	1960's
Wristwatch	1	1970's
Modern cameras (not digital)	2 - 3	1980's
CD-player	2	1980's
Desktop computer with printer	7 - 8	1990's

[*]. Indicates when the products started to be widely marketed

The conditions are, of course, much different in non-industrialized countries. Only a small minority of homes in many African countries have electric power. People living in rural areas or in slum districts in large cities in Asia and Latin America often only have a couple of light bulbs and a radio or TV, but no appliances or tools containing electric motors. The diagram in figure 2.4 shows the electric power consumption per capita in some countries [16, 17]. Sweden is obviously one of the most electricity dependent countries and it is therefore not surprising that the Swedish electrical industry is strong also from an international perspective.

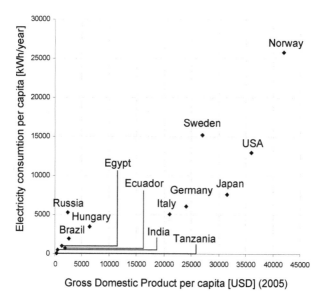

Figure 2.4 Electric power consumption vs. Gross Domestic Product

2.1.5 An anonymous product

It is certainly accurate to refer to electrical motors as our anonymous servants. They perform many different functions and work for us every day, though most of us hardly notice them, as they are integrated components in many common products. Some examples can be seen in figure 2.5. Very few private persons have ever purchased a separate electric motor. The motors are instead subject to trade between professionals, a fact that of course has impact on specified requirements and hence the development focus. Low cost solutions often prevail over performance and quality, at least for many small motors. Furthermore, only a minority of people have some knowledge about the function and design of an electrical machine. Whether this is good or bad can be up for discussion. Knowledge is important, but if it isn't sufficient enough, problems may arise when people start trying to repair faulty motors themselves.

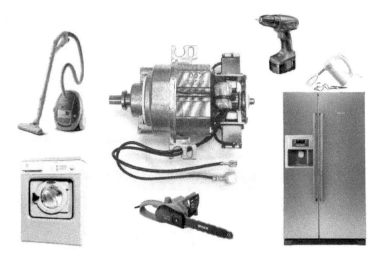

Figure 2.5 Universal motor for vacuum cleaner. This kind of motor is frequently used for various domestic appliances and hand tools. It functions both with AC and DC supply.

2.1.6   Increased use over time

Viewed from an historical perspective, the infrastructure and industrial applications of electrical machines came earlier than their domestic use. Around the previous turn of century, power plant generators supplied power chiefly for lighting, galvanic processes, electrical trams and industrial drive systems [18]. Electrical motors had not yet been introduced in private homes or in the service sector, except for some rare cases. Electrically driven lifts were installed in the late 19th century, and a few household apparateous with electrical motors had also been demostrated. [19] The development of electrical machines for private use, in spite of earlier attempts, took off after World War I with the introduction of vacuum cleaners, sewing machines and some other domestic appliances. For instance, less than 5 percent of Swedish homes had a vacuum cleaner in 1925, but this had increased to 55 percent in 1941[20]. The diagram in figure 2.6 is based on various documentation and some interviews and it gives an indication how the number of electric motors in private Swedish homes has increased during the years. Individual variations are, of course, large and the diagram illustrates what is typical for many households, not statistically correct mean values.The use of electric machines in cars has also increased rapidly, especially during recent years. Volvo's and Saab's premium models are good examples [21]. This is also included in the diagram.

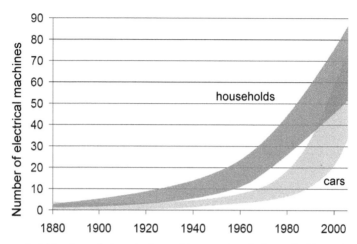

Figure 2.6 Number of electrical machines in family households and cars

New applications and an increased use of various electric apparatus have required a continuous development of rotating electrical machines. The focus of development varies, of course, depending on application, machine type and size, mode of operation, etc. as will be seen later in this thesis.

2.2   A fascinating product

2.2.1   An extreme output range

Rotating electrical machines are fascinating in several ways. The size of such a machine can be defined in different ways, but the most common, and usually most relevant measure, is its power rating. With this definition of size, these machines cover an unrivaled power range, around 15 orders of magnitude. The smallest motors used for wristwatches and other instruments are rated down to 1 µW while the largest power plant generators are well above 1 GW [22, 23]. There is no other type of machinery, which can match this. Combustion engines, for example, cover a range of roughly seven orders of magnitude. The diagram below, figure 2.7, indicates the output ranges for machines for various applications.

The practical designs are of course very different when comparing a small motor and a large generator, but the basic function is the same. It is a matter of interaction between electric current, magnetic flux and mechanical force. The function is based on the same physical laws over the whole output range.

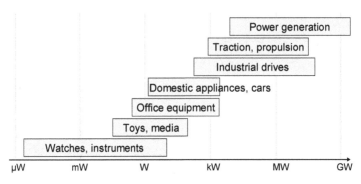

Figure 2.7 Electrical machine outputs for different applications

Very small motors used in watches, toys, CD players etc. are usually permanent magnet motors. So-called universal motors, originally intended for both AC and DC supply, are common in hand tools and various domestic appliances. For outputs in the kW range and larger three-phase AC machines dominate. Figure 2.8 shows examples of some small motors for different applications.

Figure 2.8 Micro PM motors are used for disk drives, toys, wristwatches, and many other applications

There are also other types of extremely small electric motors such as electrostatic and piezo electric, but they are not included in this study.

2.2.2 A successful product

What are the reasons behind the success of electrical motors? Other type of machines exist for producing mechanical work but all of them have much more limited use. Examples, already mentioned, are combustion engines, pneumatic and hydraulic motors, and various types of turbines. Each of them may have its advantages, but they are all niche products compared with electrical machines.

The main reasons why the electrical motor has become the worlds most widely used "work horse" are:

- The simple and clean supply of electric power
- The relatively low motor production costs
- The good possibilities to control the motors
- The high motor efficiency
- The high reliability and practically no need for maintenance
- The good adaptability to different applications
- The quite and clean operation
- The quick access to the function
- The energy conversion can be reversed

Its only real weakness is actually not its own; it is the difficulty to store sufficient energy in batteries in case of mobile applications.

2.2.3 Impact on energy balance

The electric machines perform a lot of useful work but these drive systems also consume a lot of energy. In an industrial country like Sweden, which nowadays uses plenty of electricity for heating, still roughly 1/3 of the total electric energy is used in electrical motors. A Stanford Research Institute report from 1980 estimates that as much as 2/3 of the total U.S. electricity consumption was for motor drives [24]. A recently published report claims a similar relation for Europe in 2004 [25]. One reason for the difference is probably that the domestic use of electricity in Sweden has increased much more than the industrial. Where Swedish homes and service centres consume electricity for heating, the U.S. uses it for motor-driven air-conditioning.

Part of the energy consumed is losses in the electrical motors, which typically have efficiencies of around 90 - 95 percent at nominal load, higher for large motors and lower for small ones. Sweden's total annual electric energy consumption is around 150 TWh of which approximately 50 TWh is used for motor drives. An improvement of the average motor efficiency with two percent units results in a decrease in energy consumption of around 1 TWh per year. This represents a value of approximately 300 MSEK (43 MUSD) based on average spot market prices in the Nordic countries during the years 2003 - 2004, which illustrates the importance of such a development for the society [26]. However, reduced energy consumption is more important than money. The European Union and also North American authorities have therefore adopted certain rules and laws promoting, or in the case of the U.S. even requiring, high efficiency motors [27, 28].

Another way in which development of electric drive systems can contribute to lower energy consumption is through cost effective variable speed drives. Most motors, used for fans, pumps and similar applications, have been fixed speed motors. The flow has usually been controlled by energy wasting throttling. Introduction of speed-controlled motors, regulating the speed to the actual need, improves the process efficiency significantly. This will be dealt with further in chapter 6.

2.3   The electrical machine industry

2.3.1   An important industry

Electrical machines are not only important for their users. They represent also a valuable industrial activity of significant importance for society. As an example, the total annual turnover of electrical machine production in Sweden can be estimated to approximately 3 billion SEK (430 MUSD) out of which at least 75 percent is for export. The number of employees engaged is in the order of 1600.[*]

If the development and manufacture of electrical machines is considered as a core business, there are also a number of other activities directly connected to this. There are suppliers of necessary material and components such as copper wire, electrical steel, insulation material, roller bearings and different sensors. Another category is manufacturers of complement equipment e.g. switchgear, power electronic converters, relays etc. Development, design, installation and commissioning of entire drive systems for various industrial processes require unique know-how and represent high production value and many jobs. Figure 2.9 tries to visualize how the electrical machines can be put into a larger industrial context. No quantitative analysis has been per-

---

[*]. No official statistic is available. The figures have been assembled by the author after contacts with the companies listed in table 2.1

formed, but the "secondary" areas represent, without any doubt, a very significant industrial activity.

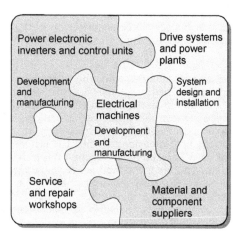

Figure 2.9 Electrical machine related industrial activities

Development of electrical machines is usually carried out by the manufacturers, while the research is a matter for both industry and technical universities. The development process is in many respects very different when you compare large tailor-made generators with small mass-produced motors. The main requirements, and the most influential factors, are not the same. This will be studied in detail in later chapters.

2.3.2   A Swedish perspective

Sweden has long held a strong position in the development of electrical machines, but mainly in that of larger sizes, > 0,1 kW. Sweden's involvement started back in the 19th century. Large power plant generators have, since then, been built by Asea, and later by ABB, the company that resulted in 1988 from a merger between Asea and BBC. ABB and the French company Alstom fusioned their power generation business areas in 1999, but this jointly owned company was completely taken over by Alstom the next year. A new company, VG Power AB, was established in 2002 by former ABB/ Alstom employees aiming at development and sales of retrofit generators and renovations. Both this company and Alstom outsource the manufacture to different sub-suppliers, e.g. to GenerPro AB, which took over the generator workshop in Västerås in 2000. This is a good example that existing knowledge and resources represent such a high value that this type of business activity survive through several structural changes, provided there is a market for the products.

ABB in Västerås is still a world-leading manufacturer of large 4- and 6-pole AC motors and generators as well as certain sizes of DC-machines and traction motors. ABB is also a very large manufacturer of standard AC motors, mainly used for industrial applications. Examples of other Swedish manufacturers of electrical motors are ITT Flygt, Danaher Motion (earlier ELMO), Ankarsrum Industries and BEVI. ITT Flygt develops and manufactures motors for its own submersible pumps, while the others sell their special products to various customers. The large domestic appliance company Electrolux was, until recently, an important manufacturer of fractional horsepower motors. In the mid 1980's, Electrolux produced annually 4.5 million motors in its factories in Västervik and Ankarsrum, primarily for its own products, but also for several external OEM customers [29]. The factory in Västervik was closed a couple of years ago and the production was moved to Hungary. In addition to the manufacturers mentioned above, there are also some other companies which make electrical motors as part of their main products. There are also some very small manufacturers of special motors, e.g. servomotors. Table 2.1 contains a list of Swedish companies engaged in development and/or manufacture of electrical machines in 2006.

Table 2.1   Companies in Sweden developing and/or manufacturing electrical machines

Name	Location	Main business	El. machine activity	Products
ABB Machines	Västerås	Electrical machines	Dev + Mnf	HV AC + DC + traction
ABB Motors	Västerås	Electrical machines	Dev + Mnf	LV induction motors
Alstom Power	Västerås	Electrical machines	Dev + Rev	Generators
Ankarsrum Industries	Ankarsrum	Electrical machines	Dev + Mnf	DC + Universal motors
Atlas Copco Tools	Nacka/Tierp	Tools	Dev + Mnf	HS PM motors
BEVI	Blomstermåla	Electrical machines	Dev + Mnf	AC machines
Bombardier	Västerås	Trains	Dev	Traction motors
Danaher Motion	Flen	Electrical machines	Mnf	AC motors

*Table 2.1   Companies in Sweden developing and/or manufacturing
electrical machines*

Name	Location	Main business	El. machine activity	Products
Emotron	Helsingborg	Power electronics	Dev	SR motors
GenerPro	Västerås	Electrical machines	Mnf	Generators + motors
Hasselblad, Viktor	Göteborg	Cameras	Dev + Mnf	Micro motors (actuators)
HDD Servo Motors	Stockholm	Electrical machines	Dev + Mnf	Servomotors
ITT Flygt	Sundbyberg/ Emmaboda	Submersible pumps	Dev + Mnf	LV induction motors
Komposit-produkter	Vikmanshyttan	Composite products	Dev + Mnf	HS PM motors
LMT	Lidköping	Grinding machines	Dev	HS spindle motors
Spintec	Eskilstuna	Spindles	Dev + Mnf	HS spindle motors
VG Power	Västerås	Electrical machines	Dev + Rev	Hydropower generators

This thesis is primarily based on Asea/ABB's development of electrical machines, because this company has been, and still is, by far the largest electrical machine manufacturer in Sweden and it has since long maintained a position in the frontline of the international development in this field.

Sweden has practically no manufacturers of really small motors such as those used in computers, instruments, toys etc. That is the main reason why this study from now on will focus on larger machines.

3   Historic review

The rotating electrical machine was invented in the nineteenth century. The different main concepts, the DC-machine and the synchronous as well as the asynchronous AC-machines, had all been demonstrated before the end of that century. It is sometimes claimed that the essential development of electrical machines was already completed short after World War I and that they therefore are fairly uninteresting with respect to current research and development. This is of course not true. These machines are far too important for modern society to be ignored with respect to R&D. The early history is nevertheless of great interest. This chapter contains a short historic review, which can contribute to a better understanding and a broader perspective on the more recent development studied in later chapters.

3.1   The scientific discoveries

The history of electromagnetism starts in Copenhagen in April 1820, when the Danish professor in physics Hans Christian Oersted (1777–1851) discovered that an electric current could cause a magnetic needle, placed parallel to the conductor, to turn. The needle turned in different directions depending on the direction of the current and also depending on the position of the needle relative to the conductor. Oersted concluded that there exists an interaction between the current and the magnet, circularly distributed around the electric conductor. He published his discovery in a letter "Experimenta circa effectum conflictus electrici in acum magneticam" to other scientists in July of the same year [30].

Oersted's experiment was demonstrated for the Academy of Science in Paris in September 1820 by the physicist Francois Arago (1786-1853). This inspired his colleague André Maria Ampère (1775–1836), professor at the École Polytechnique, to study the interaction between electrical currents. Only one week later, Ampère was able to show that two parallel conductors carrying currents in the same direction attract each other, while those carrying currents in opposite directions repel each other. Further experiments followed soon, e.g. demonstration of coils with multiple turns. The necessary foundation for the electrical motor, the knowledge concerning mechanical forces caused by interaction between electrical currents and magnetic fields (fluxes), was, in principle, laid during a few months in 1820.

The English scientist Michael Faraday (1791–1867), who started as a bookbinder apprentice and later became laboratory assistant and finally professor, has played a very important role in the early development of electro-technology. In 1821, he was able to demonstrate rotation due to interaction between electric currents and a magnet. A vertically suspended current carrying conductor, with its lower end in mercury, could rotate around a magnet protruding from the centre of the mercury bowl. He also demonstrated that he could cause the magnet to rotate around the conductor. Michael Faraday's most important contribution was made in 1831, when he discovered the electromagnetic induction, i.e. that a changing magnetic flux will induce an electromotoric force (voltage) in a surrounding coil and a current will flow if the circuit is closed. This is a fundamental phenomenon for the function of the electric generator and also for the transformer.

Figure 3.1 H. C. Oerstedt, A. M. Ampère, M. Faraday and J. C. Maxwell

The American physicist Joseph Henry (1797–1878), professor at Albany and later Princeton, seemed to have found electromagnetic induction independently of Faraday, but his first publication came half a year later than Faraday's. Henry is recognized for inventing multi layer coils in 1830, by means of which he could design very strong electromagnets.

Several other scientists made important contributions to the study of electromagnetism during the 1820's and 30's, and their achievements can be studied in literature on electrical history. The information above is based mainly of two such books [31, 32].

A complete theoretical treatise of the electromagnetic phenomena was published in 1861 and 1864 by the Scottish scientist James Clerk Maxwell (1831–1879), professor at Kings College in London. Maxwell's laws have, ever since then, been fundamental for the analysis of electromagnetic phenomena and products. Maxwell , who was a quiet and reserved person, has later become somewhat of an icon for electrical science [33].

3.2 The pioneer years

Many scientists and inventors tried to build and demonstrate various electrical machines during the 1830's – 1850's, but most of these were without any practical use. The French instrument maker Antoine Hippolyte Pixii (1808–1835), who worked for Ampère, made, in 1832, a hand-driven electromagnetic apparatus which could produce an alternating voltage. He let a horse-shoe shaped permanent magnet rotate above another horseshoe shaped iron core provided with two coils in which the voltage was induced. William Ritchie (1790–1837), a Scottish lawyer and physicist presented, in 1833, a machine with rotating coils and a stationary magnet which could produce a pulsating DC current because he had introduced special contacts for bringing the current from the rotating winding, and thus a primitive commutator was born. The English inventor Edward Clarke demonstrated three years later an apparatus, which could generate AC or pulsating DC voltage depending on the position of a contact device, used for connection between the rotating winding and the stationary circuit. Clarke built his machine for medical electro therapy, which was claimed to be a painful experience [34]. The permanent magnets used in all these machines were weak steel magnets, so they had to be fairly large. The introduction of electromagnets was a big improvement in that respect.

The first practical demonstration of an electrical motor was made by Moritz Hermann Jacobi (1801–1874). He was born in Potsdam, Germany but later became a Russian citizen and professor in Sankt Petersburg. In 1835 Jacobi published a report about an electric motor he had built the previous year. It was, in principle, an 8-pole motor consisting of four rotating and four fixed horseshoe shaped electromagnets facing each other end to end. A photo is shown in figure 3.2. In September of 1838, he managed to demonstrate that his electric motor could drive a small boat at the Neva River. He improved the design and one year later, he once again demonstrated how such a motor could be used for driving a boat, this time a larger one, with 10 – 14 persons on board. Jacobi published his results in 1840 in a report with the title "Über die Prinzipien der elektromagnetischen Maschinen" [34].

Figure 3.2 M. H. Jacobi and his motor build in 1835

Contemporary of Jacobi, the American Thomas Davenport (1802-1851), a self-educated blacksmith, performed various electromagnetic experiments.

He managed to build a motor in which the interaction between one stationary and one moveable electromagnet could bring the latter to turn half a turn. By introducing a device, later called a commutator, Davenport managed to achieve a continuous rotation. He demonstrated his motor by means of a model electric train running on a circular track. Thomas Davenport visited several famous scientists, one was Joseph Henry, in order to get support for his invention. He was finally granted a patent for his electric motor on February 25, 1837. This is claimed to be the first patent ever for an electrical machine [35].

Several attempts were made to build steam engine-driven generators. The first recorded attempt was by the English chemist John Stephan Woolrich, who built a generator for galvanisation in 1844. It was a large machine with four stationary horseshoe magnets and a disk with toroidal coils rotating between the magnet poles. The same basic idea was used by the Belgian physicist Florise Nollet (1794–1853), but he refined it and built a machine with multiple axial sections. His machine became, after his death, known as the "Alliance machine" and was, from 1857, used for generating electric power for lighthouses with arc lamps, figure 3.3. It is interesting to note, however, that the founders of the Alliance Company had hidden a strong battery, used for back-up, when the unreliable generator was demonstrated in Paris for official inspectors and even for the Emperor Napoleon III (1808-1873). This was probably not the first and definitely not the last attempt to improve test results with dubious methods. The fraud became publicly known but the company can, in spite of this, be recognized as the commercial pioneer in electric power technology [36].

Figure 3.3 An "Alliance machine", built in Paris in 1863

It is interesting to notice that all these, previously mentioned, machines are so called axial flux machines, i.e. the magnetic flux is crossing the airgap in a direction parallel to the shaft of the machine. Axial flux machines have experienced a renaissance in recent years and receive currently a lot of attention, but it is radial flux machines, which have dominated nearly since the mid of the 19th century. Both concepts are explained in the next chapter, section 4.2.

3.3 The practical breakthrough

Werner von Siemens (1816–1892), German scientist, engineer and industrialist played a central role in the emerging electro-technology. He received his technical education at the Prussian Army's Artillery and Engineering School in Berlin. It was also here, in 1847, that he established Siemens & Halske, a company for the development and manufacturing of telegraph equipment. One product, which he invented 10 years later, was a hand-driven permanent magnet "induction machine" with a 2-pole armature winding in a cylindrical rotor. This machine was a radial flux DC machine, more efficient than earlier machines due to a smaller airgap between the stationary magnets and the rotor, see figure 3.4. In 1866, Siemens' discovery of self-excitation, also known as the dynamo electric principle, led to radical improvements as permanent magnets were replaced by much more powerful electromagnets allowing the machine to generate its own excitation current instead of relying on an expensive battery for power. Siemens had found that there was enough remanence in the iron, so that the dynamo could pick-up voltage when it started to rotate. Siemens immediately realized the importance of his discovery and wrote in a letter to his brother on December 4, 1866: *"This thing has a huge development potential and can lead to a new epoch of the magnetic electricity."* (My translation) [37]. This discovery has often been referred to as the cornerstone of the power electric era. (Dynamo was the common term for DC generator.)

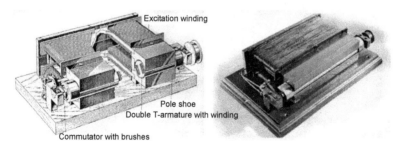

Figure 3.4 Siemens' dynamo from 1866

Siemens was not undisputed as inventor of the dynamo electric principle. Two English engineers, Charles Wheatstone (1802–1875) and Samuel Alfred Varley (1832–1921) published, independent from each other and more or less simultaneously with Siemens, similar inventions. However, it was only Werner von Siemens who realized the industrial potential in this invention [38].

The new dynamo machine had its problems. One was overheating and another was large pulsations in the generated voltage. The pulsations could be reduced by use of a so-called "ring armature" invented in 1865 by the Italian scientist Antonio Pacinotti (1841–1912) when he was professor in Bologna. He arranged a toroidal winding consisting of many coils around an iron ring, which rotated between two poles. The voltage pulsations were almost eliminated by the even distribution of the winding around the entire ring periphery. This concept was improved upon, five years later,by the Belgian electrotechnician Zénobe Théophile Gramme (1826–1901), who worked in Paris for the Alliance Company as model maker. He replaced the solid iron ring with a ring wound from iron wire, which radically reduced the eddy current losses and thus the overheating. Gramme also applied the dynamoelectric principle to his machine and started a successful commercial production. These dynamo machines were mainly used for supply of power for galvanisation and for arc lights [39].

Figure 3.5 Werner von Siemens and Zènobe T. Gramme

The ring armature had a distinct disadvantage because only a minor part of the winding was active, namely the outer part, which is passed by the magnetic flux. The invention of the drum type armature winding addressed this problem. Friedrich von Hefner-Alteneck (1845–1904), design engineer and later director within Siemens & Halske, published, in 1872, the invention of this winding, in which both coil sides are placed close to the rotor surface but under different poles. This has been the dominant type of winding for DC machines ever since and, transferred to the stator, also for AC machines. Figure 3.6 shows the principal difference between the ring armature and the drum armature windings [38].

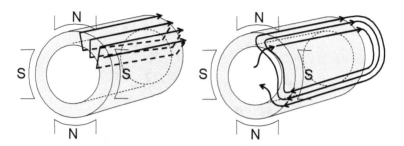

Figure 3.6 Ring wound armature (left) and drum wound armature

3.4 Early industrial development

The world saw the birth of the electric power industry in the 1870's. Earlier attempts had been far too limited. The reason was that inventors and entrepreneurs now realized that there could be a market for electric lighting and therefore established companies for manufacturing dynamos or started such activities in already existing enterprises. It was suddenly possible to generate power in larger quantities and at lower costs. This and other new applications for the use of the power were demonstrated and received great attention in newspapers and at international exhibitions. Germany and the U.S. soon became the leaders in the development of the electro-technical industry. Most of the frontline companies were strongly tied to the respective inventors.

Zénobe Gramme has been mentioned above. In 1871, he opened together with a partner, an electrical machine factory in Paris: the "Societé des Machines Magneto-Electriques Gramme". Gramme demonstrated that his DC dynamo also could work as a motor and it was even developed into a single-phase AC generator. Gramme was the first to produce a commercially successful electrical machine. His machine was considered, for several years, as the most advanced of its kind and was sold in a number of countries, even in America [39].

Siemens & Halske supplied dynamos for lighting installations as early as 1868, but very few. The machine had a solid iron armature core and therefore such large losses, that it required water-cooling. The annual sale was less than five during the first years and it took until 1878 to reach an annual sales volume above ten. The boom came two years later when production increased to 300 – 500 units per year. The technology had of course been improved several years earlier, primarily through Hefner-Alteneck's work. Figure 3.7 contains a photo of one of Siemens' mobile lighting power units consisting of steam engine and dynamo [8].

Figure 3.7 Dynamo and steam engine supplied by Siemens in 1878

Another German workshop manufacturing dynamos was established in Nürnberg in 1874 by Sigmund Schuckert (1846-1895), who earlier had worked with telegraph equipment both in Siemens & Halske and in Edison's factory. He designed his own dynamo and his company, Schuckert & Co, became an important supplier of both dynamos and other electrical products. The general economical crisis of 1901 forced the company to be fusioned with the power electric parts of Siemens & Halske [40].

The British-born American electrical engineer Edward Weston (1850-1936) was working in an electroplating industry in New York when he invented his first dynamo in 1873. He received in 1875 a very important patent for laminated cores and pole pieces, which radically improved the efficiency of electrical machines. Based on his patents, the Weston Dynamo Electric Machine Company was established in 1877. It is claimed to be the first U.S. factory of its kind. The dynamos were mainly used for lighting. Weston left the generator and lamp business in 1887 and switched to electrical instruments [41].

Charles F. Brush (1849-1929) graduated from the University of Michigan with a degree in mining technology, but his interest soon turned to electricity and lighting. Brush was not satisfied with existing dynamos and decided to develop his own. He was aware of Gramme's success in Europe and used this dynamo as a basis for his work. Brush kept the ring armature but converted the machine into an axial flux dynamo in order to increase the total airgap surface. He received a patent in 1877 and established the Brush Electric Company, which became one of the major suppliers of lighting installations until it merged with the Thomson-Houston Electric Company in 1889 [42].

Another leading American electric engineer and businessman was Elihu Thomson (1853-1937). Born in England he moved to the U.S. as a child. His first important invention, the three-coil dynamo complete with an automatic regulator, became the basis for his electric lighting system. In 1880, Thomson established together with his former colleague Edwin Houston (1847-1914), both had been teachers, a company later known as the Thomson-Houston Electric Company. The company played a successful role in the electric lighting business with an AC system. The company merged with the Edison General Electric Company in 1892 forming a new much larger company, General Electric [43].

Thomas Alva Edison (1847-1931) has been seen as the archetype of an inventor. His role in the emerging electrical industry was central. He started without much formal education as a telegraph operator, but soon became an inventor of telegraph equipment. He managed well and was able to open his own laboratory in Menlo Park, outside New York, in 1876. Two years later, Edison and his assistants, started to work on the problem of electric lighting. He was made famous for inventing the incandescent light bulb, but he in fact was not the only one, not even the first. What made Edison so successful was his system approach. His staff worked on development of all components required for a commercially viable lighting system. One of the components was the dynamo. He acquired and studied dynamos from Weston and Gramme and then started development of his own, which he completed in 1879. His dynamo was characterized by a small internal resistance, a small airgap and large bipolar electromagnets, see figure 3.8. Edison claimed that it had a better efficiency than any other generator and could be turned into a business in itself. Edison Machine Works was, therefore, established to manufacture the machines.

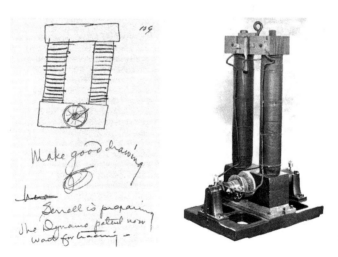

Figure 3.8 Edison's dynamo, introduced in 1879, and nicknamed "Long-legged Mary-Ann"

Edison's reputation grew rapidly, not only in America but also in Europe. He was called the "Wizard of Menlo Park" but his R&D methods were also questioned. He relied more upon extensive tests than theoretical analyses. It was also characteristic of Edison to seek publicity for his inventions, even before they were ready. His business relations with other companies, sometimes his competitors, were also somewhat complicated. In spite of this, his laboratory, which can be seen as a forerunner to later corporate research centres, served a number of companies, some of which he held shares in. These companies became, in 1889, parts of the Edison General Electric Company that three years later merged with Thomson-Houston thus forming General Electric (GE), the world's largest company of its kind [44].

The other leading American company in this business was founded by the inventor and industrialist George Westinghouse (1846-1914). Westinghouse was primarily a mechanical engineer and his first companies had a corresponding focus, but he also saw the potential for electricity. He therefore formed a company, in 1884, later known as Westinghouse Electric Manufacturing Company. In 1888, he acquired exclusive rights to Nikola Tesla's (1856-1943) patents for polyphase AC machines and systems, whereby Tesla also joined Westinghouse's company. There was, during these years, an intense fight between Edison, Westinghouse and even Thomson-Houston over the advantages and disadvantages of DC and AC systems respectively. Edison claimed that AC was too dangerous while the others stressed the better possibilities in transmission and distribution of AC. Patent rights and other commercial issues had, of course, a strong influence on the different positions. Westinghouse took another strategic step when he, some years later, acquired exclusive rights to manufacture Parson's steam turbines in America. The Westinghouse group of companies grew rapidly and had around 50 000 employees at the turn of the century [45].

Figure 3.9 C. F. Brush, T. A. Edison, E. Thomson and G. Westinghouse

The American manufacturers had a very large domestic market and did not start an extensive export of their products to Europe. They preferred instead to sell licenses or even organize daughter companies with local production, such as Anglo American Brush Co.

A German mechanical engineer, Emil Rathenau (1838-1915) managed to obtain the rights to use Edison's patents in Germany and begun in 1882 with the installation of lighting systems. The results were good and Rathenau wanted to expand into manufacturing. He realized that this could only be done if he didn't challenge the interests of Siemens & Halske. An agreement was reached and a new company, the Deutsche Edison Gesellschaft, was established with the right to manufacture light bulbs and build power supply stations, while the exclusive right to manufacture and sell Edison's dynamos was aquired by Siemens & Halske. The partners renegotiated in 1887 and the Deutsche Edison Company changed name to "Allgemeine Elektricitäts-Gesellschaft" (AEG). Siemens took over the German rights to Edison's patents, but AEG could, among other things, begin to manufacture machines up to 100 hp. The agreement between the companies was completely terminated in 1894 and AEG became a more and more powerful competitor to Siemens & Halske [46].

Switzerlands electrical industry dates back to 1876 when the Maschinenfabrik Oerlikon was established in a suburb of Zurich. Electrical machines, locomotives and machine tools belonged to the most important of its products. Only 23 years old, the engineer Charles E. L. Brown (1863-1924) was appointed head of the electrical department. Brown designed various AC and DC machines, most notably the 3-phase generator used for the famous power transmission from the hydropower plant Lauffen, located 30 km north of Stuttgart at the Neckar River, to a large electricity exhibition held in Frankfurt am Main in 1891. The Lauffen generator is shown in figure 3.10. The transmission distance was 175 km, the voltage 15 kV and the successful demonstration became a heavy argument for AC in relation to DC. Brown left Oerlikon the same year and founded, together with his colleague Walter Boveri (1865-1924), a new company, Brown Boveri & Cie. (BBC). BBC grew then rapidly into one of the major European electric corporations. Oerlikon was eventually acquired by BBC in 1967 [47].

There were of course many others who saw the possibilities in the emerging electric industry and several companies were established in different countries. Great Britain had since long been a leading industrial nation and was certainly also active in this field, but their efforts were quite fragmented. One example worth mentioning, however, is the steam turbine manufacturer C. A. Parsons & Co, which manufactured outer-pole AC generators from 1889 [48].

Another important pioneer firm was Ganz Electric Works in Budapest, the first of this kind in the Austro-Hungarian Monarchy, founded in 1878. It began with DC machines but the technical manager, Károly Zipernowsky (1853-1942), soon focused on development of the AC technology. He patented a generator in 1880, which could produce both DC and single or multiphase AC. The Ganz company became especially recognized for transformer inventions made in 1884-85 [49].

Figure 3.10 The 300 hp 3-phase generator in Lauffen, 1891

3.5 Scientific progress

Maxwell published his fundamental theories on electricity and magnetism in the early 1860's, but there is no evidence that these had any impact on the development of electrical machines until much later. On the contrary, many of the successful inventors mentioned above had no special scientific background, at least not in electrical science. They were practical and innovative persons and their designs were based on empirical knowledge. It is witnessed that they often learned through studies of each other's machines.

Electricity was instead a matter for the physicists at universities. The polytechnical institutes typically taught civil and mechanical engineering, mining technology and chemistry. In Germany, the first professor in electro-technology was appointed in Darmstadt in 1883, while the first professor in Sweden not until 1900. The electrical machine industry was clearly ahead of the academy during the take-off period around 1880. However, the generator and motor designers needed better theoretical tools and the situation improved soon. One important step was made when the English scientist and engineer John Hopkinson (1849-1898) in 1886 published a paper on "Dynamo-electric machinery", in which he presented the first rational method for calculation of the magnetic circuit in such a machine. He took thereby into consideration both the magnetic saturation and the stray flux between the poles [50]. Hopkinson eventually became professor in electrical engineering at Kings College in

London. Other key persons were the Austrian born Gisbert Kapp (1852-1922), who later was appointed professor in Birmingham and the English-man Silvanus P. Thompson (1851-1916), professor of physics and electrical engineering at London Technical College. Thompson published, in 1884, an epoch-making book "Dynamo-electric Machinery: a Manual for Students of Electrotechnics", which was undoubtedly studied by almost every electrical machine designer at that time [51].

The period from the mid 1880's until the end of the century was very essential. Most of the necessary, basic electrical machine theory was developed during these years with contribution from a number of scientists both in Europe and America. The German-born, American professor Charles P. Steinmetz (1865-1923) can be mentioned as example. He developed the analysis of AC circuits with complex numbers, he introduced equivalent circuits for electrical machine calculations and established analytical methods for determination of iron losses. Steinmetz was a very unique individual, physically crippled and almost a dwarf but intellectually a giant. He had to flee from Germany, accused of socialistic student activities, directly after he had written his PhD thesis at the university in Breslau and immigrated to the U.S. in 1889. Steinmetz worked for GE from 1893 until his death but was also professor at the Union College in Schenectady [52].

A good illustration of the position of electrical machine theory, soon after the previous turn of century, is a comprehensive series of textbooks published by professor Engelbert Arnold (1856-1911). Arnold, who was born in Switzerland, was docent at the university in Riga, then chief engineer at Oerlikon until he became the first electro-technical professor in Karlsruhe in 1894. These books, written by Arnold and some of his assistants, were the fundamental source of theoretical "know-how" for German speaking students and engineers during the first decades of the last century. Two examples are included in the reference list [53, 54].

Figure 3.11 S. P. Thompson, C. P. Steinmetz and E. Arnold

3.6 Early activities in Sweden

Sweden was in the 19th century, like most other European nations, an agricultural country, but industrialization had begun. The basis for the emerging industry was the vast natural resources, primarily the forests, iron ore and the abundant energy in rivers and streams. A rich belt of mines and steel mills named "Bergslagen" had developed across Sweden northwest from Stockholm. Sawmills had been built like a string of pearls along the Baltic coast of northern Sweden. The rivers were used both for transportation of timber logs and for water-wheels driving pumps and hammers. Mechanical workshops and textile factories were established in the cities. All these industries, but especially the steel works and soon also the pulp and paper industry, had a huge need for mechanical power. The versitale electric motor, introduced in the 1880's and 90's, met this demand and quickly replaced both water-wheels and steam engines. The prospects for the establishment of domestic manufacturing of electrical equipment was therefore bright.

The electrification in Sweden began, in 1876, with outdoor arc light installations at two sawmills. The power came from steam engine-driven dynamos manufactured by Gramme. It was also a Gramme dynamo that, two years later, supplied the first electric lights in Stockholm with power. A well-known businessman in Stockholm, Ludvig Fredholm (1830-1891) got interested in lighting systems and made, during 1881, several trips abroad in order to study different possibilities. He became convinced of the potential of electric light and managed to raise funding for a test installation at Norrbro in Stockhom, the bridge in front of the Royal Castle. Both the arc lamps and the dynamo were supplied from Brush.

Fredholm had engaged a young engineer, Georg (Göran) Wenström (1857-1927), to manage the installation and tests. In January 1882, Fredholm met, at a commercial fair in Örebro, Georg Wenström's older brother, 26-year-old Jonas Wenström (1855-1893). This meeting can be viewed as the conception of the Swedish electric industry.

Jonas Wenström was physically weak but he had a brilliant mind. Jonas, who never married, seems to have had a very trustful relation to his father, Wilhelm Wenström (1822-1901), who was a consulting engineer. Jonas Wenström had studied mathematics and science at the universities of Uppsala and Christiania (presently Oslo) and he had made study tours to Germany and France. He had become interested in electro-technology and electrical machines during the mid 1870's as can be seen from the following lines in a letter he wrote to his father on February 22, 1877: "... *Since I came here, I have thoroughly studied the subject on electrical currents impact on magnets and on each other and vice versa. Thereby even the magneto-electric machines that now exist, out of which Gramme's is the best and most practical. I have since then designed a new dynamo-electric machine, which, I think, will have several advantages before others ...*" (My translation). From this point, Jonas Wenström´s interest in electrical equipment continued and he performed more design studies, but he lacked the financial resources to realize them [55]. This changed after Fredholm's visit to Örebro.

Figure 3.12 Ludvig Fredholm and Jonas Wenström

Jonas Wenström had almost complete drawings for a new dynamo when he met Ludvig Fredholm, who already was negotiating with the Anglo American Brush Co. about patent rights for lamps and dynamos. Fredholm became interested in what Wenström showed him and Wenström offered Fredholm the patent rights for his dynamo. A prototype dynamo was then built in a forge at a grain mill. The conditions were primitive. The hand-forged sheets for the laminated core were taken from an old roof and the armature winding insulation was woven drawings, left over from Wilhelm Wenström's engineering bureau. The dynamo was a 2-pole machine with a frame, which completely surrounded the excitation winding, and a rotor with the armature winding placed in slots. Figure 3.13 shows a photo of this dynamo and it is evident why it got the nickname "the Turtle" [56].

The dynamo was ready in August 1882 and the initial tests were performed, whereupon Fredholm and Jonas Wenström signed an agreement, which guaranteed the Swedish patent rights to Fredholm. The contract would, however, only be binding if it could be proved by tests that Wenström's dynamo had the same or better efficiency than the Brush dynamo. These "bench-marking" tests were carried out in November and witnessed by Johan Cederblom (1834-1913), professor in machine technology at KTH and docent Erik Edlund (1819-1888) from Uppsala University, who was a leading electrical scientist in Sweden and later professor at the Academy of Science. They concluded, *"Wenström's dynamo gives more light per horsepower used"*. Fredholm could thus continue his efforts, which resulted in the establishment of "Elektriska Aktiebolaget i Stockholm" on January 17, 1883 [55].

Figure 3.13 Wenström's first dynamo built in 1882

This new company started its manufacturing in the attic of a mechanical workshop in Arboga, a small town between Västerås and Örebro. Ludvig Fredholm, with an office in Stockholm, had been appointed as president and Georg Wenström was employed as the manager of the workshop. Jonas Wenström remained with his father in Örebro, where he developed new improved versions of his dynamo. He designed a series comprising nine sizes ranging from 1 to 44 kW. He continued to make study trips to several European countries but also one to America. Wenström maintained extensive correspondence with leading international experts. He even exchanged letters with Edison.

Jonas Wenström belongs to a group of outstanding Swedish inventors, whose inventions have been of decisive importance for Sweden's industrial development. Contemporary with Wenström were, among others, Alfred Nobel (1833-1896), Lars Magnus Ericsson (1846-1926) and Gustaf de Laval (1845-1913), whose names are linked to inventions of dynamite, telephones and steam turbines. They became internationally famous both as inventors and industrialists and their family names remain even today as names on the industries they founded. The short-sighted, hunchback Jonas Wenström did never reach the other's social position and wealth. He remained comparatively anonymous, living with his father and sister; continuing to work on his electro-technical inventions until he catch a severe cold and passed away only 38 years old. Nevertheless, Jonas Wenström contributions to the emerging Swedish electrical machine industry can not be overestimated [57].

16 of Wenström's dynamos were manufactured the first year, 1883. Some in-

teresting technical data for these machines are given in table 3.1. The workshop in Arboga was extended after some years, but it was closed down and moved to Västerås in 1892. The workshop had then delivered 256 dynamos with a total power of 3273 hp [58]. The company had at that point in time changed name to "Allmänna Svenska Elektriska Aktiebolaget" (the General Swedish Electric Company) later known as Asea.

Table 3.1   Wenström first dynamo compared with more modern
DC machines

Machine type	B1	LD 8	LAK 80
Manufactured	1882	1932	1975
Power [W]	360	360	380
Base speed [rpm]	600	750	900
Flux density in airgap [T]	0.134	0.54	0.7
Efficiency [%]	49	64	64
Temperature rise [K]	30	55	75
Weight [kg]	166	36	19

There were other companies manufacturing dynamo machines in Sweden during the 1880's. The first was AB Hakon Brunius in Gothenburg, which started manufacturing of dynamos of Gramme's design in 1880 [20]. Another company in Gothenburg, Edwin Andrén & Co, also started to manufacture electrical machines during the 1880's. Nyhammars Bruk, an old ironwork, is a third example, which manufactured dynamos starting from 1886. These machines were designed by Johannes G. Darell (1865-1947), who had worked for Wenström during a short period [59]. All three companies finished their electrical machine production fairly soon.

Two companies, which begun to manufacture electrical machines in Sweden during the 1890's, should also be mentioned. One of them was Luth & Rosén AB in Stockholm, which started production in 1893 based on a license from Schuckert & Co in Germany. The other company was Boye & Thoresens Elektriska AB in Gothenburg, which manufactured small DC motors from 1898, but it was already then a subsidiary of Asea [59].

3.7   The 3-phase system and the induction motor

Single-phase AC generators had been built in parallel with the DC dynamo during many years. It was in principle the same machine, but with the commutator replaced by sliprings. A practical transformer had been developed and patented by the Ganz company in 1885. Lighting systems were built, either DC or AC, and various suppliers of course promoted the alternative they had access to. The AC systems were handicapped because no practical AC motor that could compete with the DC motors, existed, until the invention of the 3-phase asynchronous motor, also called induction motor. Jonas Wenström was one of those occupied with this problem. He had his concept ready in 1888 and applied for a Swedish patent on January 20, 1890. It was a patent for a complete 3-phase system consisting of synchronous generator, transformer and an induction motor. The detailed design was then ready a few months later. The motor is shown in figure 3.15. It was a 4-pole motor designed for 5 hp. Wenström received patents on this system also in Great Britain and in some Nordic countries [60, 61].

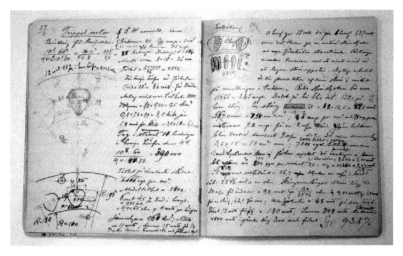

Figure 3.14 Facsimile from Jonas Wenström's notebook, April 27, 1890

There were more inventors claiming priority for the multi-phase system and the induction motor, primarily Michael von Dolivo-Dobrowolsky (1862-1919) in Germany and Nikola Tesla in America. The first observations of the phenomenon on which the induction motor is based were made as early as 1824 when Francois Arago found that a magnet needle hanging over a copper disk turned when the disk was turned. The opposite was shown one year later. The Italian physicist Galileo Ferraris (1847-1897), professor in Turin, discovered, in 1885, that he could create a rotating magnetic field by means of two alternating currents with 90° phase shift. He built a small motor with a copper cylinder as rotor, but the motor ran with 50 percent slip and the efficiency was therefore only 50 percent, so he drew incorrect conclusions and abandoned the idea of using it as an industrial motor [62]. It was instead Nikola Tesla, inventor and scientist, who applied for an American patent for a 2-phase motor in 1887. Tesla was a Serb born in Croatia in 1856. He studied at an Austrian university and worked in Prague, Budapest as well as in Paris before he immigrated to America in 1884 and began working for Edison. He left after a year due to difference in attitudes towards development methodology and to the AC/DC conflict. Tesla continued his efforts to develop a poly-phase motor resulting in the patent application mentioned above. He managed to invent both a synchronous and asynchronous motor, but he seems not to have completely understood the function of the latter. He referred the function to the appearance of poles in the rotor instead of induced currents. In spite of this, he demonstrated, in 1888, such a motor with a copper winding around an iron cylinder as rotor and, as mentioned earlier, he sold the patent rights to Westinghouse. Tesla received 75 000 dollar as down payment and a royalty as high as 2.50 dollar per horsepower of motors sold. This tells a little of how the prospects for electric motors must have been evaluated in those days. Nikola Tesla was a very eccentric person who made many important inventions, but also had several ideas, which were pure science fiction. He died, in 1943, impoverished in spite of the wealth his patents had given him [63, 64].

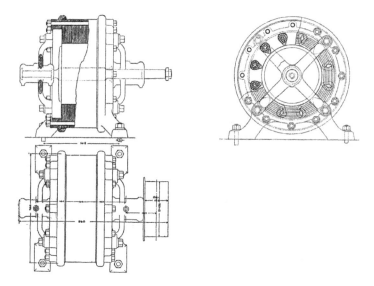

Figure 3.15 Drawing of Wenström's 3-phase motor

Two other inventors, the American Charles S. Bradley (1853-1929) and the German Friedrich Haselwander (1859-1932), also made, independently of each other, important contributions and were the first who conducted experiments with 3-phase machines. They applied for patents for synchronous 3-phase machines in 1888 and 1889 respectively. Neither of them managed however to bring their inventions to successful industrial use.

Michael von Dolivo-Dobrowolsky was the first to build a 3-phase induction motor. He was born in St. Petersburg but had to move to Germany two years after being expelled from the university in Riga due to participation in radical political activities. He completed his studies in Darmstadt and then worked as an electrical engineer at AEG in Berlin. Dobrowolsky investigated, based on what was known about Ferraris' and Tesla's results, different multi-phase concepts and came to the conclusion that a 3-phase system was most advantageous. He invented and tested both a squirrel-cage and a slipring rotor, see figure 3.16. Dobrowolsky's German patent application was filed in March 1889, 10 months before Wenström's. The important demonstration of the 3-phase system, which took place in 1891 through the power transmission from Lauffen to Frankfurt, has been mentioned above. Dobrowolsky was responsible for the development of a 100 hp induction motor, which drove a pump for an artificial waterfall at the exhibition. He was eventually promoted to technical director of AEG and appointed Dr.hon. by the Darmstadt Technical University [46, 65, 66].

Figure 3.16 Dobrowolsky's squirrel-cage motor built in 1889

It is interesting to compare Dobrowolsky's and Wenström's motor concepts. The motor used at the Frankfurt exhibition was a squirrel cage motor with six sliprings! The explanation is that the 3-phase armature winding was placed in the rotor and Dobrowolsky wanted to be able to switch between star and delta connection. His first small experimental motor had a stationary armature winding of the Gramme ring type and the short-circuited squirrel cage in the rotor. Due to the high flux leakage in the ring winding, he decided that this concept could only be used for fractional horsepower motors, so he built the next motor with a drum type armature winding placed in the rotor, which was common for DC dynamos. This was a 2 hp motor. The next step was the 100 hp motor for the exhibition in September – October 1891. There was simply no time for any further test machines [66]. A comprehensive paper on the invention of the induction motor, with emphasis on the interaction between science and engineering, can be found in the American publication Technology and Culture; see reference [67].

Figure 3.17 N. Tesla and M. v. Dolivo-Dobrowolsky

Jonas Wenström designed his 5 hp induction motor during 1890 and it was built and ready for testing in the spring of 1891. The 3-phase armature winding was of drum type, placed in the stator, and thus the squirrel cage winding was located in the rotor. There were 12 closed slots in the stator, i.e. one slot per pole and phase. Each stator coil encircled one tooth, so it was actually a kind of concentrated winding, which has become common nowadays. The rotor winding had 48 slots with one conductor per slot and short-circuit rings at each end of the rotor [61]. It is obvious that Wenström's concept, compared with Dobrowolsky's, was much more in accordance with what later has proved to be most suitable for induction motors. Without diminishing Dobrowolsky's decisive role and pioneering inventions, it is obvious that Jonas Wenström's genius and originality cannot be questioned.

One important project, for both Asea and Sweden, was the 3-phase transmission from the hydropower plant Hellsjön to the mining village in Grängesberg, a distance of 13 km. This was a Swedish equivalent to the German project mentioned earlier and that included synchronous generators, transformers and induction motors. The power was 344 kVA and the transmission voltage 9 500 V. It was carried out in 1893 and thereby opened the door for a rapid increase of AC systems in Sweden.

3.8 The Old and the New World

The early history of electromagnetism and electrical machines, briefly told above, initially had its centre of gravity in Europe, but it shifted during the last decades of the 19th century towards America. The initial steps were taken by scientists in some European universities, which is not surprising considering the fact that America, at that time, had very few scientific institutions compared to Europe. An interesting question is, what happened next? The transition took place in the 1870's. Edison had bought French dynamos from Gramme in 1878. Four years later he licensed his own design to Siemens. Young European inventors, entrepreneurs and scientists as Edward Weston,

Nikola Tesla and Charles P. Steinmetz moved to America and became very successful. Did they leave because they realized that this opened better professional opportunities or did they just follow the main stream of European emigrants? Either way, the American electric industry had obtained a world leading position towards the end of the century. A large domestic market did of course contribute, but equally important was the innovative development work carried out.

An amusing view on the differences between America and Europe was found in a one hundred years old book, "Uppfinningarnas bok" ("The book on inventions") [68]. The original text, in old-fashioned Swedish language, appears in the facsimile in figure 3.18. A free translation follows thereafter.

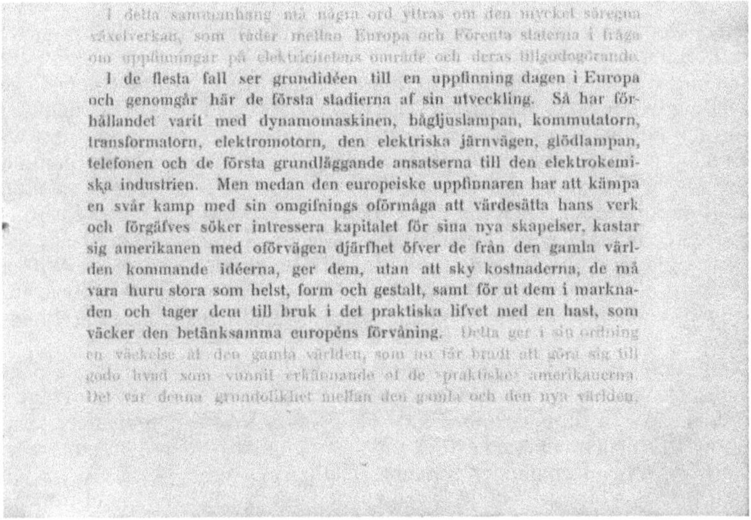

I detta sammanhang må några ord yttras om den mycket säregna växelverkan, som råder mellan Europa och Förenta staterna i fråga om uppfinningar på elektricitetens område och deras tillgodogörande.

I de flesta fall ser grundidéen till en uppfinning dagen i Europa och genomgår här de första stadierna af sin utveckling. Så har förhållandet varit med dynamomaskinen, bågljuslampan, kommutatorn, transformatorn, elektromotorn, den elektriska järnvägen, glödlampan, telefonen och de första grundläggande ansatserna till den elektrokemiska industrien. Men medan den europeiske uppfinnaren har att kämpa en svår kamp med sin omgifnings oförmåga att värdesätta hans verk och förgäfves söker intressera kapitalet för sina nya skapelser, kastar sig amerikanen med oförvägen djärfhet öfver de från den gamla världen kommande idéerna, ger dem, utan att sky kostnaderna, de må vara huru stora som helst, form och gestalt, samt för ut dem i marknaden och tager dem till bruk i det praktiska lifvet med en hast, som väcker den betänksamma europêns förvåning. Detta ger i sin ordning en väckelse åt den gamla världen, som nu får brådt att göra sig till godo hvad som vunnit erkännande af de »praktiske» amerikanerna. Det var denna grundolikhet mellan den gamla och den nya världen.

Figure 3.18 Facsimile from Uppfinningarnas bok

"Mostly, the basic idea for an invention sees daylight in Europe and passes here the first development stages. This has been the case with the dynamo machine, the arc lamp, the commutator, the transformer, the electric motor, the electric railway, the incandescent lamp, the telephone and the basic impulses for the electrochemical industry. But, while the European inventor has to struggle hard against the inability of those around him to appreciate his work, and he tries in vain to interest financiers for his new inventions, the American jumps boldly on the ideas coming from the old world. Gives them form and stature, without sparing any expenses - no matter how large, and brings them into the market and takes them in practical use at a speed, which astonishes the hesitant European."

3.9 The 20th century

It is impossible to include a review of the electrical machine development during the last 100 years corresponding to the one above for the 19th century. That would almost require an encyclopaedia. Parts of this more recent development will instead be illustrated by the cases described and discussed in later chapters. Here will only some, very brief, notes be given for the convenience of the readers, who are less familiar with the subject.

The market and the number of applications for electrical machines have grown immensely during the 20th century. They are used in every sector of modern society, as described in chapter 1. The number of motors, especially small motors, has increased more rapidly during the last decades than ever before. Many types of motors can today be seen more or less as ordinary commodities. Others are highly specific and require more special designs.

The most common machine types are the same today as 100 years ago, e.g. the induction motors, but they have gradually undergone many steps of development. A new generator or motor for a certain power is much smaller and more efficient nowadays than a corresponding old machine. The costs are also between one and two orders of magnitude lower. This depends to a large extent on more rational production methods. Line production of smaller machines began in the 1920's and more and more automatic production methods have, since then, been successively introduced. Access to better materials, especially the introduction of synthetic insulation materials in the mid 1900's, which has led to higher allowable operating temperatures, has been one important factor for development. More sophisticated and efficient cooling methods have also contributed. The power semiconductors available from the 1950's and -60's have had a tremendous impact on electrical machine development. The new converter technology opened new possibilities for variable speed drives causing some older types of motors, used for such applications, to become obsolete. Another factor of great importance has been the introduction of the powerful rare-earth type magnets during the 1980's. They facilitated development of some radically different machine concepts. The use of computers as an engineering tool has of course had a big influence on the electrical machine development as well. This started in small scale in the 1950's but soon became widely used and indispensable. Modern calculation and simulation programs are a necessary prerequisite for most machine designs today.

What has happened to the electrical machine industry during the last century? There has been one group of large manufacturers and then all the others. Most of the large companies were already established in the late 19th century and have remained in that position until now. These companies are characterized by a very wide range of electrical products. Obvious examples are GE and Siemens, which were the largest electro-technical companies in America and Europe respectively before 1900, as they still are, even if they have extended their activities into other areas. This group of large companies, Asea/ ABB has been one of them, have become more and more multinational. They have usually been both commercial and technical leaders in their spheres of interest, e.g. large electrical machines. The structure of this industry remained fairly stable for a very long time, but a radical rationalisation has taken place during the last decades. The number of independent large manufacturers has been drastically reduced.

The small and medium-sized manufacturers constitute a large and very heterogeneous group. Most of them are niche companies with limited R&D resources, but some have been successful with a certain machine concept. Companies of this type have come and gone, been restructured or purchased during the course of the 20th century.

One important country, which has not been mentioned in this review, is Japan. The Japanese electro industry grew very powerful during the last century. It became independent of licenses from western manufacturers and it started to market its products worldwide. It is today, in the case of electrical machines and drive systems, on-par with European and American companies and in some areas even ahead.

The electrical machine industry is, since the 1990's, subject to globalization. One example of this can be closure of old factories and transfer of manufacturing to so-called low cost countries. Another common step is to out-source certain parts of the production to specialist suppliers, e.g. punching and stacking of iron cores, die casting or welding of stator frames etc. A third method is what is called "branding". A company stops the manufacturing of certain machine types or sizes and buys instead corresponding machines from countries like China and Korea, but puts its own nameplate and logotype on them. Is an old industry in turmoil, or .....?

4   Technical overview

The functioning of a rotating electrical machine is not easy to understand. It requires knowledge of a few, fairly abstract, physical phenomena but also some rather straightforward practical design principles. It is not the objective of this thesis to become a textbook on electrical machine theory and design. Several good books already exist on this subject and a few of them are included in the reference list; the first two are Swedish [69, 70, 71, 72]. Nevertheless, for the convenience of those readers, who lack sufficient electrical machine knowledge, a brief and popular technical introduction is given in this chapter. The experts can leave this chapter without consideration and go directly to the next one.

4.1   Basic principles of operation

An electric generator converts mechanical energy into electrical energy and an electrical motor vice versa. Usually, the same machine can be used both as a generator and as a motor. The function, in both cases, is based on the interaction between electrical currents and magnetic flux, which are in mutual relative movement. In the case of a generator, a mechanical force brings the rotor in rotation and the flux induces currents and thus electric power. In a motor, the flux and the currents produce mechanical forces, which sets the rotor in rotation.

4.1.1   Motor operation - torque generation

The operating principle of a motor is a practical application of Lorentz' equation concerning mechanical forces acting upon current carrying conductors in a magnetic flux. The generalized mathematical expression of this law is

$$\mathbf{dF} = \mathrm{d}L\mathbf{i} \times \mathbf{B} \qquad (4\text{-}1)$$

In an electrical machine with a constant flux density (B) perpendicular to the length (L) of a conductor, the equation 4-1 can be simplified to

$$F = LiB \qquad (4\text{-}2)$$

where   F   = mechanical force [N]
             L   = active length [m]
             i   = current [A]
             B   = flux density [T]

Figure 4.1 explains the basic function of an electrical motor. The force acting

---

upon each conductor is perpendicular to both the magnetic flux and the current. Its magnitude depends on these parameters as stated by equation 4-2 and the direction is also determined by these as shown in the figure.

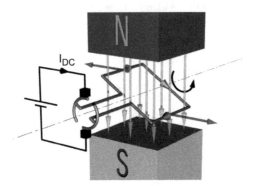

Figure 4.1 DC motor principle

The forces acting upon this simple loop result in a mechanical torque in accordance with equation 4-3, but only as long as the flux has a radial direction. It is obvious that a singular coil in an homogeneous flux, as the figure shows, will cause a pulsating torque.

$$T = DLiB \qquad\qquad (4\text{-}3)$$

where

T = torque [Nm]
D = diameter [m]

A real motor contains a large number of conductors arranged as coils usually placed in slots inside a magnetic core. It makes, of course, no difference in principle whether the conductors or the flux rotate. In the same way, it does not matter whether the rotating parts are placed inside the stationary parts or vice versa. Many motors have rotating flux and stationary coils as shown in figure 4.2.

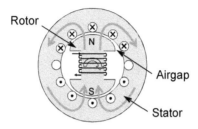

Rotor

Airgap

Stator

Figure 4.2 Motor cross section

Somewhat simplified, the torque can be determined as the sum of the tangential forces acting upon all the conductors multiplied by the airgap radius. The airgap is the small free space between the rotor and the stator as pointed out in figure 4.2.

It is convenient to define a physical entity called "linear current loading", A_S, which represents the current, evenly distributed along a unit length of the airgap circumference, and it can be expressed as

$$A_S = \frac{n_S Q I}{\pi D} \qquad (4\text{-}4)$$

where
- A_s = linear current loading [A/m]
- n_s = number of conductors per slot
- Q = total number of slots
- I = rms current [A]

The force acting on a small "unit area" of the airgap surface is, in accordance with equation 4-1;

$$dF = A_S B_\delta \qquad (4\text{-}5)$$

where
B_δ = magnetic flux density in the airgap [T]

The torque can be calculated after summation of the tangential forces on all "unit areas" constituting the airgap surface. The current through a specific winding is, of course, always the same in all series connected conductors while the flux density varies along the periphery. Assuming a sinusoidal flux distribution, the torque can be calculated using the mean value of the flux density. (These, somewhat simplified, equations are valid for DC machines, but the general conclusion is the same for an AC machine.)

$$T = \frac{D}{2} \cdot \pi DL \cdot \frac{A_S \hat{B}_\delta 2}{\pi} \tag{4-6}$$

$$T = D^2 L A_S \hat{B}_\delta \tag{4-7}$$

where $\qquad\qquad \hat{B}_\delta$ = peak value of airgap flux density [T]

<u>The torque in an electrical machine is thus proportional to the linear current loading, the magnetic flux density, and the volume, which is represented by the airgap diameter and the active length.</u>

4.1.2 Generator operation - voltage (emf) generation

The generator function can be explained with reference to Faraday's induction law, which states that a time dependent variation of a flux passing through a closed loop induces a voltage, or more correctly an electro-motoric force (emf), in the loop. Faraday's law is given in equation 4-8.

$$e = -\frac{d\Phi}{dt} \tag{4-8}$$

where $\qquad\qquad$ e $\;$ = electro motoric force, voltage [V]
$\qquad\qquad\qquad\quad$ Φ $\;$ = magnetic flux [Wb]
$\qquad\qquad\qquad\quad$ t $\;$ = time [s]

The left side of the above equation is the voltage induced in a single loop while the right is the time derivate of the flux through the loop. In case of a coil with several turns, the emf is

$$e = -N\frac{d\Phi}{dt} \tag{4-9}$$

where $\qquad\qquad$ N $\;$ = number of turns

The negative sign tells that the induced voltage will cause a current opposing the change of the flux. A simple sketch illustrating the generator function is shown in figure 4.3.

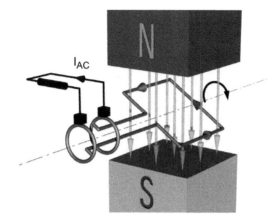

Figure 4.3 AC generator principle

It is convenient to assume a radial sinusoidal flux, which is moving (rotating) in tangential direction and passing through a coil with N turns. The width of the coil is one pole pitch, i.e. the distance along the airgap circumference from the centre of one pole to the centre of the next pole, and the length is the active length of the machine. See figure 4.4.

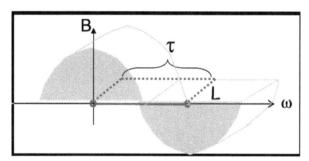

Figure 4.4 "Rotating" flux and stationary coil

The flux moves with a peripheral speed, v, corresponding to a mechanical angular speed ω_m. The flux variation through the coil depends on the number of poles, which means a full cycle for each pair of poles passing along the coil.

$$\omega_e = \frac{\omega_m p}{2} \qquad (4\text{-}10)$$

$$\omega_m = \frac{v}{R} = \frac{2v}{D} \tag{4-11}$$

where

ω_e = electrical angular speed [rad/s]
ω_m = mechanical angular speed [rad/s]
p   = number of poles
v   = peripheral speed [m/s]
R   = radius [m]

The generated emf is then, applying equation 4-9;

$$e = -N\frac{d\left(\frac{2}{\pi}\tau L\hat{B}_\delta \cos\omega_e t\right)}{dt} = \frac{2}{\pi}N\tau L\omega_e \hat{B}_\delta \sin\omega_e t \tag{4-12}$$

where

τ = pole pitch [m]

$$\tau = \frac{\pi D}{p} \tag{4-13}$$

The effective value (rms value) of the emf is then obtained by combining the equations 4-10, 4-11 and 4-12;

$$E = \frac{\sqrt{2}}{\pi} \cdot N\tau L\frac{2v}{D} \cdot \frac{p}{2}\hat{B}_\delta \tag{4-14}$$

$$v = \frac{\pi D n}{60} \tag{4-15}$$

Insertion of equation 4-13 and 4-15 in 4-14 gives

$$E = \frac{\pi\sqrt{2}}{60} \cdot nNDL\hat{B}_\delta \tag{4-16}$$

where

n   = rotational speed [rpm]

The generated emf (no-load voltage) voltage in an electrical machine is thus in principle proportional to its speed, the number of series connected turns in the winding, the airgap diameter, the active length and the flux density in the airgap.

4.2 Utilization

Usually, the development of electrical machines aims at more cost effective and less bulky machines. It is very important to design the machines so that the different materials are used as efficiently as possible. Much of the development efforts are therefore focused on solutions that improve the so-called "utilization" of the machines. This has actually been a main target for electrical machine engineers through more than a century of development.

The size of an electrical machine is usually given by its rated power. The rated power is, in the case of motors, the mechanical output power from the shaft and for generators it is often the so-called apparent output power (product of current and voltage) from the terminals. An expression for the power can be obtained as follows:

$$P = \omega_m T \qquad (4\text{-}17)$$

where $\quad$ P $\quad$ = active power [W]

$$\omega_m = \frac{2\pi n}{60} \qquad (4\text{-}18)$$

Combining equations 4-7, 4-17 and 4-18 gives

$$P = \frac{\pi}{30} n D^2 L A_S \hat{B}_\delta \qquad (4\text{-}19)$$

The power of an electric machine is thus basically proportional to its speed, the "airgap volume", the linear current loading and the flux density. It is possible to convert this and say that the power is proportional to the speed, the total armature current, integrated over the cross section, and the total magnetic flux. The expression above is a simplification in the sense that it assumes that current and flux are perpendicular to each other.

The equation 4-19 is used as base for defining a "utilization coefficient", K_u, also called Esson coefficient.

$$K_u = \frac{P_a}{n D^2 L} = \frac{\pi}{30} \cdot A_S \hat{B}_\delta \qquad (4\text{-}20)$$

where $\quad$ P_a = rated apparent power [VA]
$\qquad\quad$ K_u = factor [VA/(rpm m^3)] more often [kVA/(rpm m^3)]

It is, of course, only a matter of definition if the apparent or the active power shall be used, but the apparent power reflects better the electrical utilization of the machines.

The Esson coefficient indicates how efficiently an electrical machine is designed, but also which parameters can be used for improving the utilization. Those parameters are the linear current density and the airgap flux density. The latter is rather difficult to increase due to saturation effects in the ferromagnetic material used in the magnetic circuit. The flux density in the airgap is therefore limited to approximately 1 T. What remains is then the linear current loading, which has long been the main parameter used for increasing machine utiliszation. A higher linear current loading is obtained through deeper slots for the armature winding and/or a higher current density in the copper in this winding. The slot depth is related to the machine size and it is not so easy to play around with. This leaves the current density as the most important parameter, but an increased current density leads to higher losses and temperatures provided that the cooling is not improved. What actually has happened over the years is that new materials have been developed allowing higher temperatures and new, more efficient cooling systems have been introduced.

Typical current densities, linear current loadings and utilization factors vary quite a lot between different machine types and sizes. The diagram in figure 4.5 shows some representative examples on utilization factors. It is also common to use other key parameters for comparing the utilization of the machines. Most common are power density, kW/kg, and torque density, Nm/kg.

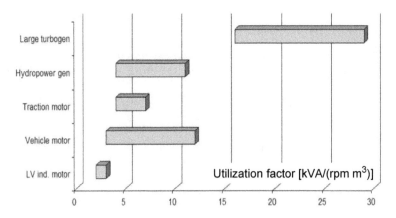

Figure 4.5 Comparison of utilization factors based on continous rating

4.3   Basic concepts and main components

4.3.1   Radial and axial flux

The topology of a rotating electrical machine can be arranged in different ways, but it is possible to identify a few main concepts. The most common is the radial flux layout, which means that the magnetic flux is crossing the airgap in the radial direction. There is also the axial flux design in which the flux passes through the airgap in a direction parallel to the shaft of the machine. Figure 4.6 illustrates the principal difference between these two concepts. A third solution is the so-called transversal flux design, but that will not be covered here. It can be studied in several other papers and publications, e.g. in a KTH licentiate thesis from 2001 [73].

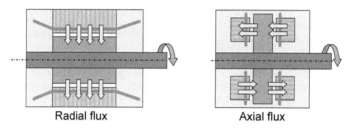

Radial flux   Axial flux

Figure 4.6 Basic layouts of radial flux and axial flux machines

Radial flux machines are by far the most common, and the rest of this chapter will therefore focus on them. Some reasons for this dominance are that the radial machines, having the same loading over the whole airgap surface, become in most cases compact, are suitable for both large and small machines, high and low speeds, and can operate with small airgaps. Axial flux machines are limited, due to mechanical reasons, but can be suitable when extra short axial length is required.

4.3.2   Pole arrangements

The machines can be built as inner pole or outer pole machines. The DC machines, which will be dealt with in section 4.8.1, have always been provided with rotating armature windings, while the poles carrying the excitation winding are located in the stator. This means that they are outer pole machines. AC machines have usually had inner poles although there are many exceptions. Figure 4.7 shows both these topologies.

Figure 4.7 Inner pole hydropower generator, 64.5 MVA, from 1964 and outer pole DC motor, 200 kW, manufactured around 1905

Another important difference is between salient pole machines and cylindrical machines as illustrated in figure 4.8. Some types of machines must be one or the other depending on their function. Others, e.g. synchronous machines, are built with cylindrical rotors for very low pole numbers and salient rotors for higher pole numbers.

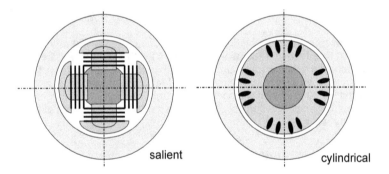

salient   cylindrical

Figure 4.8 Salient and cylindrical rotor machines

4.3.3   Inner and outer rotor

From a principal point of view it is possible to change position between the stator and rotor and build an outer rotor machine. This is not common, but it can be suitable for certain applications, especially in case of PM machines. Wheel motors in an electric vehicle are good examples. The main advantage is that the airgap diameter, and thus the torque, can be increased within the same outer dimensions. See figure 4.9.

Figure 4.9 Inner and outer rotor machines

4.3.4 Main components

It is common to talk about the active part of an electrical machine, which consists of the magnetic circuit and the windings, i.e. those parts which are conducting the magnetic flux and the electric currents. A three-phase winding in an AC machine is normally placed in slots in a laminated stator core. The core consists of teeth and back. Depending on the type of machine, there can be different rotor windings or none at all. Rotor windings in cylindrical rotors are located in slots, while salient pole rotors have a winding consisting of series connected coils wound around each pole core. The non-active parts are also very essential and they comprise mainly the structural parts such as stator frame (stator housing), end-shields, bearings, shaft etc. Figure 4.10 is a guide for identification of essential parts in common electrical machines.

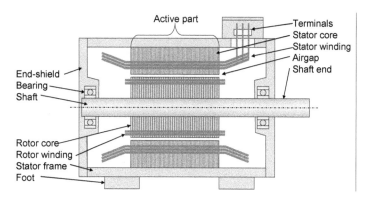

Figure 4.10 Main parts of a typical electrical machine

4.4 Windings

A 3-phase winding, usually in the stator, can be made in a lot of different ways. Many of them can be rather complicated. Nevertheless, the scope of this text is not to go into such details, but to give a simple description and to introduce certain basic terminology. It is at this stage sufficient to discuss 3-phase slot windings.

Each phase consists of a number of coils and each coil can comprise one or more turns. They can all be connected in series or in a few parallel groups depending on the specified voltage. The coils belonging to the different phases are equally distributed in slots around the airgap periphery. Each coil has two coil sides located at a certain distance from each other, different for different types of winding. There can be one or two coil sides per slot depending on whether it is a single-layer or double-layer winding. Each coil can also be divided into the active slot portion and the non-active coil ends. The coil ends can be arranged in various ways for connecting the individual coils to each other. The phases have to cross each other, resulting in end regions that will look like a basket of coils. The most common winding configuration is the so-called distributed winding, which can be concentric or lap type. The latter is illustrated in the winding diagram figure 4.11.

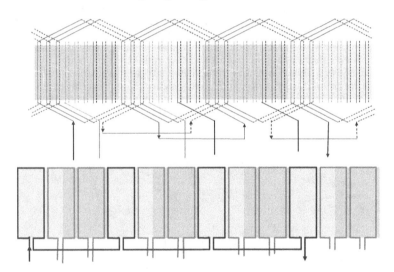

*Figure 4.11 Winding diagrams: distributed 4-pole lap winding (upper)
and concentrated 8-pole winding (lower)*

Figure 4.12 a) End windings in a large 4-pole motor and b) PM motor with concentraded winding

A different type of winding, also shown in the figure, is the concentrated winding in which each coil surounds only one tooth. Figure 4.12a shows a photo of the end windings in a large machine, while 4.12b contains a photo of a motor with concentric winding. Smaller machines usually have single-layer windings placed in semi-closed slots in which each conductor is made up of a number of thin, circular copper strands. These windings are known as mush windings or random windings, because the individual strands can come in any arbitrary position when a coil is inserted through the narrow slot opening. Large machines, on the other hand, have nowadays double-layer windings in open slots. Each coil has a rectangular cross-section and they are pre-formed as, so-called, diamond coils. The conductor also consists of a number of strands, but now of rectangular shape. It was earlier common to use single-layer concentric windings in semi-closed slots for large machines. Those coils had rectangular sections and were pre-formed as hairpins, which were pushed axially into the slots. They were abandoned primarily due to difficulties to rationalize the production process. Figure 4.13 shows cross-sections of these common types of slots including slot and strand insulation.

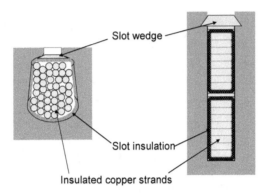

Figure 4.13 Semi-closed slot with mush winding and open slot with pre-formed coil sides

The reason why conductors are divided in parallel strands is to avoid eddy current losses due to skin effect and proximity effects, which occur in conductors placed in an alternating magnetic flux. The higher the frequency, the smaller the individual strands must be. Another phenomenon, which can cause extra losses in the winding, is currents circulating through parallel strands depending on differences in voltage. This can be avoided by transposition of the strands, i.e. letting series connected strands take different positions in the slots they pass through, so that all parallel circuits become equal. This is done systematically in larger machines, while for smaller machines it is sufficient to let this be done randomly. Figure 4.14 gives a simple example on a controlled transposition. (Another method is illustrated by figure 7.14.)

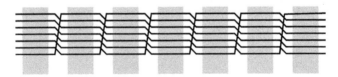

Figure 4.14 Transposition of parallel strands. The strands shift position after having passed axially through the stator core a number of turns.

4.5 Important materials

The development of electrical machines has been very dependent on materials available. Essential are of course the "active" materials such as electrical conductors and the ferromagnetic materials used for the magnetic circuit. The development of insulation materials has also had a great impact on electrical machine development. Various structural materials for taking mechanical forces, constituting enclosures or interfaces to other equipment, have also influenced machine development.

4.5.1 Conductors

Copper is by far the most important electrical conductor used in electrical machine windings. Aluminum is also common, but more or less only for rotor windings in induction machines. There have been attempts to replace Cu with Al in stator windings, but without major success. Bronzes and steel are used in certain details, e.g. sliprings, while carbon is the base material for brushes. Low-temperature super conductors of niobium-titanium have been used in several experimental machines since the 1960's, but never obtained any practical use in electrical machines. The high-temperature super conductors, which can be cooled by liquid nitrogen (77 K = - 196 $^{\circ}$C) instead of liquid helium (4 K = - 269 $^{\circ}$C), were discovered in 1986 and have recently been used in some prototype machines [73]. The global production of copper, for the electrical industry, was 11 million tons in 2005, out of which approximately 2.4 million tons was used for motors and generators. The price for copper has increased much during the last years and is subject to large fluctuations. For instance, the copper price was 1360 USD/ton in March 1999 while it was 8700 USD/ton in April 2006. Such variations will, of course, complicate optimization of electrical machines and similar products. In spite of the increased copper price, the Swedish company Elektrokoppar, which produces both copper and aluminum wire, claims that copper will remain dominant in electrical machines also in the future.

4.5.2 Magnetic materials

There are basically two types of magnetic material used in electrical machines, soft and hard. The soft materials constitute the stator and rotor cores and they are chiefly various grades of so-called "electrical steel sheet". However, solid steel can also be used where there is no alternating flux. Soft magnetic composites, consisting of sintered or pressed, insulated iron powder, have also become common for small machines operating with high frequencies. The hard materials are the permanent magnets.

4.5.2.1 Electrical steel

The essential requirement on the electrical steel sheet is that it shall be an efficient conductor for the magnetic flux, i.e. it shall have a high permeability, and it shall have low losses when subjected to alternating flux. It should also be easy to punch. Rolled steel sheet has been available from the infancy of the electrical machines, but development of special grades suitable for this application began in the 1910's. Characteristic of this kind of sheet is that it is alloyed with silicon, which increases its resistivity. A higher content of silicon reduces the specific losses, but the sheet becomes at the same time more brittle and more difficult to punch. The sheets are also provided with some heat resistant surface insulation, which prevents short-circuits between the stacked sheets. The type of steel sheet used in modern electrical machines is cold rolled and usually non-oriented and the thickness is normally 0.35, 0.5 or 0.65 mm. Thinner sheets means lower losses but also higher costs. The Swedish rolling mill Surahammars Bruk has recently developed its process and is now able to produce 0.2 and even 0.1 mm cold rolled sheets. The reason for this development is that many modern electrical machines operate at such high frequencies that it can be advantageous to use very thin lamination in spite of the high cost. Figure 4.15 presents magnetization curves for high quality electrical steel used for larger generators in the 1940´s and 1990´s respectively. A magnetization curve shows the relation between flux density and magnetic field strength. Table 4.1 gives some other data for the same sheets. It is obvious that the improvements are rather moderate except for the costs [74, 75]. Surahammars Bruk has recently reported on development of lean grades of non-oriented electrical steel which means that a low loss level can be obtained with steels that are less alloyed. This has been achieved through improvements of the manufacturing process [76].The diagram in figure 4.16 illustrates that there has been a rapid development during the last years. It is interesting to note that Swedish steel was once considered state of the art in the field of electrical machines. The British professor John Hopkinson wrote in 1886 in his pioneering report on Dynamo-Electric Machinery, published by the Royal Society, about some core materials "… *[which] have a magnetic permeability but little inferior to the best Swedish charcoal iron"* [77]. Asea considered the supply of electrical steel so critical that it acquired its own steel work, the several hundred years old Surahammars Bruk, in 1916. It remained a part of the Asea Group for 70 years.

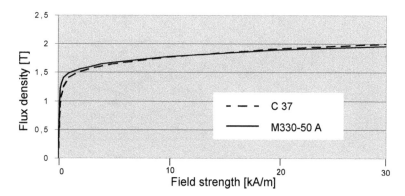

Figure 4.15 Magnetization curves for electrical steel

Table 4.1   Data for high grade electrical steel

Grade	Year	Silicon [%]	Thickness [mm]	P_{10} [W/kg] Typical / guarantee	P_{15} [W/kg] Typical / guarantee
C 37*	1950	3.5	0.5	1.50 1.65	3.30 3.60
Ck 33**	1970	2.8	0.5	1.25 1.45	2.90 3.30
M330-50 A**	2005	2.8?	0.5	1.29 1.35	3.03 3.30
M270-35 A**	2005	2.8?	0.35	1.01 1.10	2.47 2.70

*. Hot rolled
**. Cold rolled

The world production of electrical steel was, in 2005, close to 11 million tons, which corresponds to one percent of the total steel production. Seven of the top ten producers were companies located in the Far East [78]. Most of the electrical steel is used for rotating electrical machines according to information from Surahammars Bruk.

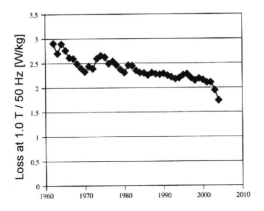

Figure 4.16 Recent reductions of electrical steel losses

There has been, during later years, a development of so-called amorphous steel, which has radically different properties; very low losses but also low saturation level [79]. So far, these materials have had no practical impact on normal electrical machine manufacturing and are therefore left without further comments.

4.5.2.2 Soft magnetic composites

Soft magnetic composite (SMC) or iron powder is another modern material which has one distinctive advantage over the electrical sheets, namely that it can be used even in case of three-dimensional flux. It consists of small, insulated iron particles so that the material becomes fairly isotropic, somewhat depending on the pressing method. This property facilitates development of quite unconventional machine topologies. The permeability is, however, significantly lower than for normal laminated cores and the losses are higher, at least for normal frequencies. The use of this type of material is hence limited to small machines working at high frequency. A couple of interesting examples are machines developed in two recent PhD projects, one at KTH and one at LTH [80, 81]. The development of soft magnetic composites started in the 1980's and the leading manufacturer is the Swedish company Höganäs AB [82].

4.5.2.3 Permanent magnets

Permanent magnets have been used since the very beginning of electrical machine development. A permanent magnet has a very different magnetization curve in relation to soft magnetic materials. It is characterized by a high remanent flux density and a large coercivity force as shown in figure 4.17.

The former is the flux density that remains when there is no external magnet-ization left and the latter is the negative magnetization (field strength) required to reduce the flux density to zero. A permanent magnet has its working point in the second quarter of the magnetization curve, preferably where the product of flux density and field strength has its maximum. This represents the energy density of the magnet and can be considered as a measure of how powerful the magnet material is.

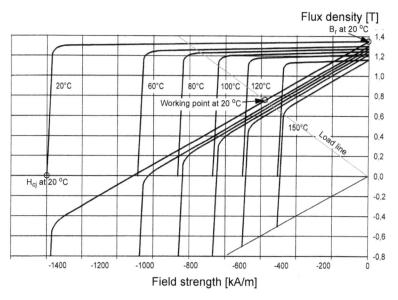

Figure 4.17 Magnetization curves for NdFeB PM material

The first permanent magnets were from magnetite, a mineral found in nature, and they were used for magnetizing iron compass needles. The permanent magnets used for electrical machines were initially steel magnets, which had been magnetized by means of current carrying coils. These steel magnets were weak and it was therefore necessary to make them large in order to obtain a sufficient flux. The AlNiCo magnets developed in the 1930's represented a big step in remanence flux density, but their coercitivity force was small so they have not been used much for electrical machines. Cheap ferrite magnets became available in the 1940's and have found a lot of applications from loudspeakers to refrigerator stickers. Ferrite magnets are used for creating the magnetic flux in billions of small motors, but they are not powerful enough for the sizes dealt with in this study.

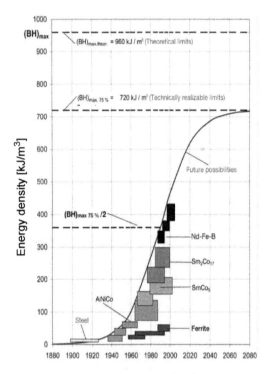

Figure 4.18 PM material development

The introduction of so-called "rare earth metal magnets" during the 1970's and 80's opened the possibilities for development of new types of electrical motors. First came the samarium-cobalt (SmCo) magnets and later those from neodymium-iron-boron (NdFeB). These types of magnets are either pressed or sintered and machined to final shape and dimensions. They are very powerful, but have also some disadvantages. One drawback is that they are mechanically brittle, another is their corrosiveness, which necessitates some sort of surface coating. NdFeB magnets are especially temperature sensitive and are irreversibly demagnetized above certain temperatures. These types of magnets are quite expensive compared with ferrite magnets, particularly those of SmCo. Even if Nd and Sm belong to the rare earth metals, the availability is not critical. China and, to some extent, Australia have ample resources and are major producers of the bulk material. The costs have decreased over the years but no further drastic reductions are expected. Figure 4.18 illustrates the development of PM materials and table 4.2 compares some of their key parameters [83, 84].

Table 4.2   Key data for PM materials

PM material	B_r [T]	H_{cj} [kA/m]	BH_{max} [kJ/m³]	Max temp [°C]	Material cost [%]
Ferrite	0.4	250	40 - 60	250	10
AlNiCo	1.2	75	50 - 75	500	50
SmCo	0.9 - 1.1	600 - 2000	150 - 250	300	150 - 200
NdFeB	1.1 - 1.4	1000 - 2500	250 - 400	100 - 200	100

4.5.3   Insulation materials

Insulation material is used to separate parts of the machine with different electric potential and to prevent harmful currents to flow. Ideally the insulation should be infinitesimally thin because it steals space from the active copper and iron and it constitutes an obstacle for the heat dissipation, especially from the windings. Most critical, in both respects, is the insulation inside the slots. Real insulations, however, require a certain space as the examples in figure 4.13 in section 4.4 show. The insulation is subject to combinations of electrical, thermal, mechanical and sometimes even chemical stresses. These can cause a long term deterioration of the material or, if very high, an instant breakdown. The insulation is usually the most temperature sensitive material in an electrical machine and determines the maximum allowable operation temperature. The development of more durable insulation materials has therefore been instrumental for the development of more compact, reliable and cost effective machines.

Early insulations were made from natural materials such as cotton, silk, rubber and cellulose fibre. Mica was introduced very early for high voltage slot insulations and it remains an excellent choice for this application. Shellac, a resin obtained from tropical trees with help from a lice, was used as an impregnation varnish for a long period, from the early 1900's. Enamel came into use as insulation on thin conductors in the 1920's. High voltage slot insulation, made from asphalt bitumen impregnated cotton ribbons coated with mica splittings, was developed during the same period.

A big step was taken by the introduction of synthetic insulation materials during the post World War II era. Glass fibre is an example used both for spinning around thin conductors and as tape with or without mica. Polyester and later epoxy resins replaced the old impregnation varnishes. Other new materials were polyamide and polyurethane used for different types of tapes and varnishes. Well known trade names are Nomex and Kapton. The insulation materials are divided into a number of temperature classes according to table 4.3 [85]. The diagram in figure 4.19 shows the development of dielectric strength of some common high-voltage insulation materials. The use of certain modern insulation materials is complicated by the fact that they can be hazardous for both workers and the environment.

Table 4.3   Temperature classification for electrical machine insulation systems, IEC Publication 85

Designation	Temp. limit [°C]	Examples of materials
Y	90	Non-impregnated textile, paper, wood, rubber
A	105	Shellack impregnated textile, paper, polyamide
E	120	Asphalt impregnated textile, paper, polyethylene
B	130	Asphalt or polyester impregnated mica, glass
F	155	Epoxy impregnated mica, glass, polyurethane
H	180	Silicon impregnated mica, glass
C	> 180	Silicon impregnated mica, glass, ceramics

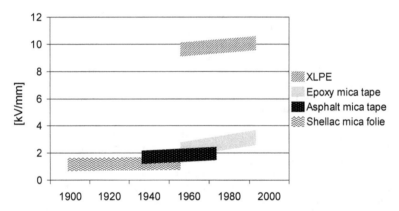

Figure 4.19 Allowable electric stress on HV insulation materials

Supply of insulation material was strategic for leading electrical machine manufacturers and could even be seen as a competitive advantage. Asea decided therefore in 1915, when World War I made the import of such material difficult, to build a factory for insulation materials. The company's central laboratory was established more or less at the same time and R&D concerning insulation materials has remained one of the laboratory's prioritized areas since then [86]. Today, these types of materials are acquired from specialized external suppliers.

4.5.4   Structural materials

An electrical machine can not be built and operate without some structural components that complement the active parts. Examples of such components are stator frames, end-shields, shafts, bandages and retaining rings, various supports and covers. Important functions of these components are to provide necessary mechanical strength and stiffness or to transfer mechanical forces. Another function can be to constitute some sort of enclosure or protection. It is obvious that the active parts of a machine are also used in these respects, e.g. the rotor core must transfer the torque to or from the shaft and withstand the centrifugal stresses.

The requirements of the structural materials can be very different from component to component. High mechanical strength is often, but not always, required. It is in some cases necessary to use nonmagnetic and even electrically insulating components. Other properties can be related to the production process, e.g. weldability. The most common materials for the non-active parts of electrical machines are various types of steel and aluminum. Plastic materials, both thermoplastics and laminates, are also frequently used. Composites containing glass or carbon fibres are other examples, which have good or excellent mechanical strength respectively.

4.6   Losses

Electrical machines, especially the larger ones, usually have very high efficiencies. The losses, nevertheless, represent a considerable waste of energy and are sources for temperature rises, which limit the performance of the machines. It is therefore a common development target to reduce the losses and, hence, increase the efficiency. Equation 4-21 gives the basic relations.

$$\eta = \frac{P_{output}}{P_{input}} = \frac{P_{input} - P_{loss}}{P_{input}} \qquad (4\text{-}21)$$

where η = efficiency
P_{output} = output power [W]
P_{input} = input power [W]
P_{loss} = total loss [W]

The total loss is the sum of a number of different losses described below.

4.6.1   Mechanical losses

There are, in principle, two types of mechanical losses in electrical machines; friction losses and "pumping losses", i.e. the power required to circulate a coolant or lubricant through a machine. The friction losses consist of bearing losses, air friction losses and in some cases brush friction losses. The bearing losses depend on the type of bearings, whether rolling type bearings or sleeve bearings, but also on the load, the speed and the viscosity of the lubricant. Equations 4-22 and 4-23 below indicate the influence of these main parameters, but further details can be found in numerous engineering handbooks [87, 88].

Losses in a rolling bearing:

$$P_b = \mu F v \qquad\qquad (4\text{-}22)$$

where
μ = friction coefficient
F = bearing load [N]
v = peripheral speed of inner ring [m/s]

Losses in a sleeve bearing:

$$P_b = 0,5d^2 l\sqrt{p\eta n^3} \qquad\qquad (4\text{-}23)$$

where
d = shaft diameter [m]
l = bearing lengh [m]
p = specific bearing load [N/m^2]
η = dynamic viscosity [Ns/m m^2]

When a rotor rotates in air or an other medium, there will be certain losses due to the friction between the rotor and the air. It is obvious that a smooth cylindrical rotor will cause much lower air friction losses than a rough, salient pole rotor. The size and speed of the rotor also have a big impact on the air friction losses. Equation 4-24 shows the dependence in the case of a simple cylindrical rotor. More information can be obtained in special literature. [88, 89]

$$P_{af} = kAv^3 \qquad\qquad (4\text{-}24)$$

where
k = air friction coefficient [Ws^3/m^5]
A = cylinder area [m^2]
v = peripheral speed [m/s]

Brush friction losses only occur in machines with sliprings or commutators. These losses are more or less proportional to the peripheral speed, total brush area, brush pressure and the friction coefficient [88].

The friction losses are rather limited in most machines but can have a significant impact in larger machines and especially in high speed machines.

4.6.2　Iron losses

Iron losses are, by definition, concentrated to the ferromagnetic components in the active part of the electrical machine. These losses are caused by alternating or rotating magnetic fluxes and can be split in two different kinds of losses, hysteresis and eddy current losses respectively. Both of them require attention.

4.6.2.1 Hysteris losses

A ferromagnetic material consists of very small, so-called domains, which can be considered as individual magnets (magnetic di-poles) depending on a uniform direction of the electron-spin in the atoms constituting a domain. When subject to an external magnetic flux, the domains turn in the direction of this flux. If it is an alternating flux, the domains keep turning back and forth which requires a certain amount of work, expressed as the hysteris losses. This phenomenon also explains the saturation effect, because eventually there are no more domains to turn. The hysteris losses depend on the material itself, the frequency of the flux and the flux density. Equation 4-25 is an approximate expression of the specific hysteris loss.

$$p_h = k_h f B^2 \qquad (4\text{-}25)$$

where
$\quad p_h$ = specific hysteris loss　[W/kg]
$\quad k_h$ = material constant
$\quad f$　 = frequency [Hz]

4.6.2.2 Eddy current losses

The cause of the eddy current losses can be traced back to Faradays' induction law (equation 4-8). If an alternating flux is passing through a piece of metal, it will induce an emf and hence a current, analogous to a closed loop. The current density becomes higher close to the surface as shown in figure 4.20. These types of induced currents are called eddy currents and are quite prohibitive for the use of solid iron in case of alternating fluxes. It is therefore necessary to use laminated cores with insulation between the sheets, which prevents the circulation of large eddy currents.

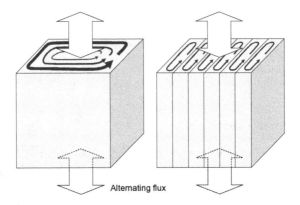

Figure 4.20 Eddy currents through a solid core and in a laminated core

The eddy current losses in a core depend on the frequency, flux density, thickness of the sheets and resistivety of the material approximately as indicated in equation 4-26.

$$p_e = \frac{k_e f^2 B^2 t^2}{\rho} \qquad (4\text{-}26)$$

where

p_e = specific eddy current loss [W/kg]
k_e = constant
t = sheet thickness [mm]
ρ = resistivity [Ωm/mm²]

Iron losses in a certain part, e.g. the back of the core, are obtained by multiplying the specific losses ($p_h + p_e$) by the weight of that part yielding the total iron losses to be the sum of the losses in relevant parts, usually the teeth and the core back. It should be noticed that iron losses are influenced by manufacturing related factors such as quality of the insulation varnish, sharpness of the punching tool, residual stresses etc. This is taken into account by the partly empirical constants k_h and k_e.

4.6.3   Resistive losses

The resistive losses are often referred to as ohmic losses or copper losses, which can be misleading in case of an aluminum winding. Nevertheless, these losses are quite simply those caused by the winding resistance and the current passing the winding. The relation shown in equation 4-27 is well known. It is common to use the DC resistance for calculating these losses and refer any increase, depending on AC currents, to stray losses (see 4.6.4). It is also important to consider the influence of temperature on the resistance. This is quite significant in the case of copper and aluminum.

$$P_{res} = RI^2 \tag{4-27}$$

where P_{res} = resistive loss [W]
                    R = resistance [Ω]

4.6.4 Stray losses

Stray losses are a complicated matter and difficult to predict with high accuracy. Most of them are caused by eddy currents in various parts of an electrical machine. They can be attributed to skin effects and proximity effects in AC windings as well as circulating currents due to leakage fluxes. The iron losses described above are usually referred to no-load conditions while extra iron losses caused by leakage fluxes or harmonics are considered as stray losses. Alternating leakage fluxes in non-active components as stator frames and end shields will also induce eddy currents and consequently cause stray losses. The same can occur in active components, which in principle only should carry a direct flux, e.g. the rotor surface in a synchronous machine. The main source for most of these leakage fluxes are the armature current and stray losses (P_{stray}) are therefore highly dependent on the load but also on the frequency as indicated in equation 4-28.

$$P_{stray} \sim f^2 I^2 \tag{4-28}$$

Design measures that can reduce leakage fluxes, eddy currents and harmonics are important in order to avoid large stray losses.

4.6.5 Total losses

The total loss in an electrical machine is the sum of those described above, i.e. the mechanical, iron, resistive and stray losses. It is common, at least for machines working at a fixed speed, to group them in no-load and load losses respectively. The no-load losses can be measured when the machine runs at nominal speed and voltage and without any load. They consist mainly of mechanical and iron losses. The load losses are thus resistive plus stray losses, which both are proportional to the square of the armature current. This is, however, not completely stringent for all those types of machines that need an excitation current or no-load current for creating the no-load flux. It is nevertheless a common and frequently used approach.

The losses and the efficiency of a machine are usually calculated at rated load. It is obvious that the no-load losses reduce the efficiency for low loads.

One example of an efficiency curve for a normal machine is given in figure 4.21 a. The efficiency is usually higher for larger machines as illustrated by figure 4.21 b, which shows typical efficiencies at rated load for some common machine types. The relationship between the various losses in a large synchronous generator and a standard induction motor is presented in the diagrams in figure 4.22.

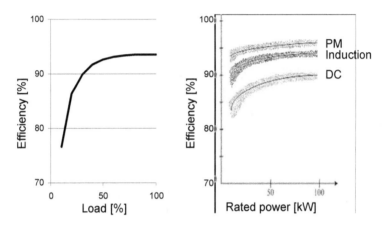

Figure 4.21 a) Efficiency vs. load for common motor. b) Typical efficiencies at rated load for common motor types

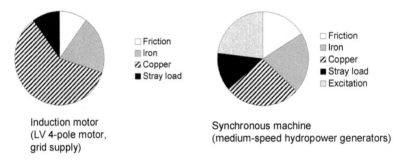

Induction motor
(LV 4-pole motor,
grid supply)

Synchronous machine
(medium-speed hydropower generators)

Figure 4.22 Examples of loss distribution in synchronous generator and induction motor

Figure 4.23 contains a diagram, which is a good illustration of the power flow and losses in an induction motor. It shall be observed that the resistive losses are divided into two parts, one for the stator winding and one for the rotor winding. Corresponding diagrams can be drawn also for other machine types.

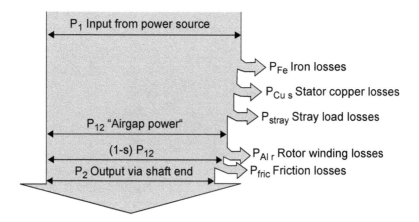

Figure 4.23 Diagram showing the power flow in an induction motor

4.7 Cooling

The losses in an electrical machine produce heat that can become harmful if not dissipated. Too high temperatures can lead to a rapid ageing or even immediate breakdown of various materials, but can also result in high mechanical stresses due to differential thermal expansion. It is therefore necessary to provide the machines with adequate cooling, at least for machines of sizes covered in this thesis. The larger the machines, the more efficient the cooling must be. This is illustrated in figure 4.24, which shows that the heat dissipation per area unit to the ambient air increases for a larger machine, assuming the same loss density per unit volume. This is why large machines have more sophisticated cooling systems than smaller ones.

Heat can be dissipated through thermal conduction, convection and radiation, or most commonly through a combination of these, along with the transportation of a cooling medium. Radiation does not contribute much to the cooling process, while conduction and convection play highly important roles. Conduction means that the heat is passing through a piece of material. The temperature drop depends on its dimensions and on the thermal conductivity of the material. Metals are good heat conductors while electrical insulation materials have poor thermal conductivity. It is therefore difficult to transport the resistive losses from a winding through its insulation to surrounding parts. Convection is the transfer of heat from a solid material to a gaseous or liquid cooling medium. In this case, the temperature drop is for a certain heat flow depending on the available cooling surface and the convection coefficient. The latter can vary quite a lot and it is influenced by the type of coolant, its speed and whether the flow is turbulent or laminar. Finally, the heat trans-

---

port capacity, by means of a circulating coolant, is determined by the mass flow rate and the heat capacity of the coolant.

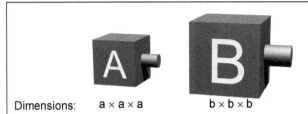

Dimensions: $a \times a \times a$ $b \times b \times b$

Assume same specific load, i.e. kW/m^3 (or kW/kg) and same efficiency. The relation between the losses is

$$\frac{P_{lossA}}{P_{lossB}} = \frac{P_A}{P_B} = \frac{V_A}{V_B} = (a/b)^3$$

The relation between cooling surfaces is

$$\frac{A_{coolA}}{A_{coolB}} = (a/b)^2$$

The relation between temperature rises assuming same coolog method is then

$$\frac{\Delta t_A}{\Delta t_B} = \frac{P_{lossA}}{A_{coolA}} \times \frac{A_{coolB}}{P_{lossB}} = a/b \qquad \text{therefore,}$$

Larger machines require more effective cooling systems!

Figure 4.24 Cooling of small and large machine

The equations 4-29, 4-30 and 4-31 below can be used for calculation of temperature drops for conduction, convection and coolant flow respectively.

$$\Delta T_{cond} = W \frac{t}{A\lambda} \qquad\qquad (4\text{-}29)$$

$$\Delta T_{conv} = \frac{W}{A_{conv}\alpha_{conv}} \qquad\qquad (4\text{-}30)$$

$$\Delta T_{flow} = \frac{W}{Qc_{th}} \qquad (4\text{-}31)$$

where

ΔT	= temperature drop [K]	
W	= heat flow [J/s]	
t	= thickness (length) of the conducting piece [m]	
A	= area of the conducting piece [m^2]	
λ	= thermal conductivity [W/m K]	
A_{conv}	= cooling surface [m^2]	
α_{conv}	= convection coefficient [W/m^2 K]	
Q	= coolant flow rate [kg/s]	
c_{th}	= thermal capacity [J/kg K]	

The cooling of a machine can be arranged in many different ways. The early machines were open and the losses were dissipated to the ambient air without any particular means. The next step was to introduce forced cooling, i.e. let a shaft driven or separate fan blow ambient air through the machine. Hazardous environments required closed machines which initially were cooled by natural convection to the ambient air. This was not very effective and these closed machines were therefore provided with shaft mounted fans blowing air along cooling fins on the outside of the stators. Even more effective systems were introduced with forced inner cooling and built-on heat exchangers in which the circulating air could transfer the dissipated heat to a secondary coolant, usually water or air. Figure 4.25 gives a few examples on cooling arrangements.

It has been mentioned that the insulation is a barrier for the heat transfer and in order to avoid this, concepts have been developed where the coolant is flowing inside the insulation, in direct contact with the conductors. Such solutions are expensive and are only used on very large machines where it is necessary. This is usually referred to as "direct cooling" while most other machines use "indirect cooling" for their armature windings. Figure 4.26 shows the principal difference between direct and indirect cooling of some common windings. The most efficient and common coolant for directly cooled windings is de-ionized water, but oil and gases can also be used.

It is possible to consider different media as coolants in electrical machines. Most common is, of course, air. Water is often used as a secondary coolant in closed machines but, as mentioned above, also as a primary coolant in directly cooled windings. Hydrogen, at a certain overpressure, is an efficient coolant used in large turbogenerators and synchronous condensers, but this requires that such a machine is hermetically closed. Oil is used both as a lubricant and a primary coolant for sleeve bearings but it can also serve as an alternative coolant in some other cases. Table 4.4 contains a comparison between some different cooling media and shows that water is the most efficient of them.

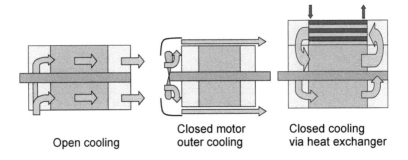

Open cooling   Closed motor outer cooling   Closed cooling via heat exchanger

Figure 4.25 Basic cooling concepts

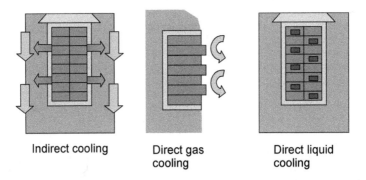

Indirect cooling   Direct gas cooling   Direct liquid cooling

Figure 4.26 Indirect and direct cooling of windings

Examples of machines using special cooling principles such as heat pipes etc. exist, but such machines are not dealt with in this study.

Table 4.4   Comparison of properties for cooling media

Fluid	Spec. heat	Density	Vol. flow	Heat abs.
Air	1.0	1.0	1.0	1.0
Helium	5.25	0.138	1.0	0.75
H2, 1 bar	14.35	0.07	1.0	1.0
H2, 4 bar	14.35	0.28	1.0	4.0
Oil (Transil)	2.09	848	0.012	21
Water	4.16	1000	0.012	50

4.8 Overview of electric machine types

Induction motors, salient pole generators, traction motors, compressor motors, reluctance machines, universal motors, outer pole machines, PM motors, high speed generators, single phase motors, hydropower generators, transversal flux machines, wheel motors, shunt motors, closed machines, class F motors etc. etc. The list of electrical machine types could cover a page or more. Isn't there some sort of systematization? No, not really. Even if some official classification of certain types of machines exists, the nomenclature is not always clear. A specific machine can be identified with different terms, e.g. pump motor, induction motor, flange motor, squirrel cage motor, closed motor, low voltage motor, aluminum motor. There are obviously many ways to group electrical machines. Common ways are according to application, topology and mechanical design but the most important grouping is with respect to operating principle. Application is self-evident and topology has been dealt with in section 4.3 above. This short overview will therefore focus on function and some important characteristics. A diagram in the form of a family tree is presented in figure 4.27 as an attempt to provide a simple overview.

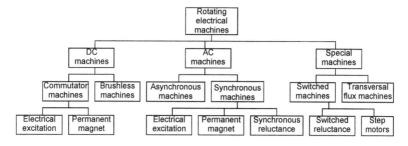

Figure 4.27 Electrical machine "family tree"

A brief explanation of the function, some characteristics, advantages and disadvantages of the most important electrical machines are presented in the next pages. Further details and other machine types are included and explained in later chapters when necessary.

4.8.1 DC machines

The classical DC machines are outer pole machines with an armature winding in the rotor connected to a commutator. The armature current is led to or from the rotor through brushes. The currents in the individual rotor coils are AC currents and the main flux becomes an AC flux inside the rotor. The rotor core must therefore be laminated. The current through the armature winding creates a flux, which is perpendicular to the main flux. This phenomenon is called armature reaction and it also exists in other types of machines. The resulting flux will have a slightly different direction and the brushes are not any longer in the neutral position; see figure 4.28 a. This could cause harmful sparking and that is why the brushes were turned a certain angle on old DC machines. The problem was that the armature reaction is proportional to the load and the brushes must be adjusted accordingly. The only way to do this on very old dynamo machines was to have one man allocated just for this job [90]. The commutation theory was developed during the 1890's by professor E. Arnold in Karlsruhe and others. The real improvement came with the introduction of the commutation poles around 1900 [91]. The armature current passes through the commutation winding in such a direction that it is counteracting the armature reaction at all loads allowing the brushes to remain in the neutral position as shown in figure 4.28 b.

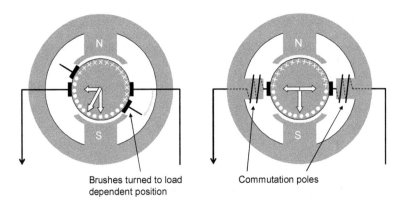

Brushes turned to load dependent position

Commutation poles

Figure 4.28 a and b   Cross sections of DC motor without and with commutation poles

DC machines have traditionally been named after how their excitation has been arranged. The excitation winding could be in parallel or series with the armature winding or both. The machines are then referred to as shunt, series and compound machines respectively. The excitation current was controlled with resistors and each of these machines had somewhat different characteristics. Series motors had a large starting torque and were therefore used for traction applications. A fourth type of DC machine, the separately excited, has become dominant since the introduction of power electronic rectifiers. All four types are schematically shown in figure 4.29. Small DC machines can also be excited by permanent magnets.

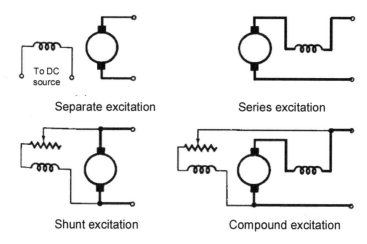

Separate excitation   Series excitation

Shunt excitation   Compound excitation

Figure 4.29 DC-machine excitation principles

DC motors have been the most common choice for variable speed drives for at least ¾ of the previous century. Their main advantages are excellent speed controllability and good overload capacity. Disadvantages are that they have several windings and a commutator making them comparatively large, expensive and less efficient than most AC machines. The speed range is limited due to commutation. The commutator and the brush-gear require more maintenance than other machines usually do. Their share of the market has decreased over the years, but the development is not over. ABB in Västerås introduced a new DC motor generation only a few years ago. A sketch of such a motor is shown in figure 4.30.

Figure 4.30 Modern DC motor from ABB

4.8.2 Synchronous machines

Synchronous machines are in some respects similar to DC machines. Both are, for instance, magnetized by a DC flux. If the commutator was replaced with sliprings, a DC machine could, in principle, be converted into a synchronous machine. A practical difference is, however, that the latter normally are inner pole machines while the former are outer pole machines.

A synchronous machine usually has a 3-phase winding in the stator. As shown in the simple 2-pole machine in figure 4.31, the phase windings are geometrically displaced at 120°. When the 3-phase currents pass through their respective windings, each of them create an mmf, which together result in a rotating mmf and flux with constant amplitude. Figure 4.31 attempts to give a visual explanation of this.

The rotor poles are attracted by the rotating flux causing the rotor to rotate synchronously with this, hence the nomenclature synchronous motor. The angle between the poles and the stator flux will increase if the mechanical load is increased. If it gets too large, the motor will fall out of synchronism. The machine can also be operated as a generator, in which case the flux from the rotor poles induces emfs and currents in the 3-phase winding. The rotor is then rotating ahead of the resulting stator flux. Figure 4.32 illustrate both these situations.

The explanation above refers to a 2-pole machine, but it is generally applicable independent of the pole number. A 2-pole machine rotates a complete turn, i.e. 360° mechanically, for one electrical period, which coincides with 360° electrical degrees. In case of higher pole numbers, the relation between mechanical and electrical angles can be determined by equation 4-32.

$$\alpha_m = \frac{\alpha_e}{p/2} \qquad (4\text{-}32)$$

where α_m = mechanical angle
α_e = electrical angle

This means that a machine fed with a fixed frequency will rotate slower the higher its pole number is.

The poles can either be electrically excited by a DC current or by permanent magnets. It is also possible to let the machine be excited only by the stator currents, but that requires that it is a salient pole design. Such a machine is called a synchronous reluctance machine and it works simply because the salient poles are attracted by the rotating flux. All three types of machines, the electrically excited, the PM and the synchronous reluctance are illustrated in figure 4.33.

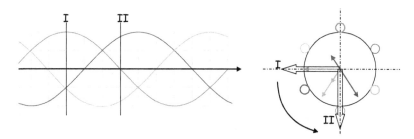

Figure 4.31 Creation of a rotating flux by a 3-phase winding

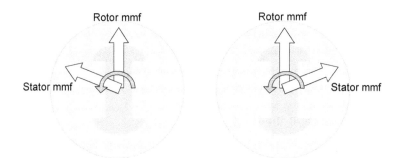

Figure 4.32 Pole angles for motor and generator operation respectively

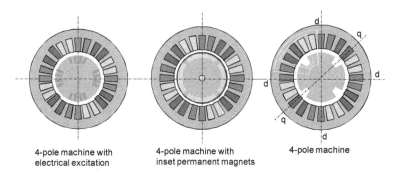

4-pole machine with
electrical excitation

4-pole machine with
inset permanent magnets

4-pole machine

Figure 4.33 Sketch showing cross sections of synchronous machines

The electrically excited synchronous machine is first of all used as a generator, from small diesel-driven standby generators to large power plant units. It
is also used as motor, but mainly for larger outputs, i.e. in the MW range. The
advantages of such synchronous machines are excellent possibilities to con-

trol voltage and reactive power, high efficiency and an extremely wide power range. Disadvantages are that they need a special exciter to provide the current for the excitation winding, they are difficult to start in motor operation and they are relatively expensive. The excitation current can either be provided from a separate exciter, static or rotating, via brushes and sliprings or from a so-called brushless excitation system, which will be described in section 7.8.1.

PM machines are mainly used as inverter fed motors, e.g. as servomotors, but have in later years become a viable alternative as motors in vehicles and as generators in wind mills. PM machines are characterized by high power density and torque density as well as high efficiency, because there are no excitation winding losses. On the other hand, they can be somewhat difficult to control, certain magnets are sensitive to high temperatures and the most powerful magnets are expensive. Some inverter fed PM motors are called "Brushless DC motors" but are in fact also synchronous motors.

Synchronous reluctance machines are not widely used, even though they are simple and have good efficiency. This is probably because they can not compete with induction motors due to inferior starting properties.

4.8.3 Induction machines

Synchronous machines can only operate at a speed which is exactly the same as the speed of the rotating flux. Asynchronous machines can, at least theoretically, operate at any speed except that of the rotating flux. Synchronous and asynchronous machines have, in principle, the same type of stators but different rotors. The functioning of an asynchronous machine is based on the interaction between the rotating flux and the rotor currents, which are induced by this time varying flux. That is why the machines are often called induction machines.

The stator currents create a rotating flux in the same way as described above for the synchronous machines. This flux will also pass through the rotor, which usually is equipped with some type of slot winding, and will induce currents in this winding. The rotor winding acts, in principle, as a secondary winding in a transformer and the rotor currents have the same frequency as the stator currents as long as the rotor stands still. The current carrying conductors will be subjected to forces as described in section 4.1 and also illustrated in figure 4.34. The forces result in a torque that brings the rotor into rotation.

The speed difference between the flux and the rotor diminishes as the rotor accelerates. The frequency decreases in the rotor as does the induced emf. The rotor currents will decrease too, but not in a linear way. If the rotor could reach the same speed as the flux, no currents would be induced because the flux is then constant in relation to the rotor.

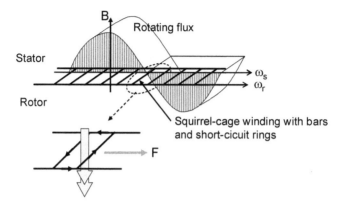

Figure 4.34 Sketch showing the basic function of an induction machine.

An induction motor accelerates, therefore, to a speed that is slightly lower than the synchronous. This speeds is determined by the intersection between the motors torque characteristic and the load torque, see figure 4.35. If the machine is forced to run above the synchronous speed, it will operate as a generator. The difference between the synchronous and the rotor speed is defined as slip and it is often given in percent or per unit in accordance with equation 4-33.

$$s = \frac{n_s - n_r}{n_s} \qquad (4\text{-}33)$$

where

s = slip [p.u.]
n_s = synchronous speed [rpm]
n_r = rotor speed [rpm]

The maximum torque for a specific induction motor is influenced by its inductance, but the slip at which the maximum torque occurs depends on the rotor winding resistance. An increased rotor resistance means that the maximum torque is reached at a lower speed. This is also illustrated by figure 4.35.

There are two main types of rotor windings used in induction motors, squirrel cage windings and 3-phase slot windings combined with sliprings. The squirrel-cage consists of a number of un-insulated circular, rectangular or otherwise profiled conductors placed in chiefly axial slots in the rotor core. The conductors are inter-connected by short-circuit rings at both ends of the rotor. The profile, and even the material of the conductors, have impact on the torque characteristic and on the starting current.

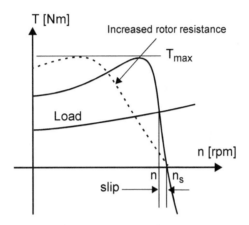

Figure 4.35 Torque characteristic for induction machine

In the case of a slipring motor, the rotor winding is connected in series with an external resistor, which is successively reduced so that the maximum torque can be maintained over a larger speed range. In addition to this, the start current is limited compared to the corresponding current in a squirrel cage motor, which explains the extensive use of slipring motors long ago, when the grids were weaker. Figure 4.36 contains a connection diagram of a slipring motor.

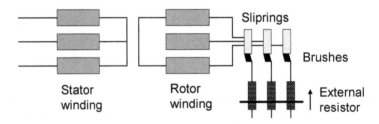

Figure 4.36 Connection diagram for slipring motor

The induction motor is the most common electrical machine for industrial applications, especially the squirrel cage type. The squirrel cage motor is a robust and cost effective product with fairly good efficiency. It can easily be started by connection directly to the line, at least in modern strong grids. A disadvantage is that it needs reactive power for its excitation and motors with high pole numbers tend to have a low power factor. Induction motors are nowadays frequently used in variable speed drives in which they are fed from power electronic inverters.

4.8.4 Switched reluctance machines

Switched reluctance (SR) machines are similar to induction machines in the sense that the flux is produced by the stator currents, but they are very different in other respects. Characteristic for an SR machine is that it has salient poles both in the stator and rotor but with different pole numbers. The stator poles are provided with coils while the rotor just contains laminated steel or, in some cases, SMC. Figure 4.37 shows a sketch of a 8/6-pole motor.

Figure 4.37 SR motor with 8 stator and 6 rotor poles

The working principle for an SR motor is the following; a current is turned on through two diametrically opposite stator coils, creating a magnetic flux through these poles. A pair of adjacent rotor poles is pulled into alignment with the flux, whereupon the current is turned off and two other stator poles are magnetized. These poles will then attract another pair of rotor poles, so that the rotor keeps moving. A continuous rotation is thus achieved by successive switching of the currents in different stator coils. The term "reluctance motor" comes from the necessary provision that the flux crosses a very different reluctance ("magnetic resistance") when stator and rotor poles are aligned with each other compared to when they are facing pole gaps.

The SR motor is, from a design and manufacturing point of view, very simple and cost effective. It has good efficiency because there are practically no rotor losses other than the friction losses. It is robust and easy to cool. It is, in spite of this, not widely used. The reasons for this are that the motor is unable to function without some sort of inverter, the torque contains pulsations and the motor has a reputation of being noisy. It is suitable for variable speed drives but has difficulties to challenge the speed controlled induction motor, which has a huge advantage in its large production volumes for fixed speed drives.

5 The development process

Development is a word which is used in many different contexts. The focus of this study is on product development, though related processes, such as market and production development, will also be discussed. Electrical machine development processes contain several common elements independent of the type of machine. This chapter will deal with the general structures of the electrical machine development process as a background for the specific studies presented in later chapters. The first parts of this chapter should be familiar to readers with industrial experience, but they are included for readers with mainly academic background.

5.1 Definitions of "development"

The word "development" is common and it is used in many different contexts. We speak of "global development", "developing countries", "historical development", "weather development", "development of a disease", "personal development", "development of an exposed film", "technical development", "development of a new car model", "development department", "development of an idea", etc, etc. The list of examples could go on. Most people are familiar with these different expressions, and the use of the word "development" does not generally cause any misunderstandings. It is, nevertheless, important to use relevant definitions, especially in an academic text.

Dictionaries give several definitions of the words "develop" and "development". Examples can be found for instance in Webster's New World Dictionary [92]. The following definition taken from the Swedish "National Encyclopedia's Wordbook" is relevant for this thesis, even though the word "complex" can be questioned [93]:

"Development, a process during which something is changed and usually becomes more complex or valuable. " (My translation)

A few more specific definitions of "technical development" can be found in Mats Fridlund's study of the development cooperation between Asea and Vattenfall, documented in his Ph.D. thesis, "Den gemensamma utvecklingen" ("The mutual development") [94]. He gives a few different definitions of "development". The first one is "technical innovation activity". The second definition can be described as "progress", e.g. technical progress. A third definition is used on a macro level in the meaning "modernization of societies". The first two definitions are used in my thesis and the use of each is hopefully evident from the context.

Development, in accordance with the first definition, includes both product and production development while activities leading to increased know-how and new engineering tools reside in the borderland between the first and second category. Market development does not fit directly under the heading "technical development", but there are certainly strong connections.

In his thesis, Mats Fridlund also introduces three types of "construction work", namely "technological construction work", "social construction work" and "cultural construction work". It can be disputed whether "engineering work" would be a better expression, but most important is, of course, the interpretation. The "technological construction work" is traditional engineering activities, such as performing calculations and simulations, designing products and tools, testing prototypes and products. The "social construction work" includes creation and use of alliances with other actors of importance, such as colleagues at other departments or externally with suppliers, consultants, customers, scientists and others. Finally, the "cultural construction work" performed by engineers connects their development work to a wider public through representational activities, presentations, publications, exhibitions, etc. All these three categories of "engineering work" will appear in the following chapters.

5.2 Research and Development

R&D is a widely used notion, but it is not always clear what is included when the term is used. Many companies present facts and figures on R&D in their annual reports. Often normal design work on customer orders is included, even if it does not contain any element of development. Companies consider it an advantage to show a high portion of R&D out of the total costs.

Research can be defined as a systematic work in order to find and establish new knowledge in a certain field. It is often divided into "basic research" and "applied research". However, the OECD countries agreed in 1970 on a common terminology in which research also includes "development work". Basic research is, according to OECD's definition, a systematic and methodic search for new knowledge and new ideas without focus on any specific application. Applied research is defined as a systematic and methodic search for new knowledge and new ideas but focusing on a specific application. Finally, development work is defined as an activity which systematically and methodically utilizes research results and scientific knowledge in order to create new products, processes and systems or significant improvements of already existing ones [95]. This definition of development is clearly more stringent than the one used in many companies' presentations.

This thesis deals with "Development of Electrical Machines" in a wide sense, but the word "development" is mainly used in accordance with OECD's definition. Some activities can and will be referred to as applied research and others as normal order related engineering.

5.3 Product development

Industrial products are, with very few exceptions, developed and manufactured to give a certain financial profit. This is a fundamental prerequisite, in principle, for all product development. Beside this, the variations in objectives, methodology, complexity, size of effort, etc. are practically unlimited. The development of a new car model is, of course, very different from the development of a pair of pruning shears. Other examples could be pharmaceutical drugs, computer programs, large transformers, shoes, ice cream with new flavors, steel alloys, TV sets and so on. Products are very different and thus also the development projects.

In spite of what has been said above, it is possible to identify some major steps which are common in development projects, even if all steps are not always explicit. The main steps are the following:

- Start: The initiative for a development project can come from different actors. A manufacturer notices that his current product is not competitive any longer and needs to be replaced. An OEM customer turns to a supplier with a request for a new product. An inventor comes up with an idea for a new product. New legislation makes it necessary to replace an old product that does not meet present rules. A company has decided to enter into new markets. A decision has to be made in each case by relevant persons and necessary resources have to be allocated.

- Pre-study: It has become common practice to begin development projects with some sort of pre-study. Such a study can range from a brief review of a project proposal to an extensive activity including market analyses, concept design of several alternatives, experiments, sourcing analyses, writing of a complete business plan and a detailed plan for the main project.

- Evaluation and decision: A final decision is taken based on the results of the pre-study.

- Specification: The importance of this phase is strongly emphasized in modern project management courses. It often involves participation of sales engineers, customers or marketing consultants, along with production specialists. The specification work usually becomes extra extensive in products containing a lot of software development. Not only the product but also verifying tests must be specified.

- Design: This is a core activity in product development projects requiring a great deal of creative work. It includes detailed layout design, technical calculations and simulations, choice of material and components, verifying experiments, and detailed design of prototypes and test rigs. In case of software development, it includes programming. Calculation of production costs and quality assurance activities are also included in this phase.

- Prototype manufacture and testing: This part of the work can vary widely. Many products require extensive testing of several prototypes and sometimes also official approval through certification. In other cases, it can be sufficient to make and test one prototype for verification of the specified performance. There are also several examples where the prototype phase is omitted and the first products are made directly for customers.

- Documentation for series production and marketing: This phase is perhaps not considered as the most glamorous part of a product development project, but it is absolutely necessary. The documentation consists of manufacturing drawings, purchasing specifications, quality assurance documents, brochures, catalogues, pricelists, manuals, maintenance instructions, spare part lists, etc, as relevant from case to case.

- Production preparations: This is a very expensive part of some projects , especially in case of mass-produced products where it can be necessary to invest in new production lines, including buildings and machine tools. On the other side, there are projects where existing resources can be used and the production preparation is limited to a few man-weeks required for production planning.

- Product release: This point is relevant for series manufactured products and, in principle, represents the end of a development project and the beginning of regular sales and manufacturing. A development project can contain a number of product releases covering different models, sizes or even markets.

The above list is typical for development projects within the mechanical and electro-technical industries. The steps are listed in chronological order but there is often certain over-lapping. The companies try to shorten development times as much as possible. Much iteration can be required, especially during the design and prototype steps. The projects are usually followed up through regular progress reports and review meetings. Figure 5.1 describes a product development project in graphical form.

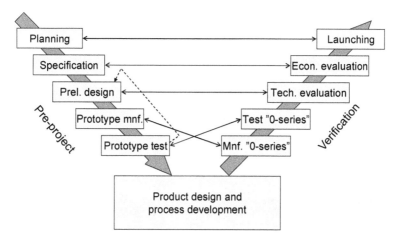

Figure 5.1 Diagram showing a typical product development project.

5.4 Market and Production Development

5.4.1 Market development

"Market development" can be interpreted in different ways. It can be a report of how the sales of a company's products have changed in a certain region during recent years. It can also be a planned activity aiming at entering into new markets or increasing shares in existing markets. It is the latter which is of interest in this thesis, especially its connection to the product development.

Market development can be an integrated part of a product development project, but it can also be a separate effort to increase sales of already existing products. It is not unusual that a market expansion requires improved or modified products, so it will quickly influence on the product development. The so-called Ansoff matrix in figure 5.2 illustrates, in a simple way, the options companies have when they are planning business expansion through market and/or product development.

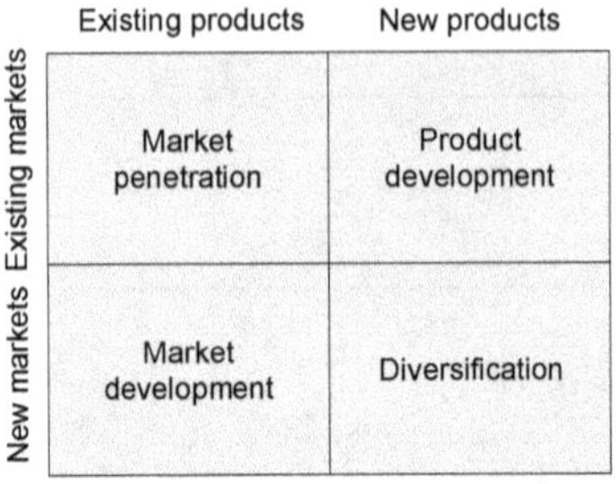

	Existing products	New products
Existing markets	Market penetration	Product development
New markets	Market development	Diversification

Figure 5.2 Expansion through market and/or product development

Examples of activities normally included in market development are market studies, creating a market strategy and a market plan, making sales documentation, training of the sales force, meetings with potential key customers, or participation in exhibitions and trade fairs. Market development can also mean establishment of new subsidiaries or branch offices, engagement of agents and distributors, along with entering into license agreements for local production in some markets.

5.4.2 Production development

Production development usually goes hand-in-hand with product development. New designs often require altered production methods. Conversely, the introduction of improved manufacturing technology also needs modified designs, or can open possibilities for new and better solutions. There are numerous examples of this and some of these examples will be dealt with in later chapters. Production development is not only a matter of better manufacturing operations, but also improved factory layouts and logistics resulting in shorter lead times and reduced amount of parts in production.

Investments in factory buildings, production lines, machine tools and other tools are not considered as development from the accounting point of view, but they are strongly linked to both product and production development. The necessary engineering for these investments is normally part of production development. The investments are usually expensive and, therefore, have a huge long-term impact on the product development. The design engineers try to re-use the existing tools as much as possible for new or modified products.

It is also a matter for production development to perform analyses whether various components, which are parts of the products, shall be made in-house or outsourced to suppliers. Basically, it should be a question of simple cost analysis, but this decision is often affected by other aspects, such as reliability of supply, lack of resources, need for confidentiality, etc. Outsourcing has become more common during the last two decades and it has usually influence on the product development. A supplier can have different production equipment and methods, which have to be considered.

The quality of a product depends both on design and manufacturing, and quality assurance measures must, therefore, be part of both product and production development. In the latter case it is primarily a matter of choosing reliable methods, using accurate tools, keeping the workshop clean, along with intergration of appropriate control operations and tests in the production. This is done in close cooperation with the product development engineers, or it is even specified by them.

5.5 Electrical Machine Development and Design

5.5.1 Standardization levels

There is a big difference between a motor for a vacuum cleaner and a generator for a hydropower plant. The first is small and it is produced in hundreds of thousands. The second is large and only one or a few units are manufactured. It is possible to find everything in between. Some manufacturers have therefore introduced "standardization levels" as a mean for a systematic approach in the development work, especially regarding rules for possible variations and documentation principles.

0. This level represents individually designed machines, which require new development in several respects. Most of the development is made on customer order. Examples can be a new type of wind power generator, a wheel motor for a vehicle or a machine which requires extensive extrapolation.

1. The next level contains machines with standardized design principles, which are applied on "tailor-made" machines. Many alternative solutions exist and dimensions can be chosen relatively freely. The design principles have been developed earlier but the specific designs are made on customer orders. Examples are hydropower generators and traction motors.

2. This level includes machines that have been developed as series, but with freedom for customers to specify ratings and choose between many options concerning cooling forms, etc. The cross-section dimensions are strictly standardized while the active length can be varied. The necessary order design is very limited. Examples are large industrial motors and certain turbogenerators, i.e. machines in the MW range and larger.

3. This is a level which consists of standard machines documented in catalogues, but with a number of options and also possibilities for minor customer-specific modifications. They are manufactured on

customer orders and can in most cases be specified without involvement from a design department. Examples are many induction and DC motors for industrial applications.

4. This level refers to standard machines produced in series or batches according to standard documentation. Frequently ordered sizes and models are kept in stock. A few options specified and documented in catalogues are allowed. Examples of such options are small, foot or flange mounted, induction motors for which different pole numbers and different voltages are available.

5. The highest level of standardization concerns completely standardized machines. They are mass-produced without any options for variations. Example: vacuum cleaner motors.

The above structure is representative for the normal business situation. Many manufacturers are of course prepared to make exceptions if, for instance, an important OEM customer asks for large numbers of a special machine.

It is evident, from the list of standardization levels above, that development of electrical machines is carried out both in specific development projects and as part of the design on customer orders. As expected, the lower the level of standardization, the more complex is the development carried out on customer orders.

5.5.2 Development for series production

The most common objective for development of electrical machines is cost reduction. This is valid independent of machine type and size. A provision is of course that the machines meet the required performance and posess high quality and other properties that make them competitive.

Machines in the range below 100 kW are usually manufactured in large numbers. Even machines up to 1 MW can often be considered as series manufactured. One example of this is traction motors. The machines can be highly standardized and sold to a wide market according to the manufacturers catalogue, or they can be specially made for one large customer to his specification. The development projects could be somewhat different in these two cases but the basic approach would be similar. It is a matter of developing a new machine and verifying it through prototype tests.

The development of an electrical machine, or a series of machines, starts from a specification of the required performance (rated power, torque, speed, voltage, etc.) and design (cooling and protection forms, mechanical arrangements, etc.). In the case of a series, the structure and the number of variants have probably been thoroughly analyzed in a pre-study. Standard motors have to comply with official standards which have an important impact on the development. Customer specific machines are often subject to very special requirements, e.g. due to mechanical integration with other equipment.

The type of machine to be developed is well known in the case of a series, i.e. 3-phase, squirrel-cage induction motors in a certain range. The situation is not always the same for customer specific applications, for which it can be necessary to first choose a machine type. Should it be an induction or a PM motor or perhaps a switched reluctance?

Experience, from previously manufactured machines, is a very important source of know how for development of new ones. Old test results help the electrical design engineer to decide on initial dimensions as input for his calculations. He then tries to find the best solution for the active part of the machine which fulfills the required performance. The calculation is an iterative process that can be more or less automatic depending upon available computer programs. Numerical optimization methods are sometimes used as well. The diagram, in figure 5.3, illustrates the sequence in which dimensions, windings, machine parameters and performance are calculated. Depending on the type of development project, there can be restrictions on dimensions because it is necessary to use existing, expensive punching tools.

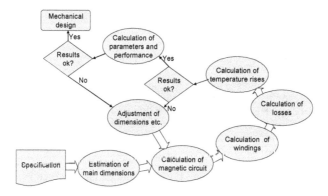

Figure 5.3 Electrical and thermal calculation procedure

In principle the mechanical design follows after the electrical and thermal calculations, but it is, in reality, necessary to have some overlap. This part of the development process includes layout design of the whole machine, detailed design of the different parts, choice of materials for non-active parts, stress analyses, bearing calculations, cooperation with manufacturing specialists, drafting of necessary drawings, making purchasing specifications and cost calculations. The mechanical design is a diversified activity and normally needs much more manpower than the electrical design.

The next step is the manufacture and testing of one or more prototype machines. The results show if any modifications are necessary and whether new prototype tests must be performed. When the test results are satisfactory the extensive work of documentation necessary for regular sales and production remains. This documentation phase is less comprehensive in a customer specific project. Figure 5.4 contains a block diagram showing the steps in a typical standard motor development project.

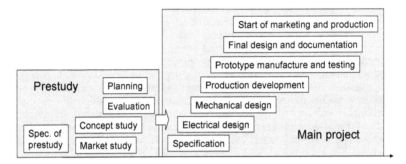

Figure 5.4 Main steps for development of a standard motor series

5.5.3 Development of large machines

The development of large machines is, in some respects, different from what has been described above even though there are also many similarities. A fundamental difference is that large machines are, in principle, always built on customer orders and prototypes are seldom used. The first machine with certain performance and design becomes a "prototype" before it is shipped to the customer.

New generations of large industrial motors and certain generators are developed in projects similar to those for smaller machines and only minor design work remains to be performed on customer orders. Other large machines, e.g. hydropower generators, are subject to a more continuous developing process. This means that a new type of rotor spider can be introduced at one time, a modified stator coil-end support the next time, and a new lubrication system for the bearings at yet another time. The development of such new components is often combined with the ordinary design work on actual customer ordered machines. If the developed component turns out well, it becomes part of the design standard for the machine type concerned.

The background for the development of a component or system is usually the need for reduced costs or improved quality. New concepts can be needed as preparation for machines with ratings outside the familiar range. Development of new systems, e.g. an insulation system, or new concepts such as direct water-cooling are carried out in separate projects. Tests are performed in different test rigs, but full-size prototypes are almost never built. That comes with the first customer order.

The customer's specification is the basis for the design of a particular machine. The work for large machines starts with a concept layout along with electrical, thermal and mechanical dimensioning. Available experience from previous machines is important as input for the initial calculations. These analyses, together with layout design and cost calculations, are often made in order to enable the manufacturer to make a quotation. If it leads to an order, the same steps are repeated but more in depth.

The activities in the above paragraph can be described as electrical, magnetic, thermal and mechanical design of the machine as well as the determination of its characteristics and properties including the expected manufacturing costs. The problem facing the design engineers is, other than the physics and mathematics involved, a number of restrictions, requirements and other constraints that must be taken into account. These conditions can be combined in four groups, namely:

- Site and environmental constraints: These constraints are predominantly applicable for large machines. One example can be the maximum outer dimensions due to transport limitations such as railway gauges. Another can be restrictions on cooling due to no access to fresh cooling water.

- Official standard restrictions: A typical example is allowable temperature rises.

- Customer requirements: Obvious examples are rated output and speed, but can also be maximum synchronous and transient reactances.

- Manufacturers design standard: Here restrictions in concepts and dimensions due to existing tools, accepted cooling principles and available insulation systems can be mentioned.

These conditions can be looked at as boundary conditions surrounding an area within which the final design solution must be found; see figure 5.5. The whole procedure can be seen as an optimization with constraints, provided a suitable target function is defined. A common target function is the sum of the manufacturing costs for the machine and the capitalized value of the losses. Sometimes other factors such as weight, volume or maintenance costs can be included.

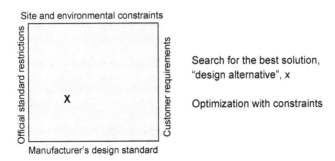

Figure 5.5 Boundary conditions for design of large electrical machines

The calculation and layout design are followed by detailed design and drafting, purchase of material and components, production preparations and manufacturing including specified quality control. Many machines are tested in the manufacturers works before shipment, but some are too large. These are instead transported in pieces, which are assembled on site and the acceptance tests are then performed in the customer's plant. The fact that large machines are designed, manufactured and installed before they are tested constitutes of course a risk, but this is accepted as normal practice. This will be dealt with further in chapter 7. The development sequence for a large, tailor-made machine is illustrated in figure 5.6.

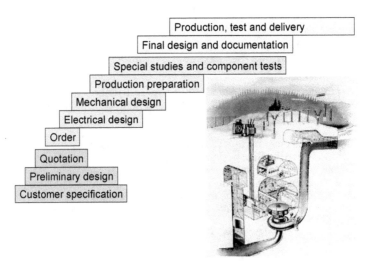

Figure 5.6 Development /design steps for a large machine, e.g. an hydro-power generator

5.5.4   Engineering tools

For a very long time the slide rule and the drawing board have been the classical tools used by engineers developing electrical machines. Until the late 1950's, when the first computers were installed, all calculations were done manually. The early computer programs followed more or less exactly the same routines as those used in the manual calculations. The main advantages of using computers were increased capacity and that they allowed the engineers to calculate and compare several alternatives with greater ease. The drawing board remained, until the 1980's, even if there were earlier attempts to rationalize the drafting work, for instance by letting computers calculate measurements. These measurements were then printed in separate lists and attached to manually drawn standard drawings, which were not to scale. The introduction of CAD (Computer Aided Design) had a rapid impact. Computer workstations, with interactive 3-dimensional design programs, have now completely replaced the drawing boards.

Some CAD programs can directly generate geometries for calculation and simulation programs and they are also directly connected to systems for CAM (Computer Aided Manufacturing). This allowed for the automatic programming of NC machines (Numerically Controlled machine tools) and even robotics. Figures 5.7 and 5.8 gives examples of different generations of drawings.

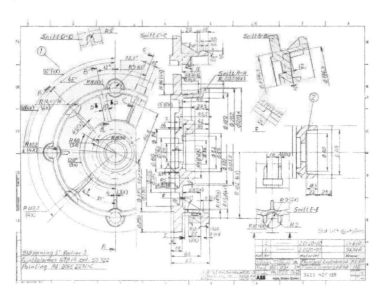

Figure 5.7 Manual drawning of induction motor end-shield

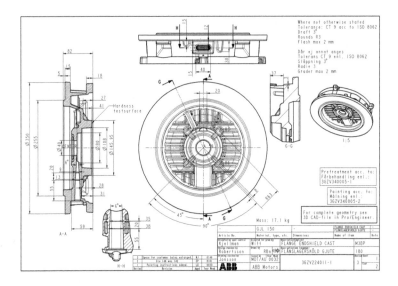

Figure 5.8 CAD drawing of induction motor end-shield

Electrical, thermal and mechanical calculations were computerized very quickly, essentially during the 1960's. As mentioned above, computerization was initially a help in increasing the calculation capacity, but the development of more powerful computers soon opened possibilities for new and better calculation methods. Numerical analyses were introduced as a complement to existing analytical methods.

FEM (the Finite Element Method) has become a very powerful tool for simulation of different fields, both vector and scalar fields. FEM was first used for mechanical stress analysis in the aircraft industry [96], and the same application was first used also in the electrical machine industry. Early examples from Asea are from the years around 1970 [97]. Simulation of electrical and thermal fields soon followed while electro-magnetic FEM analysis came late in 1970's, at least within Asea. The latter application was interactive from the beginning, which was not the case with the earlier ones.

Magnetic fields are complicated to simulate, primarily due to non-linearities that depend on saturation in the iron. It was impossible to attack this problem with analytical methods. It is still not easy, but modern 2- and 3-dimensional FEM programs are very helpful. An example of an old-fashioned magnetic field analysis is a manual flux plot for determining the commutation flux in a DC machine, figure 5.9. It requires almost an artistic talent of the engineer to draw the quasi-quadrates in such a plot.

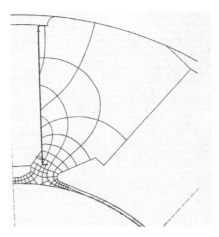

Figure 5.9 Manual plot of flux between commutation and main pole in DC machine

Figure 5.10 shows a flux plot from an FEM simulation. It shows the flux distribution that also exists in the non-linear parts of the magnetic circuit. It is possible, through such a simulation, to obtain also a number of other results such as inductances, flux harmonics, magnetic forces, etc. However, a deeper look at this topic is outside the scope of this thesis, but further description can be found in numerous books and papers, for instance in those included in the reference list [98, 99]. FEM simulation is mainly used as a development tool, for new standard series as well as for special customer ordered machines. It is usually not included in routine calculations.

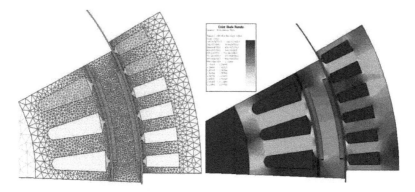

Figure 5.10 Mesh for FEM analysis and resulting flux density in a 4QT PM machine

Interactive simulation programs solving differential equations have also become widely used for the analysis of transient phenomena in complex systems. A good example is ABB's program SMT for transient analyses of synchronous machine systems. This program was developed by Claes Ivarson (1946-2004), senior scientist at ABB Corporate Research. The program can, for instance, simulate shaft oscillations at various transient conditions in a system consisting of an electric grid, a large synchronous motor, and a mechanical load, e.g. a compressor.

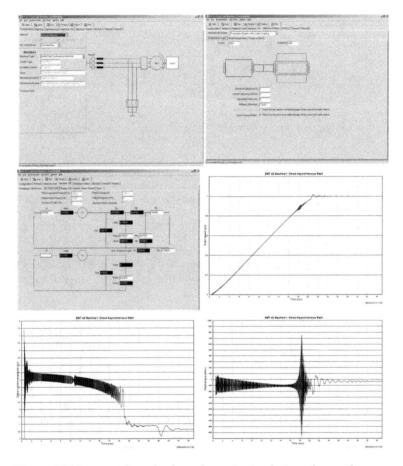

Figure 5.11 Input and results from dynamic simulation of a synchronous motor asynchronously starting a compressor

Such a simulation can show the risk for resonance between the electrical and the mechanical parts of the system and the size of the torque peaks that can occur. This type of simulation is primarily a tool used in connection with quotation and order design as conditions vary from one installation to the next [100]. Figure 5.11 shows an example of a simulation of this type.

Optimization has already been mentioned above. A traditional method of optimization has been to calculate a number of alternatives and then compare the results. How close to an optimal solution it is possible to reach depends on the skill and experience of the design engineer. However, the use of numerical optimization routines, which search for a minimum or maximum of a specified target function, is also common in electrical machine development and at order design of tailor-made machines. Such an optimization is always restricted by a number of conditions and the target function normally has a large number of variables, 10 - 20 [101].

5.5.5 Know-how

Electrical machines are fairly complicated products and both theoretical competence and practical experience are needed for their development. Knowledge from a number of scientific disciplines such as electro-magnetic theory, thermo dynamics, solid mechanics, fluid mechanics, control theory and others are important. There are, in the same way, many practical areas from which a development team must have know how and experience. Examples are application know-how, CAD, quality assurance, manufacturing, testing and so on. The diagrams in figures 5.12 and 5.13 give a good idea of what is required of scientific and practical know how. This will be more evident in later chapters.

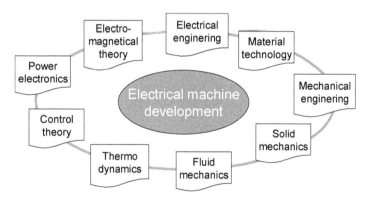

Figure 5.12 Scientific know-how

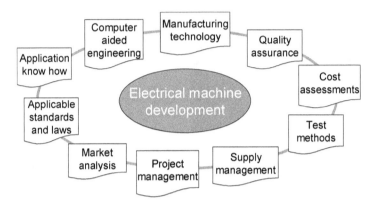

Figure 5.13 Practical know-how

Leading electrical machine manufacturers such as ABB, earlier Asea, have for the most part the know how required for the development of new machines in-house. ABB has departments for development and design of different machine types and the personnel in these departments are usually very experienced and qualified. Nevertheless, some problems need special competence and that is why the company's corporate research unit often becomes involved. Suppliers can also be a source of know-how; a good example of this is bearing technology.

Other manufacturers are also of interest in this respect as know-how can be transferred directly through license contracts or agreements allowing mutual exchange of technical information. A competitor's technology can also be studied through purchase of his products for inspection and bench marking testing or through a study visit to some location where his machines are installed. Technical conferences, trade fairs and publications can, of course, also be important sources of know-how.

Several technical universities play an important role in electrical machine research, especially in the development of tools for theoretical analyses. Their contribution to product development has been linked not so much to new concepts and technical solutions, but instead to the education of graduated engineers and researchers who later become key persons in the industrial development.

6 Induction motors

Large steps and dramatic technical breakthroughs do not characterize the development of standardized induction motors. It has rather been a matter of evolution characterized by steady progress. The importance of these motors can, however, not be overestimated. They have always been one of the base products within the electro-technical industry and the study of their development is therefore well worth including in this thesis. There is reason to believe that not only pure technical factors, but also factors related to market and production, have had major influences on the development of induction motors. The non-technical factors are, of course, equally important and interesting as the technical. Induction motors are used more and more in variable speed drives, and this application is therefore given considerable attention in this chapter.

Induction motors have been a key product for Asea, and later ABB, and the company has always maintained a leading position in this field. The study is therefore, to a large extent, focused on this company, but it also includes some information on and comparisons with, other manufacturers, especially those which have been, or still are, part of the Swedish electro technical industry.

6.1 The electrical workhorse

Electrical motors, larger than 1 kW, are mostly 3-phase induction (asynchronous) motors and are mainly used for industrial and similar applications. There are a few industrial drives, which require motors up to 60 MW or higher, but induction motors are usually limited to around 15 - 20 MW [102]. Larger sizes are always synchronous motors which are often preferred even in the range 5 - 20 MW. The main reasons for this preference are that synchronous motors have better efficiency and lower reactive power consumption, which can justify their higher price.

Induction motors are used for driving pumps, fans, compressors, machine tools, transport equipment, paper machines, printing presses, textile machines etc. The list of applications could go on. Figure 6.1 shows, as an example, large fans driven by induction motors. Motors above 1 MW are almost always high-voltage (HV) motors with rated voltages higher than 3 kV, while motors below 500 kW are usually low-voltage (LV) motors with voltage ratings below 1 kV. This study focuses on the latter.

Figure 6.1 Induction motors driving large fans, a common industrial application.

The basic reasons why the induction motor has become the preferred industrial workhorse are as follows:

• The easiness to supply the motor with 3-phase power

• The possibilities to start the motor directly on line without expensive extra equipment

• The robustness and high reliability of the motor

• The comparatively low costs for the motor

There are a number of different types and sizes of induction motors, which usually are strictly standardized, though there are some exceptions. Induction motors are usually manufactured in large quantities, especially the smaller sizes. Examples of types or variants are motors with squirrel-cage rotors and those with wound rotors (slipring motors), open-ventilated and closed motors, foot mounted and flange mounted motors, single-speed and two-speed motors, 3-phase and 1-phase motors, motors with cast iron housing and those with aluminum housing.

Most of the induction motors have been, and still are, fed directly from a fixed frequency grid (50 or 60 Hz), which means that they run with approximately constant speed, determined by the pole number. However, an increasing number of motors nowadays are inverter supplied and operate with variable speed. This will be dealt with in more detail in the latter part of this chapter.

6.2 Half a century of expansion

6.2.1 Asea's first induction motors

As mentioned in the historical review, section 3.7, the induction motor was invented around 1890. One of the key persons, in its initial development, was the Swedish inventor Jonas Wenström, whose achievements led to an early introduction of such motors in Sweden. His first induction motor, which is shown in figure 6.2, was a 4-pole motor with a short-circuit rotor and a sort of concentrated 3-phase winding in the stator. It was built and tested in 1890-91, but its performance was not good. However, the motor was successfully redesigned, one year later, by a young engineer, Ernst Danielson (1866-1907), who later became Asea's chief engineer and technical director. Asea delivered this motor to a Swedish steel mill, Fagersta Bruk, in late 1892 [103]. Ernst Danielsson's contribution to the initial development was of vital importance. He had studied mechanical engineering at KTH, worked a period for Asea in Arboga and then moved to America, where he was employed by GE from 1890 until 1892, when he returned to Sweden.

Figure 6.2 Jonas Wenström's original induction motor from 1891
(a drawing is displayed in figure 3.15)

A remarkable example of technical self-confidence and commercial risk exposure is Asea's delivery of generators and induction motors to a rolling mill[*] at Hofors Bruk in Sweden two years later. The company had, in the contract, accepted high penalties for delays and, on top of that, liability for damages in the order of half the share capital in case the equipment had to be rejected. Siemens & Halske had also been invited to make a bid but had declined due to feared technical difficulties. Nonetheless, Asea's delivery,

---

*. Wire rod mill

which included four induction motors rated 200 hp at 450 rpm each, met the requirements with good margins. This became an excellent reference for the company and increased its efforts to develop and market AC machines in relation to the DC machines, which had dominated so far [104].

It was obvious, from the beginning, that some sort of standardization of the motors was desired. If a motor-series is considered to exist when a certain type of motor is made in more than one size and for several customers, Asea actually introduced its first series of induction motors in 1892-93 i.e. right from its inception. The performance was now in order but the motors were too cumbersome. A comprehensive redesign was made at the end of the 1890's by the electro-technical manager Arvid Lindström (1866-1944), who in 1904 became KTH's first professor in theoretical electro-technology and later also in electrical machines. It is interesting to read the following lines in Asea's History 1883–1948 concerning this modified motor series: "*Practically the same core dimensions were maintained, but the rated outputs could in some cases be increased up to 75 percent through better cooling. The mechanical strength was also improved. This so called B-series comprising 11 types RSB-RAB was built for rated outputs between 1 and 120 hp.*" (My translation) [105]. This is an interesting remark about a very early shift of motor generations. It would be more difficult to increase the utilization of that order today.

Figure 6.3 Ernst Danielsson, Arvid Lindström and Emil Lundqvist

The first motor series was not suitable for mass production, but this problem was addressed in 1905-06, when Asea's chief engineer Emil Lundqvist (1872-1942) initiated a complete redesign in order to reduce manufacturing costs. New core dimensions were introduced, which allowed the inner parts of the sheets that remained after punching large cores, to be used for smaller motors. The same cross section, for more than one core length, was also introduced [106]. Production cost obviously became a key issue for the development engineers already during the infancy of the induction motors.

Emil Lundqvist, who had studied mechanical engineering at Chalmers, traveled to America in 1893 with a recommendation letter addressed to Nikola Tesla in his pocket. This helped him to gain employment at Westinghouse, where he stayed for three years. He then moved to Germany and later on to South Africa before he received a position as chief engineer at Asea in Västerås. Both Ernst Danielsson and Emil Lundqvist are good examples of young engineers, who attained knowledge and experience through a few years work in the leading American companies. They were not the only ones. Emil Lundqvist left Asea after a few years and later he became president of Stora Kopparbergs Bergslags AB, Sweden's oldest industrial enterprise.

OUTTRÖTTLIG OCH PÅLITLIG, AN-
VÄNDBAR FÖR ALLA KRAFTÄNDAMÅL,
ARBETSFÖR I ALLA LÄGEN.

Figure 6.4 "That one is my best journeyman".
Front-page illustration in Asea motor brochure from 1913. Asea's logo-
type was the swastika until this had become a symbol for the Nazis.

6.2.2 Asea's motor series until the mid 1940's

Induction motors have been built both as slipring motors and squirrel-cage motors. Slipring motors offered a possibility to limit the start currents and they were therefore more common at that time, when the power supply was weaker. Motors with squirrel-cage rotors, which are more robust and cost-effective, dominate nowadays. A short description can be found in section 4.8.3.

Motor manufacturers like Asea have had several types of induction motor series at the same time, e.g. for slipring motors and for squirrel-cage motors. Other important variants have been motors with cast-iron housing and with aluminum housing, open ventilated and closed motors, motors with sleeve bearings and those with ball-bearings. Asea had more than 20 series of induction motors during the period of 1910-45. Many existed, as already mentioned, in parallel with each other. It is outside the scope of this study to give a detailed overview of the development during this period, but some observations can be mentioned.

Practically all catalogue-listed motors were initially open ventilated, i.e. cooled by ambient air passing through the inner parts of the motors. Figure 6.5 shows examples of such motors. Both Asea's History 1883-1948 and a paper concerning "Development of the Induction Motor" written by Hans Hedström (1911-1981), Asea's chief engineer for smaller machines, give the impression that closed motors were introduced first in the early 1930's. Closed motors were a special type series, in which the air passes on the outside of the stator, driven by an external shaft-mounted fan, developed due to the need for safer motors in dusty industries [107, 108]. In spite of this, an Asea leaflet from 1905 informs buyers that all motors can be obtained partly or completely enclosed, in which case they must be de-rated to 70 % of nominal output. Another leaflet from 1908 presents even a waterproof version of induction motors in the range 0.75 – 55 hp [109].

Figure 6.5 Asea slipring motor type M10 from 1911 and
slipring motor type MK from 1923

The early motor series had sleeve bearings, usually placed in the end shields, but pedestal bearings were also used for some of the larger motor types. Asea built its first motors with ball bearings in 1910 and this type of bearing soon became standard when the company developed its new induction motor series. One advantage of ball bearings was to get rid of the oil lubrication and another was that the ball bearings required much shorter axial length than the sleeve bearings. Asea used rolling type bearings, on a large scale, earlier than its competitors and it is claimed that the reason for this was a close cooperation with SKF, which rapidly had become a world-leading company for such bearings. SKF was founded in 1907 by the Swedish engineer Sven Wingquist (1876-1953), inventor of the self-aligning ball bearing. Good relations with competent suppliers were important for the product development also in those days.

Various kinds of 3-phase windings were used such as single layer concentric windings with end turns in two or three tiers and double layer distributed windings. Most of them were made as so-called mush windings (random windings). The slot insulation was first made from presspan combined with an impregnated cotton fabric, but the latter was replaced by micanite in the mid 1920's making the insulation less sensitive to moisture. Insulation of the individual copper strands was improved at the same time, from double layer cotton spinning to a varnish protected by single layer cotton spinning. The squirrel-cage windings in the rotors were made of copper bars placed in circular slots and soldered to short-circuit rings of copper wire. This solution was soon replaced by solid short-circuit rings soldered or welded to the copper bars. See figure 6.6. Asea made its first attempts to improve the starting capability of the squirrel-cage motors in 1924 by introduction of deep bars in the rotor winding.

Figure 6.6 Squirrel-cage rotor with copper bars and welded short-circuit rings, 1933-34

Stator frames and end-shields were made from cast iron from the very beginning. This was a traditional production method used for all kinds of machinery, but in the case of induction motors, it is still used. The "cast iron motors" are very robust and also less noisy than other designs. Nonetheless, Asea launched its first series of "aluminum motors" (type MBB) in 1945. Figure 6.7 shows a sketch of an MBB motor. Not only the stator frames and end shields were made from cast aluminum, but soon also the complete rotor winding. This meant of course a considerable reduction of the weight but also a more rational rotor production [106, 107].

Figure 6.7 Asea's first "aluminum motor", MBB, introduced in 1945

All product development is, in a broad sense, market driven. The products must meet customer needs and the prices must be competitive. A more specific definition of market driven development could be that certain customers or industry branches request special designs, which require development of new machine designs. The development of closed and later even explosion proof motors are good examples of this. Not only mines and chemical plants but also textile industry and grain mills have needed such motors due to hazardous atmosphere.

6.2.3 Sales volume

It has already been mentioned that Asea or "Allmänna Svenska Elektriska Aktiebolaget", as it was named, delivered its first induction motor in 1892. This was the same year as the company moved from its original hometown of Arboga to Västerås, which remains the main center for ABB in Sweden. The company built both AC and DC machines for the expanding Swedish industry. Machine number 1000, an induction motor, was supplied in 1896, but actually to an English customer. The export market was clearly attractive more or less from the beginning. Machine number 100 000 was delivered in 1916 and 1 000 000 was reached in 1939. The diagram in figure 6.8 shows

Asea's accumulated electrical machine production until 1948. The vast majority of these, close to 90 percent, were induction motors, so the diagram gives a good picture of how sales and manufacture of such motors developed during that period [110]. Figure 6.9 presents some additional information concerning the first decades. Growth was very rapid from 1895 when the 3-phase AC machines had become accepted. The most important customers belonged to the Swedish base industry, chiefly the steel and mining companies and the pulp and paper mills [104].

Millions of units

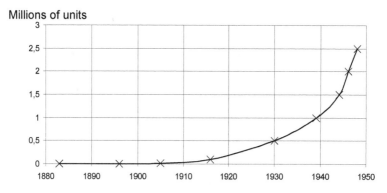

Figure 6.8 Accumulated numbers of electrical machines delivered by Asea until 1948

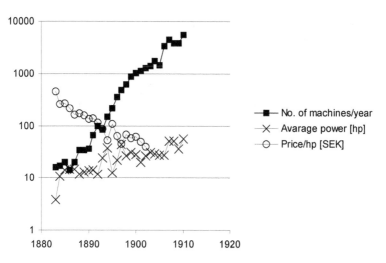

Figure 6.9 Asea's initial electrical machine production

6.2.4   Motor factories

Around the previous turn of century, Asea's smaller electrical machines were manufactured, not only in the main factory in Västerås, but also by Boye & Thoresons Elektriska AB in Gothenburg, The Fuller-Wenstrom El. Mfg. Co. in London and Norsk Elektrisk AB in Christiania (present day Oslo). It was, at least to some extent, a matter of induction motor production. Asea had, from time to time, major owner interests in these companies. Motor production in Gothenburg was soon closed, but production continued until 1916 in Oslo and much longer in England. The development and design of the machines was concentrated in Västerås. The establishment of all these factories could partly be motivated by the need for extra production capacity, but production in the Norwegian and British companies was more a question of securing presence in the local markets [110].

The electric motor business was considered to have a bright future and it is therefore not surprising that a number of other Swedish companies started similar manufacturing in competition with Asea. One of these companies was, as mentioned earlier in section 3.6, Luth & Rosén AB in Stockholm, which started manufacturing of motors in 1893 based on a license from Schuckert & Co, which in 1903 became part of Siemens & Schuckert. Luth & Rosén was initially a large supplier of AC and DC machines but focused later on geared motors and mechanical products. Asea showed interest in acquiring this company a number of times and did eventually do so in 1930, in part to avoid large foreign competitors from taking control of the company and thus securing a strong position in Sweden. Not long after Asea acquired Luth & Rosen, its electrical machine production in Stockholm was closed and transferred to Västerås, while the mechanical manufacturing remained in the Stockholm factory.

Another important manufacturer of electrical machines was Nya Förenade Elektriska AB (NFEA) in Ludvika, which was established through a fusion of three smaller companies; Holmia, Adolf Ungers Industri AB in Arbrå and Magnet in Ludvika. The two latter companies produced both generators and motors. Asea bought NFEA in 1916, primarily to increase production capacity, moved all manufacture of electrical machines to Västerås and turned the Ludvika works into a transformer and HV switchgear factory. There is no evidence that this acquisition had any direct influence on Asea's motor technology. Neither NFEA nor Luth & Rosén were financially strong enough for further independent expansion.

There were also three other companies, which began electrical machine production early last century. The first was Motorfabriken Eck, later renamed Elektriska AB Morén, which was established in Partille outside Gothenburg in 1907. Next, and by far the most important, was Elektromekano (Svenska Elektromekaniska Industri AB) in Helsingborg, which built generators and motors even up in the MW range. The third was AB Härnöverken in

Härnösand, which manufactured motors with gears. The latter two commenced with motor manufacture during the second part of World War I, which were commercially good years for most of Swedish industry. Asea officially took over these three companies during 1946-47, but continued their production for a number of years. Asea's acquisition of these companies was essential to increase its ability to meet the very large demand for electrical equipment after Word War II, but there were also other reasons. The real acquisitions had been made many years earlier, but were arranged as secret option agreements to avoid accusations of a domestic monopoly. Each company had continued to operate independently with technical development and production, but in accordance with Asea's marketing strategy [111].

6.2.5 Elektromekano

Elektromekano serves as a good example of this development period of the Swedish motor industry and Asea's relations to its smaller domestic competitors. Elektromekano evolved out of an old mechanical workshop in Helsingborg, which decided to start manufacturing of electrical machines and transformers in 1916. Its first years were characterized by rapid expansion and positive results, but the company soon suffered from the depression after World War I, and went bankrupt in 1923. After being restructured by Swedish banks, Elektromekano managed to establish certain cooperation with the large German companies AEG and Siemens. Elektromekano's financial results remained poor and domestic competition from Asea and Luth & Rosén, which had formed a cartel, was fierce. The president of Elektromekano tried to find a strong purchaser and he succeeded finally when the large telephone company LM Ericsson (LME) took over in January of 1931. At the time, LME belonged to Ivar Kreuger's (1880-1932) business empire. Kreuger, a world-famous Swedish businessman, controlled approximately 70 % of the shares. Asea had, much earlier, made half-hearted attempts to gain control of Elektromekano, but was now concerned that this company had become part of a much more powerful group. The situation changed drastically, however, in March of 1932, after Ivar Kreuger committed suicide in his apartment in Paris. He had engaged himself and his companies in large-scale, dubious financial arrangements and had no possibility to meet his obligations. Bankruptcy followed. This is known as the "Kreuger crash" that had huge consequences, not only in Sweden, but also for large banks and industries in foreign countries [112]. Because of this turn of events, Asea could now easily acquire Elektromekano, the only remaining domestic competitor, in the spring of 1933. This transaction was, however, not completed. Asea's management converted the purchase to an option, in which LME officially remained as owner, but received the agreed price as a loan from Asea with the shares in Elektromekano as security. The reason for this arrangement has already been mentioned above. The company officially became a member of the Asea group in 1946 [111].

A brilliant Danish engineer, Jens Lassen la Cour (1876-1953), served as president for Elektromekano from 1921. la Cour is in this context an interesting person. He had studied mechanical engineering at ETH in Zurich and was then employed as assistant to the famous electrical machine professor Engelbert Arnold at the Technical University in Karlsruhe in Germany. la Cour was co-author of professor Arnold's first books on electrical machine theory and later became the main author for subsequent books, even while holding leading functions within Swedish industry [113, 114]. In 1907, after three years as chief engineer in a Scottish company, la Cour moved to Sweden and into a position as technical director of Asea. The company lacked sufficient competence for its rapid technical expansion so the president, the legendary J Sigfrid Edstöm (1870-1964), who knew la Cour from his time at ETH, succeeded in employing him. la Cour made many significant contributions to Asea's technical development, e.g. the introduction of the latest electrical machine theories as the basis for the development of new machine series. He also invented and patented a 3-phase commutator motor (see section 6.7.3). la Cour left Asea in 1914 after a conflict with Sigfrid Edstöm, which infected the relations between Asea and Elektromekano for a long time. la Cour, and some other technical managers, criticized Edstöm for being too ignorant in technical matters to be able to lead the company in the right direction [115]. la Cour became director of Norsk Hydro but participated in the establishment of Elektromekano in 1916 and moved to this company in 1918. Elektromekano's development into a technically successful supplier of machines, transformers and other electro-technical products can, in great part, be attributed to Jens Lassen la Cour [116].

Figure 6.10 Sigfrid Edström and Jens Lassen la Cour

Elektromekano had a wide product program including various kinds of AC and DC machines. Deliveries of synchronous generators rated up to 4000 kVA indicate the capacity. Its induction motor series ranged from approximately 0.1 to 1000 kW. The company had a strong position as a supplier to Swedish industry and public utilities. A brief review of Elektromekano's order files, now stored at ABB's central archive in Västerås, shows that the domestic market was by far its most important, but it had some export to the Nordic countries as well, especially Denmark. After it became part of the Asea Group, there was also some export through Asea's foreign companies.

The matter of Asea's dominance in the Swedish market was politically sensitive during the period between 1945 – 55. Some members of Asea's board, specifically Marcus Wallenberg (1899-1982) and Ruben Rausing (1895-1983), wanted to avoid an Asea monopoly while part of the executive management saw a possibility for structural rationalization. They therefore had different opinions on whether Elektromekano should be closely integrated or not. The company was left with the freedom to operate very independently until the beginning of the 1960's when Asea's new president, Curt Nicolin (1921-2006), started to re-organize the company into product divisions. Elektromekano manufactured rotating machines under its own trade name until 1962 and, as a division of Asea, it continued to build small motors in the Helsingborg factory until 1967. Elektromekano had, for almost half a century, been Sweden's second largest manufacturer of heavy electrical equipment, and it had successfully carried out the development of its electrical machines and other products without external support [117, 118].

6.2.6 Asea's foreign motor production

It has been mentioned above that Asea had certain induction motor production in England, but this was not the only country. Poland had, since long, been an important market for Asea, but the government wanted local production, so the Swedish company acquired, in 1937, a majority share in a large Polish motor factory, Polskie Towarzystwo Elektryczne S.A., located in Warsaw. It was operated by Asea until 1944, when the factory was totally destroyed by retreating German troops. In 1944, Asea purchased a factory in Persan outside Paris for manufacturing of motors and other electro-mechanical products. The company also built a new factory for smaller motors in the vicinity of Melbourne 1946-48. Asea's motor production in England was closed in the mid of the 20th century, while the factory in Australia continued until the early 1990's, and that one in Persan was sold in 1998. The French factory developed and manufactured DC- and 3-phase commutator machines while the Australian unit built small induction motors designed in Sweden [119].

The largest, and most fascinating production site, was nevertheless Asea's Russian motor factory. Asea had, as early as 1893, established a representative in St. Petersburg. Russia was, together with Great Britain and the Nordic countries, a prioritized export market. Asea's sales to Russia increased during World War I, because Siemens and AEG were prevented to deliver equipment to an enemy nation. Russia was the largest foreign market during the war representing 29.5 percent of the company's total invoiced export during 1914-17. It was against this background, that Asea, in 1916, started to build a motor factory in Yaroslavl at the Volga River, 300 km northeast of Moscow. Then came the Russian October revolution on November 7, 1917, which caused a long interruption. Production, on a small scale, first started in 1927 while full production commenced in 1929-30. Asea's large motor factory in Yaroslavl, displayed in figure 6.11, was then one of the most modern in the world [120].

Figure 6.11 Asea's motor factory in Yaroslavl, USSR around 1930

The initial production was based on the supply of semi-finished components from Västerås until all equipment had been installed and a complete production line was available. The product range was limited to LV motors in the power range 0.2 – 500 kW. The factory in Yaroslavl had more than 1000 employees, around 50 from Sweden, but it did not perform any product development, even if it had a calculation and design office and also a laboratory. The production in the Yaroslavl factory grew rapidly up to approximately 30 000 units per year or more than 55 000 hp during the last accounting year 1931-32. This can be compared with an annual production of roughly 50 000 machines for the Västerås factory during this period.

It was considered remarkable that the Soviet authorities did not confiscate Asea's properties in Yaroslavl, and even allowed the company a concession to complete it and start production. The reason could have been that the communists gave electrification high priority – Vladimir Lenin (1870-1924) had said, "Communism is soviet power and electrification of the whole nation" - and they wanted access to modern technology. Nevertheless, it gradually became more difficult to run the factory and Asea therefore initiated negotiations with the Soviet government. These negotiations resulted in the Soviet state acquiring the factory in 1932 for an amount close to 8 MUSD. The investment had paid off very well for Asea [120]. An interesting side note to this history is that the factory remains an electrical motor factory, ELDIN, now in private Russian hands.

6.2.7 Production development

Cost effective production has been fundamental to machine development almost from the beginning. A good example of its importance is given in a letter from Asea's founder and first president, the technically very interested business man Ludvig Fredholm. He writes in 1888 a letter to Jonas Wenström making suggestions for the design of *"an extremely cheap machine"* and Fredholm asks *"could you not cast the magnetic yoke and the magnetic poles in one piece and then put the pre-fabricated coils around the latter?"* (My translation) [121]. Another interesting example is from a letter, which Jonas Wenström sent home to his father while he was on his American journey in 1888. He had just visited Westinghouse in Pittsburg when he wrote: *"This concerns the manufacture of the armature core. They have found that paper between the sheets is completely unnecessary. The oxide coating provides quite sufficient insulation between them. They have come to this important news through experiments and have found that the temperature rise is not higher – rather lower – without paper instead of with. On the other hand, they get room for more iron and the manufacture becomes of course cheaper. But in order to make full use of this, the sheets must be treated like theirs. They have first of all their excellent, accurate presses, which punches the segments without any single burr being detected. It is generally considered, here in America, that the machine tools are a main item, which must not be overlooked. Therefore, they don't ask how much they spend on this, as long as it gives a good result. Secondly, they use much thinner sheets than what we usually do, ¼ mm, I think. ...... And thirdly, the sheets (punched) are carefully annealed in special ovens, ... The visit to Westinghouse gave me a lot of good information."* (My translation) This can perhaps be considered as an example on early, legal industrial espionage, but more important is how much emphasis was put on production issues [21].

A further example is taken from Asea's annual report for the year 1929, four decades later [122]. This report includes a clear description of the situation and the main direction for the ongoing development. *"The activities during the year have continued along the same guidelines as in recent years. The demand on the company's products has remained lively, both from our own country and from abroad, and the company's turnover has increased compared with previous year. However, the price level for electrical products does not at all correspond to what a product based on such a comprehensive technical scientific research as well as many years practical experience ought to be worth. The competition within the electrical power area remains strong, and in order to successfully meet this, the company is forced to continue the modernization of design and production work, which requires building of new workshops and the introduction, in larger scale, of automatic and labor saving machines and equipment. The work during the past year has been characterized by intensive activities in this direction, and it is the board's intention to continue with this in coming years according to the existing plan."* (My translation)

Figure 6.12 Asea's winding workshop for smaller induction motors in 1935

Production development aims primarily at reduced manufacturing costs, shorter lead times and often also improved quality. Other objectives, at least nowadays, can be avoidance of operations, which could risk the health of the workers or have a negative impact on the environment. This is of course valid even for much of the product development, but it is, in the case of production development, achieved through introduction of new or improved manufacturing processes. There is often a strong interaction between these two kinds of development but not always. The years before World War I were a period of establishment of new factories and increase of production capacity. Motor manufacture was in many respects manual. Existing machine tools were driven by overhead belt transmissions. Introduction of rolling bearings in lathes and milling machines during the 1910's improved precision and thus product quality. The large belt transmissions were replaced by individual electric motors for driving the machine tools during the 1920's. Multi spindle machines also became available, which improved machining accuracy as the number of set up operations could be reduced. Line production was introduced for smaller motors at the end of this decade. Figure 6.12 shows line production in Asea's stator winding workshop. The examples above have certainly rationalized the production and improved product quality, but none of them led to radically new product designs [121].

Electric arc welding came into use in the early 1930's and it definitely changed many designs as lighter, but still stronger, welded parts, made from forged or rolled steel, could replace cast iron components. This was important for larger machines but not so much for smaller motors. Something that really changed the motor design was the introduction of die casting (pressure casting) of aluminum. It has been mentioned above that Asea's first motors with aluminum stator housings and squirrel-cage rotors were presented in 1945.

6.2.8 Competition

The domestic competition, up until the end of World War II, has been described above. One by one the competitors were taken over by Asea, which now had grown into a fairly large electro-technical company also from an international perspective. Its major competitors were AEG, Siemens and BBC, at least if the two World War periods are excluded. Especially the first two had a strong position in case of common induction motors, also in the Scandinavian market.

Initially, AEG was very successful and became a leading manufacturer of induction motors under Dobrowolsky's technical leadership. However, the company did not manage to keep up with Siemens and it run into financial difficulties. AEG negotiated in 1925 with Siemens about a merger, but Siemens terminated the negotiations due to the sensitive monopoly situation it would create. AEG continued on its own, but always as number two in Germany [123].

Siemens has, in a broad sense, always been Asea's main competitor, but this situation was not reciprocal. A comprehensive history of the Siemens company, covering the period up until 1945, does not even mention Asea. In the second volume of that history, the invention of squirrel-cage windings with deep bars is mentioned as a typical example on how a leading manufacturer can transfer its research results into commercially viable products. In 1919, the Siemens scientist Reinhold Rüdenberg (1883-1961) published the principle of the "eddy current" rotor winding. Rüdenberg was later promoted to electro-technical chief of the company but had to leave Germany in 1935 due to his Jewish background. The author of the Siemens history has, slightly chauvinistically, claimed that such practical implementation of research results makes the large difference between a leading company like Siemens and other European companies, which mainly act as followers [124]. Another novelty from Siemens was the totally enclosed motor with external shaft-mounted fan and axial cooling fins, which the company introduced in 1927.

In the following three years, the share of enclosed motors increased then from 10 to 40 percent of the company's total motor production. It can be noted that Asea was roughly four years behind Siemens with both these designs [125].

Both World Wars changed the relations between Asea and its large German competitors, at least temporarily. Asea had the advantage of less competition in certain markets and it had, with few exceptions, its production facilities intact. Siemens "assets of substance", on the other hand, were reduced with roughly 40 percent during the first war and around 80 during the second [126].

Asea and BBC do not seem to have been tough competitors in the field of standard induction motors. BBC focused on development of motors suitable for specific applications; especially can be mentioned textile machinery, an area where the Swiss industry was particularly strong. BBC introduced squirrel-cage motors up to 3 kW with die-cast aluminum windings already in 1935, almost 10 years ahead of Asea [127].

6.3 Standardization

Standardization is of immense importance for the industrialized society. It reduces costs for industrial products, it facilitates both domestic and international trade, it provides interchangeability of products and it establishes rules with respect to quality and safety. Early, but initially not very successful, attempts were made at the time for the French revolution, aiming at a common standard for measures [128].

Organized international standardization actually started within the electrical area. In 1906, the International Electrotechnical Commission (IEC) was founded at a meeting in London. "Rating of Electrical Machinery" was one of four technical committees formed already before World War I. A Swedish national committee for standardization was established in 1907 and it is evident that electrical machines were early on its agenda, because that technical committee has been designated number "2" out of approximately one hundred. Terminology was the first.

IEC's first publication concerning electrical machines, which was titled "Rating of Electrical Machinery", was issued in 1911, and it consisted primarily of extracts from existing rules in various countries. It was in total six countries. Sweden was one of them. The others were Belgium, France, Germany, Great Britain and the U.S. The publication contained rules concerning rating, name-plates, overload, heating including temperature limits and testing, over-speed tests, efficiency measurements, and dielectric tests [129]. Since then, IEC's general standard for electrical machines has been developed in several steps. The latest version is IEC 60034-1, Rotating electrical machines - Part 1: Rating and performance, Edition 11, 2004, which is applicable for most types of machines of the sizes treated in this study.

For LV asynchronous motors there are, however, also standards on ratings and dimensions and the background is as follows. As previously mentioned; motor series were introduced by manufacturers at a very early point of time. The reasons are obvious; the costs for engineering, tools etc. could be spread over several units, thus reducing the costs per motor. This was initially strictly a manufacturer standard. As motors become more widely used, the customers began to purchase motors from different suppliers and soon the customers wanted to be able to replace motors from one supplier with those from other suppliers. This led to standardization of dimensions such as shaft height, shaft end diameter and length, footprint etc. as well as standardized ratings (output, voltage and speed). These kinds of standards were introduced by the international organization IEC in 1959 [130], but in America by NEMA (National Electrical Manufacturers Association) already some years earlier. They have of course been subject to a number of revisions during the years and the IEC standard applicable today is "IEC 60072-1 Dimensions and output series for rotating electrical machines from 1991" [131]. Asea's Hans Hedström was, as representant for Sweden, a strong advocate for the IEC standardization of dimensions and outputs.

These types of extensive standards will, by definition, impose restrictions on development engineers. An important question is, how serious these limitations are? One possible indication, out of many, could be a comparison of the development of specific output and torque for such standard motors with other types, which haven't been subject to this type of standard, e.g. larger motors and traction motors. The diagram in figure 6.13 seems to prove such an effect. To compensate for this, manufacturers of standard motors chose sometimes the next smaller physical size for a certain rated output and provide the motors with adapters in order to maintain the correct shaft height and footprint.

A recent kind of standard requirements are the rules for minimum efficiency introduced both in Europe and in the USA. The background is of course that industrial motor drives consume a lot of electric energy and an increase of motor efficiency contributes to lower energy consumption. The European Union has introduced three efficiency classes for LV asynchronous motors ranging from 1 to 90 kW [132]. The suppliers are supposed to state which classes their motors belong to. Some suppliers, e.g. ABB and Siemens, have chosen to work with two motor series with different efficiencies and different costs in order to meet various demands and competition [133, 134]. The American rules are stricter. A law enforced by the Department of Energy under the Energy Policy Conservation Act [28] requires that all standard LV motors from 0.75 to 150 kW sold in the USA must be certified that they meet the efficiency requirements. This must be verified by tests carried out by accredited laboratories. In order to fulfil the efficiency requirements, the suppliers have to reduce motor losses by using more active material, higher quality electrical steel sheet, but also through optimization with low losses as target function. It is evident that high efficiency motors will be more costly. The diagram in figure 6.14 shows the European efficiency classes as well as the American requirements.

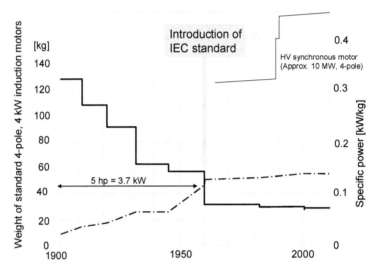

Figure 6.13 Performance development for smaller induction motors

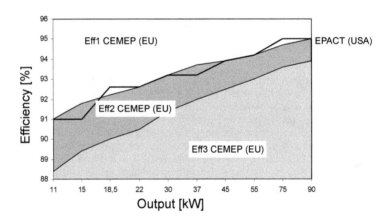

Figure 6.14 Efficiency requirements for 4-pole induction motors

6.4 Shift of generations

Standard products, such as LV asynchronous motors, are normally developed in generations. A new series replaces an older one, when its customer attraction and profitability diminish or when other motives exist. Examples of such motives can be availability of new materials or technical solutions, industry structure rationalization, new legislation etc.

Asea and later ABB have, as mentioned earlier, had several series of induction motors in parallel and all of them are not replaced by a new product generation simultaneously. On the contrary, it is done successively in order to obtain a fairly even load on development resources and key functions for production and marketing. It is, nevertheless, possible to identify some major shifts in generations of LV induction motors, and it is of interest to study the reasons behind these, the objectives for the new series, and how the development was performed.

6.4.1 New Asea motor series during the 1950's

Two of Asea's old induction motor series were replaced in the mid 1950's. One was the "MK series", which had been introduced in the 1920's. These motors were open-ventilated. The active parts had been improved during the years while the outer parts had remained unchanged. This series of motors had now become too large and heavy. The second series consisted of the "MR motors" developed in the 1930's. These motors were closed and provided with shaft-driven fans for external rib cooling. The utilization was still reasonably good but the motors had concentric stator windings, which resulted in large coil ends. Other components also needed to be modernized. It was therefore necessary to replace both these old type series.

An open-ventilated motor was most cost effective for a certain rated power because the cooling air passed in direct contact with the active components where the losses are created. The closed motor, on the other hand, had the advantage of being well protected and therefore more reliable. The cost difference was not that large for small motors and Asea was consequently only manufacturing closed motors up to around 10 hp. These were the MBB motors introduced in 1945 as mentioned earlier in section 6.2.2. Above this size, it was considered important to have both open and closed motors as well as short-circuit and slipring motors. Four closely related type series were hence developed as shown in table 6.1.

There had been attempts to agree within IEC on an international standard for footprints, shaft heights etc. but they had all failed. Asea decided, therefore, to follow the American NEMA standard for these new motors, expecting this standard to become generally accepted. This assumption proved later to be wrong.

Table 6.1   Asea's new induction motor series introduced in the 1950's

Type	Rotor	Cooling	Power [kW]*
MBB	Short-circuit	Closed	0.2 - 7.5
MAC	Slipring	Open	7.5 - 130
MBC	Short-circuit	Open	7.5 - 130
MARC	Slipring	Closed	7.5 - 130
MBRC	Short-circuit	Closed	7.5 - 130

*. 4-pole motors

Much emphasis was put on the insulation system. The stator winding was a distributed mush winding with one coil side per slot. Asea introduced class E insulation allowing 15 °C higher temperatures than the previous class A. The copper wire was insulated with a polyurethane varnish, later replaced by polyester varnish. Extensive laboratory tests were performed to prove dielectric strength, moisture resistance, ageing and sensitivity to vibrations. The rotor winding in slipring motors had the same design and insulation as the stator windings. The short-circuit rotors had windings with high copper bars in order to improve the starting properties. All motors were provided with rolling type bearings and, in this case, several tests were also carried out to verify lubrication methods and sealings. Figure 6.15 shows photos of two of the motors developed [135].

Figure 6.15 Open ventilated slipring motor type MAC and closed squirrel-cage motor type MBRC introduced in the mid 1950's

A new series of larger induction motors was developed during the late 1950's. These motors covered the output range 150 – 1500 hp and were designated MAD and MBD for slipring and short-circuit motors respectively. They were built both as low-voltage and high-voltage motors. The smaller sizes had the same type of windings as described above. They could be open-ventilated or provided with ducts that led the cooling air to and from the motor. They replaced an old series, MA, and it was primarily access to new materials that led to this development [136].

6.4.2   Asea's first IEC standard motors

Elektromekano in Helsingborg had been a member of the Asea Group since 1946 but operated very independently with its different product series. The company had its own series of small induction motors, called KR, which competed with Asea's MBB motors. This situation formed, of course, a basis for rationalization. The efforts to establish an international dimension standard for electrical motors had finally been successful. As a result of this, IEC published its first recommendations on such a standard in 1959 [130], and both Sweden and several other European countries prepared approval of these rules as national standards. The Asea Group had thus at least two strong incentives for the development of a new generation of small induction motors, i.e. up till 10 hp. The responsibility to develop and manufacture this new series of motors, the "M-motor", was allocated to Elektromekano. This seems as a good decision considering the performance of the two old series, which were to be replaced. Table 6.2 shows that Elektromekano's motors were, in general, smaller than Asea's for the same output. The weights are difficult to compare as some had aluminum frames and others cast iron. The decision to let Elektromekano develop and manufacture this motor series could also have been a compensation for the Helsingborg factory, when it finally had become a victim for structural rationalization (see section 6.2.5).

The development of the M-motor was in several ways extensive, and not only because of the new dimension standard. New materials such as cold rolled electrical steel, giving a better fill factor, and polyester-based insulation materials, allowing higher temperature rises, were used. Die cast aluminum was used for the stator housing, the end-shields and for the squirrel cage winding in the rotor. The shaft mounted fan was made from plastic. Improved production methods facilitated closer tolerances, which could be used for reduction of the airgap. All of this contributed to smaller and lighter motors as shown in table 6.2.

Application of new methods for calculation and design also had a big impact. This was the first motor series developed within Asea using computer calculations as an engineering tool. The same analytical equations as used in manual calculations were applied, but the computer allowed many more alternatives to be calculated and evaluated. This helped the development engineers to find more optimal solutions than previously. It was thus possible, even in spite of much higher utilization, to maintain the same efficiency as earlier. A cross section of the new M-motor is presented in figure 6.16 [137].

It can be noted that the size of a standard type electrical motor was, from this point on, specified by the shaft height.

The larger induction motors mentioned in section 6.4.1 did not follow the new dimension standard even if it was also intended to cover these sizes. The experience from the smaller M-motors with aluminum housing was positive and therefore inspiring an extension to larger sizes. This was an important reason for developing a series of larger M-motors, but not the only one.

Table 6.2   Shaft heights and weights for 4-pole induction motors for Asea's new M-series, old MBB-series and Elektromekano's KRB-series

Power [hp]	Shaft height [mm]			Weight [kg]		
	M	MBB	KRB	M*	MBB	KRB
0.25	63	80	70	5	6*	8
0.5	71	90	80	7	9*	11
0.75	80	100	80	8.5	12*	12.5
1	80	100	85	10	14*	16
2	90	112	100	16	30.5	30
3	100	125	115	22	46	40
5	112	125	125	31	58	60
7.5	132	140	145	42	82	85

*Aluminum

Figure 6.16 Asea's M-motor, its first according to IEC's dimension standard, 1961

Customers required low prices and short delivery times and especially the latter led to the need for new solutions. The series was to include slipring and short-circuit motors as well as open-ventilated and closed designs. With four different pole numbers and two standard voltages, the number of variants, which had to be kept in stock, would be far too high. The company decided instead to chose a modular concept in which a limited number of components were available from stock and they could then be quickly assembled according to specific customer orders.

The cooling system was the key to the modularization. The open-ventilated motors were partly cooled in the same way as the closed motors, but had an additional, parallel airflow through the inner parts of the motor. This was achieved through combination with different end shields and it increased, in some cases, the output by 50 percent compared to the closed version. Figure 6.17 shows photos of a closed and an open variant. This new series covered the output range 7.5 – 55 kW. The chief engineer Hans Hedström wrote in his annual report for 1964 that a cost reduction of 15 percent had been obtained for the open-ventilated motors [138].

Figure 6.17 Open ventilated foot motor and closed flanged motor from the late 1960's

Surprisingly enough, Asea had made the decision to have three different temperature classes for these motors, namely 120, 130 and 155 also known as class E, B and F. The copper wires always had class F insulation and the same was valid for the impregnation varnish, but the slot insulation could be chosen by the design engineer according to the thermal load. This does not make sense, especially taking into consideration the modularized structure discussed above [139].

A cooperation agreement was signed in 1965 between Asea and the Danish motor company Thrige-Titan A/S concerning development and manufacture of small induction motors. One result of this was that Thrige-Titan obtained the responsibility for development of a new type of small motors in the output range 0.12 – 3 kW. This series received the type designation MT and it was, to a large extent, based on a concept from Asea for a replacement of the small M-motors. The Danish company took over the implementation and a new factory for mass production of the MT-motors was built in Odense [140].

All development activities were not focused on motors in accordance with the new standard. There was also optimization and redesign of some older types. An interesting result is presented in the M-division's* annual report for 1968. Several variants of larger motors had been recalculated and optimized by means of a computer program and cost reductions of 10 – 15 percent had been achieved compared with the manual calculations. However, the report claims that the resulting dimensioning does not seem to vary much between the two methods. No explanation is given, but the program worked with 11 variables and small changes, in each of these, could obviously yield substantial cost differences [141].

6.4.3   Development during the 1970's

The president of the M-division, Lage Becker, wrote in a PM to the Asea management in August 1970; *The development of new machine series depends primarily on the possibilities to reduce the production costs. It is therefore not possible to state a common period of time during which a [machine] type shall be manufactured.* (My translation) [142]. Cost reduction was the primary goal for the development projects. This has also been confirmed in interviews with different key persons, e.g. Jan-Christer Zanders, manager for Asea's induction machine design office during the 1960's and 70's.

Several existing motor series were redesigned or even replaced. Computer programs were used for better optimization of the motors but also for rationalization of the calculation and design work. Class F insulation was introduced as standard for all machine types. Aluminum had long been used as conductor material in squirrel cage rotor windings. Copper was an excellent conductor for the stator windings but it was also expensive. A comprehensive study of the possibilities of using aluminum in the stator windings was therefore started in 1969.

In 1970-72, the development focused on a new series of open-ventilated motors in the 75 – 280 kW range. It was decided to use aluminum in the stator windings and in the short-circuit rotors but to maintain copper in the rotor windings for slipring motors. It is more difficult to cool the latter. The main project in 1973-74 was development of a series of closed motors rated 55 – 315 kW. It was planned to use aluminum stator windings also for these. Finn Möller, who was the M-division's manager for development and design, pays

---

*. Asea's M-division had been established in 1962 with responsibility for marketing, development and manufacture of asynchronous and synchronous machines with a maximum stator core diameter of 1000 mm.

a lot of attention to the aluminum windings in every one of his annual reports from 1970-75. Tests of pre-series seem to have confirmed the expectations and it is also stated that several hundred motors were supplied to customers. A short comment in the annual report from 1975 says that the recent price trends for aluminum and copper have not been favorable for aluminum [143].

At an earlier time in this development, in September 1971, the development manager Wolfgang Krecker (1927-1994) wrote a PM concerning the development of the closed motors. He gave examples on the costs for the conductor wire for certain motors, e.g. 27 kg Cu at 10.37 SEK/kg = 280 SEK compared with 9.6 kg Al at 9.10 SEK/kg = 88 SEK. It is evident that such a substantial cost saving was desirable. It was also concluded in the same document that the efficiency would be affected, and that the specified target, to have at least the same efficiency as Siemens and AEG, had to be revised [144].

The 1976 annual report for the division does not mention aluminum windings at all. It seems like an issue that just disappeared. Finn Möller has, however, mentioned in an interview that aluminum windings continued to be used in certain types of medium sized motors, specifically the MBM series. Certain types of MBM motors had, in spite of this, copper windings in order to fulfil the performance requirements. It is interesting to note that Asea considered it worth mixing copper and aluminium windings in the same series and the same production line. The aluminum windings were used for approximately 10 years until the production of these motor sizes was transferred to Finland and also requests for higher efficiency made the aluminum windings obsolete. The technical result had been good and there were no complaints from the customers. The 1976 annual report tells instead that the development of new designs would focus on continued mechanization of manufacturing operations [145]. This is an important remark because it reflects which area the main efforts were made within, for development of new induction motor series. Introduction of new machine tools and production methods often required new or modified designs. The product and production development had to go hand in hand. Some examples can be given.

More powerful die casting presses were installed, which enabled the designers to abandon cast iron stator frames and end shields and instead chose aluminum also for larger sizes. Numerically controlled multi-operation machine tools improved the machining accuracy allowing the tolerances to be reduced. Automatic winding machines came into use and replaced most of the manual work, see figure 6.18. All windings were not suitable for such an automatic process, however, which imposed restrictions on the stator winding layout. The first industrial robotics appeared in the workshop, but they were mainly used for lifting and moving components and had, at that time, no impact on the motor design. Asea was on its way towards one of the best mechanized motor factories in the world (comments from customers during the 1990's.).

The oil crises during the 1970's increased the awareness of the need for energy conservation and higher motor efficiency. Some customer even applied a loss evaluation, but most of them were not prepared to pay extra for high efficiency motors. According to minutes from a meeting in February 1977, representatives from the M-division claimed that they did not believe in a split up of the standard motor market into a high and a low efficiency group [146]. Less than a year later, Finn Möller wrote, in his annual report, that two motor series would probably exist, one more expensive with higher efficiency, and one low price with lower efficiency [147]. It took a long time until this happened, but it is now a reality.

Figure 6.18 Automatic winding machine in Asea's motor factory, 1970's

Another important issue was noise. The customers required quieter motors and Asea had, in this respect, some difficulties compared with those competitors that had kept their cast iron motors. Every annual report from the development and design department during the second half of the 1970's explicitly addressed this as a problem of growing importance.

The development of new motor series had been the responsibility of the functional line organization. Each main department carried out the part of the work, which belonged to their usual tasks. A new principle, with a special project organization that would be responsible for the complete project until commercial production started, was introduced in 1976 [145].

6.4.4   Competition from Eastern Europe

Asea's traditional main competitors were Siemens, AEG and BBC. A strong competitor in the Nordic market was also the Finnish manufacturer Oy Strömberg Ab. The only Swedish manufacturer of standard motors, besides

Asea, was AB Elmo in Flen. These were all market economy companies with a fairly similar cost structure; prices and performance did not vary too much.

A new type of competition grew strong during the 1970's. Manufacturers from the German Democratic Republic, Czechoslovakia and Poland increased their market shares in West European countries considerably due to a low price policy. Asea estimated that the Swedish import from Eastern Europe in 1977 represented a market share of approximately 50 percent for induction motors in the 0.75 – 45 kW range. The Swedish manufacturers claimed that the East European companies sold their motors at prices below the costs for material and that they thus were dumping the market. Elmo wrote a letter to the Swedish Department of Trade, which initiated a study that was completed in October 1977. The study showed that the average price for East European motors rated 3 – 7.5 kW was 5.20 SEK/kg while the corresponding figure for Swedish motors was 17,31 SEK/kg. At large, the prices for imported, low price motors were only 37 percent compared with corresponding motors from Western Europe. An example from Asea was a standard 55 kW motor, for which the material costs were 7.53 SEK/kg. The corresponding complete motor could be purchased from Eastern Europe at a price level 5.70 SEK/kg [148].

According to Finn Möller, Asea bought some East European motors and tested them. The conclusion was that they did not have the same quality as Asea's and Siemens' motors had. Especially the efficiency was much lower. Demanding customers realized this and preferred motors from leading manufacturers. The cost for the motors was usually only a small fraction of the investment in an industrial plant and failures, which could stop or disturb production, were very costly. Möller claimed that the competition was bearable but irritating. However, competition has often triggered development of new product generations, and it did so also this time. The development of Asea's next series of standard motors will be dealt with in the next section.

6.4.5   Pre-study, pre-project and main-project

The introduction of a new motor series is the result of a long development process usually beginning with a pre-study. Discussions concerning a new series of smaller induction motors, in this case up to 45 kW, had started in 1977 and a pre-study was launched in the autumn of the following year. The purpose of this study was to decide on design principles and performance requirements, but also to propose a pre-project. The pre-study, which took approximately half a year, concluded that a new series should be based on conventional design principles. Other concepts had been studied by the Central Laboratory but were not considered realistic. The necessity to improve performance was a major argument for replacement of the old series, especially with respect to torque characteristics and noise. Efficiency and power factor were considered more or less acceptable as they were. Bench marking had been done against eight different competitors. Some experimental motors were built and tested in order to investigate how the cooling was influenced by various stator house designs.

The pre-study showed that optimization of the active parts would not give any cost reductions, because the possible improvements had to be used for better performance. Other changes and production development could hopefully give close to 8 percent lower costs. The engineering resources, required for the main-project, were estimated to 26.5 man-years. The new series was to consist of six standard sizes and the generation shift was intended to be completed during 1981-84, but, in reality, it took more time than projected. A pre-project was proposed that required 2 – 3 man-years of work. The results of the pre-study were reported in May of 1979 [149]. It is worth noting that the pre-study did not leave several alternatives to be further investigated within the pre-project. The recommendations were very specific. The annual report for 1979 mentions that the need for slipring motors, in this lower power range, had almost disappeared. The power grids had become strong enough for direct starting of short-circuit motors. It was a similar situation for open-ventilated motors of this size. The vast majority of motors were closed.

The recommended pre-project started in July 1979 and was completed in April the following year [150]. It included primarily the following activities:

- Market analysis and sales prognoses.
- Performance requirements and detailed design solutions.
- Detailed analysis of manufacturing operations and cost calculations.
- Investment analysis, especially concerning tools.
- Manufacturing and testing of some prototypes
- Detailed planning of the main-project
- Study of consequences of production in Asea factories in foreign countries.

An important conclusion of the pre-project was that improvements and cost-reductions, which could motivate a shift of generation, must be achieved through many small changes in design and production. No radically new solutions had been found. The new generation would have the improved performance required from the market and, at the same time, the production costs would be decreased by 5 – 6 percent. The pay-off time for the project was calculated to be 3 -4 years. 2/3 of the project costs would cover investments in tools and 1/3 engineering and experiments.

The pre-project report stated that improvement of torque characteristics, efficiency, etc. could be theoretically analyzed with good accuracy by means of better calculation programs, while reduction of the noise level still had to be treated empirically. The report was presented to the board of the M-division in June 1980 and an approval, to start the main-project, was obtained. An important comment from the board was that operation with frequency control should be taken into account (see also section 6.7.7) [151].

This new generation of smaller motors was designated MBT and the only version developed was a closed short-circuit motor, see figure 6.19. The MBT series would consist of six frame sizes, the shaft heights 112, 132, 160, 180, 200 and 225 mm, cover the output range 4 – 45 kW and include pole numbers from 2 to 8. All motors should have class F insulation, but would only be utilized in accordance with the class B standard. This was a fairly common praxis. Both customers and suppliers did not fully believe in using the total, allowable temperature range, so this was a way to get some extra safety margin, but at the cost of a somewhat too large motor.

Figure 6.19 Cross section and photo of MBT motor, 1982

The main project started directly and the shift of generation was to take place successively, frame size by frame-size, from the end of 1981 until the end of 1984. The long time span was determined by the engineering and testing capacity. The target was to improve performance according to the technical specification and to reduce production costs by 5.5 percent. Reduction of the noise level had been identified as a very important issue. Asea's target was that the motors should be second to none in this respect. Later tests proved that changed combinations of the number of stator and rotor slots had been effective in this respect. The planning was based on an annual production volume of close to 100 000 motors from 1984 [150].

It has been mentioned above that no radically new design concepts had been identified which could be implemented in the new series. It is significant that almost half of the cost reduction was to come from cheaper ball bearings and simplified bearing installation. A number of small improvements should make up for the rest.

A steering committee and a project manager were appointed. A binder placed in ABB Motors archive in Västerås contains protocols from 55 steering committee meetings. These documents have been reviewed and have given clear information about the focus and the execution of the project. The steering committee consisted of the division manager and a few persons representing the design and production departments. The marketing department had no member in this group. The steering committee had the responsibility both for management of the project and follow up on its progress.

A team of six persons worked within the "MBT project"; electrical and mechanical design engineers and production specialists; not all of them full time. The major steps to be carried out were electrical dimensioning, mechanical design and drawing, manufacturing process development including quality assurance operations, acquirement of tools, manufacturing and testing of prototypes and pre-series, and preparation of various types of documentation [152].

The different motors in a standard series like the MBT have to be defined in a product catalogue. This means that the electrical engineers have to calculate the performance for a large number of specific motors. The standard series, covered by the six frame sizes, comprised 10 output ratings at four different pole numbers. The motors had different ratings for 50 and 60 Hz and could be obtained for different voltages such as 400 or 690 V. They could also be wound as two-speed motors, which meant that the stator winding could be connected for two different pole numbers (see section 6.7.3). All this required calculation of hundreds of motors to determine start and maximum torque, currents at different operation modes, power factor, efficiency and machine parameters as resistances and inductances. All these calculations, including estimation of temperature rises, were made with ABB's analytical program "OSKAR", successively developed since the early 1970's [153]. The mechanical design engineers had a more limited number of variants to cope with. Most of the frame sizes could comprise two different core lengths and each motor could be built either foot- or flange-mounted as can be seen from the catalogue [154]. The mechanical analyses required for this range of motors are rather limited. The relation between electrical and mechanical engineering, in this kind of development project, was, according to the protocols mentioned above, approximately 1/4 while in the case of larger machines it is often in the order of 1/10.

It has already been mentioned that the executive management had required that frequency converter operation had to be taken into account. The protocols from the steering committee as well as interviews do not reflect that this received much attention, if any at all. Nonetheless, a technical specification for the development of the larger frame sizes, MBT 180 – 250, written in October 1984 specifies: *"Frequency converter supply: Standard motors shall be able to meet the operation requirements, i.e. even continuous frequency converter supply within the frequency range 0 – 60 Hz. The possibility should be taken into account (and the need investigated) to operate up till 7000 rpm"* (my translation) [155]. According to a steering meeting protocol from June 1985, the design manager Finn Möller reported that YT (Asea's converter department) had decided to force the development of a new frequency converter and *"it is of utmost importance that we keep up"*. The protocol also referred to tests planned at Corporate Research with an existing experimental motor [156]. Only a minor fraction of the manufactured motors were for variable speed drives and it thus did not make sense to let these influence the design and costs for the standard motors. The simple solution was just a down- rating of the output for converter fed motors (see section 6.7.9). The frequency converters represented a much higher cost, so a slightly more expensive motor made little difference.

Asea Motors carried out the MBT project mainly with internal resources. Some assistance was, especially in the beginning, received from Corporate Research. This was mainly a matter of theoretical and experimental investigations focused on noise and cooling. The archived documentation contains numerous reports of which one is listed as an example in the reference list [157].

Noise has been mentioned as a major problem. It had been observed that the rotor slot configuration had a big influence on the "electro-magnetic noise" and a comprehensive study of this phenomenon was initiated separate from the MBT project. This study was carried out by Ingemar Olofsson, later technical manager and then division manager for the LV motors. He extended the study to comprise all aspects of noise in this kind of motors [158].

During the actual period, i.e. the first part of the 1980's, Asea also had factories for production of induction motors in Spain, Australia and Mexico. All of them were supposed to manufacture the new MBT motors, at least certain sizes. They did not participate directly in the development project but received various proposals from Västerås for comments. They also made the necessary drawings for their own production.

6.4.6   The manufacturing process

The competitive strength for standard induction motors depends more on the manufacturing process than the product design. The design does not differ much between manufacturers all over the world. It is easy to buy these kinds of motors from other manufacturers and take them apart to study them in detail. This is most likely done by every manufacturer. The performance and quality of the motors can nevertheless be very different. One reason, for these differences, is the quality of the material used. More expensive materials, e.g. electrical steel, can have lower losses and result in motors with higher efficiency. Another reason is the manufacturing process which is much more difficult to copy than the motor itself. A rational production usually means low manufacturing costs, which improves competitiveness. Asea's motor factory introduced numerically controlled machine tools as well as robotics as early as the 1960's. The production process also has a huge impact on the product quality. The treatment of the material, the quality of the tools, the kind of processes used, the tolerances achieved, the quality assurance applied and the skill of the workers. All of these factors are very important.

The development of standard motors cannot be understood and discussed without some knowledge about the manufacturing process. This has, in recent years, been influenced by outsourcing and other globalization efforts, a matter that will be further analyzed in section 6.8.3. Asea and later ABB still had, during the 1980's and 90's, a highly integrated production process for induction motors, starting with almost raw material and finishing with complete products. It had not always been that way. Lage Becker mentioned in an interview that the integration level was increased during the 70's in order to reduce lead times by avoiding disturbances in supply of components. The diagram in figure 6.20 illustrates the major steps in the production process, applicable to the size of motors represented by the MBT series. The diagram actually reflects the situation, as it was around 1990, when the concept of winding loose stator cores had been introduced instead of the previous method of winding cores, already cast into the stator housings.

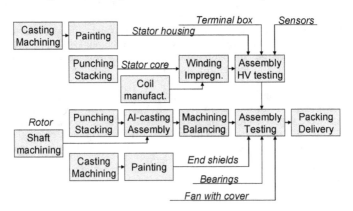

Figure 6.20 Simplified production flow diagram for standard induction motors

The input for the production is raw material such as aluminum and processed materials in the form of cold rolled electrical steel, enameled copper wire, insulation folios and resins. Standard components, like ball bearings and temperature sensors, but even specially designed plastic fans, were manufactured by specialized suppliers.

Stator frames and end shields were made in a die casting line provided with robots for handling. The necessary machining was made before the pieces left this line.

Figure 6.21 Punching tool, stator core stacking, insertion of coils, end winding bandaging, stator impregnation, rotor balancing

The punching process had been radically improved during the 1970's through the installation of new presses and investments in new tools. This was essential both for production cost and for quality. Stator and rotor segments were punched in the same operation and the stacking of the cores was also integrated in this process station. The stator cores were then transferred to the winding line for automatic insertion and forming of the pre-wound coils. The winding operation ended with impregnation and high-voltage testing. The introduction of automatic winding has radically reduced the amount of labor required for this process. A standard frame size 132 motor today needs around 60 minutes for manual winding but only 6 minutes for automated winding. The wound stator cores were finally pressed into the stator frames and terminal boxes were mounted.

The rotors were, in the meantime, manufactured in a separate line. The rotor cores were provided with squirrel-cage windings from aluminum in a die cast process. Shafts had been machined in an automatic lathe and were then pressed into the cores. Final machining of the rotor surface and balancing were also done by automatic processes.

Final assembly, testing and packing were still done manually. Figure 6.21 contains a collage of photos from the motor production.

6.4.7   Large induction motors

In 1980, Asea got a new president, Percy Barnevik, who started to decentralize the company during the early 1980's and Asea Motors became a subsidiary to the mother-company Asea AB in 1984. Asea Motors had two divisions in Västerås, one for DC- and large AC-machines and one for somewhat smaller AC-machines. The latter had three production lines; the A-, B- and C-workshops. The A-workshop built motors in frame sizes 112 – 225 mm, B manufactured the sizes 250 – 355 mm and C sizes 400 – 630 mm.

The MBT series, described in the previous section, was manufactured in workshop A. In some respects, especially technical, motors belonging to workshops B and C were more challenging and it is therefore of interest to include some comments on these larger machines.

The largest motors were the MBR and MAR series developed during the late 1970's and early 80's. These series covered the frame sizes 400 – 630 mm and outputs up to 5000 kW. These machines were available with low- and high-voltage windings, squirrel-cage (MBR) as well as slipring (MAR) rotors and several cooling arrangements such as open ventilated or enclosed with built-in heat exchangers. The design concept was modularized allowing a lot of variants to be built with a limited number of components. Figure 6.22 shows photos and cross section of an enclosed MBR motor with an air/air cooler at the top [159].

*Figure 6.22 Stator on base frame, rotor bars and short-circuit ring, exte-
rior and cross section sketch of MBR motor with built-on air/air cooler*

Major technical issues for the MBR motors were losses, noise and starting
properties. An important change, for this size of motors, was that uninsulated
aluminum bars replaced insulated copper bars in the squirrel-cage windings;
a solution which later gave rise to problems in motors with skewed windings.
These problems were addressed through extensive studies and even development of a dedicated computer program [160]. The large motors required
much more theoretical investigations than the smaller sizes, for which different prototypes could be manufactured and tested with much greater ease. The
number of graduated electrical engineers in the design offices, for the three
product lines, gives an indication of this. Table 6.3 shows these numbers in
the years 1979 and 1984.

*Table 6.3   Electrical enginers in Asea's development and design
department for AC motors, MK*

Office	1979	1984	Comments
MKA	3	1.5	Motor size 112 - 225
MKB	3	1.5	Motor size 150 - 355
MKC	4	4	Motor size 400 - 630
MKX mm	5	2	Development

The medium sized MBM motors, part of them with aluminum stator windings, were mentioned in section 6.4.3. An improved version of these motors was developed during the mid 1980's. The primary goal was to improve the efficiency and this was achieved through replacement of aluminum with copper in all stator windings, improved quality of the core lamination, more aluminum in the rotor cage windings and smaller fans. The diagram in figure 6.23 shows how the efficiency has been successively improved when new generations, of this range of motors, have been developed.

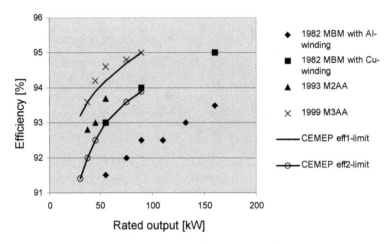

Figure 6.23 Efficiency development for ABB standard motors

All the electro-magnetic analyses performed, up to the end of the 1970's, had been analytical, when the first attempts to use FEM were made. Asea Motors had just employed Chandur Sadarangani, who had developed a FEM program at Chalmers as part of his PhD work. The program, which could solve two-dimensional, non-linear, static problems and perform linear time variant analyses, was implemented in a mini computer and was used, in the PhD thesis, for calculation of fluxes in a hydropower generator and eddy current losses in its stator winding [161]. A corresponding program had been developed within Asea and it was initially utilized for transformer analyses. This program was now used by Sadarangani for calculation of stray losses in rotor conductors in squirrel-cage motors [162]. This was probably the first time FEM was used for electro-magnetic studies of rotating machines within Asea*. Chandur Sadarangani later became professor of electrical machines and power electronics at KTH.

---

*. Lage Becker and Erik Morath had used numerical methods for calculation of the commutation flux in a DC machine in the early 1960's, probably using the finite differentiation method.

The electro-magnetic FEM analyses required powerful computers and it was first during the second half of the 1980's that workstations, with sufficient capacity, were available for interactive simulations. This kind of FEM was therefore not used much in electrical machine design until Asea, in 1987, established a central unit "Rotating machines" for development of applied simulation tools based on the company's own "Ace program". Asea Motors was one of the major participants in this activity, which actually was a unique initiative from five different electrical machine units to start a joint effort [163].

6.5 From Asea to ABB

6.5.1 Asea acquires Strömberg

The European heavy electro-technical industry suffered from overcapacity from the 1970's. Asea, which consider itself as the smallest of the large world leading electrical companies, could not finance increased development costs just through organic growth. The management saw an integration of the Nordic companies as a suitable way to increase the business base. It was against this background that Asea, in June 1986, announced that it had acquired the Finnish company Oy Stömberg Ab with 7 000 employees and later the following year the Norwegian A/S Elektrisk Bureau with 9 000 employees. Asea had held, since the 1960's, a minority interest in the Danish motor manufacturer A/S Thrige-Titan, but this activity was integrated with "Asea Danmark" in the early 1980's resulting in a total work force in the order of 3 000 persons. All these acquisitions were contributing significantly to Asea's growth from 40 000 employees in 1980 to 73 000 in 1987 [164].

Strömberg had a strong position as an electrical machine manufacturer, at least technically. It built standard type induction motors in Vaasa in north-western Finland and larger machines in Helsinki. The company's motors had a good reputation among its customers, but its economic results were poor. A working group was appointed, directly after Asea's acquisition, with the task of investigating the electrical machine activities in the two companies in order to propose structural re-allocations. The result of this was that Asea's production of larger induction motors (the B- and C-workshops) was transferred to Helsinki while all Strömberg's DC machines were taken over by Asea. This was to some extent a painful process, especially for Asea Motors, which considered that they lost much more than they gained through this product exchange. Asea Motors' management protested in vain that the Västerås factory was more rational and profitable and the products at least as modern as Strömberg. The preceding year, Asea Motors AB result of operations (before depreciations and financial items) was 7.4 percent of the revenues [165]. Everything related to the B- and C-workshops was transferred to Waasa and Helsinki respectively. The decision was probably based more on political than strict economic reasons. Asea could not acquire Strömberg and close factories and drastically reduce the workforce. It can be asserted, however, that this decision contributed to a radical shift in the relationship between Sweden and Finland with respect to electrical machine development and manufacture. Today more than 50 percent of ABB's worldwide electrical machine business emanates from Finland. Asea, and former Strömberg, took the first coordination steps, but any effective joint development did not start until a much larger merger became official.

6.5.2   Asea and Brown Boveri form ABB

The largest international merger in Europe, up until that point, was announced at press conferences in Stockholm and Zurich on August 10, 1987. Asea and BBC were to join and form a new company, ABB, beginning its operation on January 1, 1988. The new company would be one of the largest electrical industries in the world with 170 000 employees. The resources for development and manufacture of electrical products and systems would become very large. Electrical machines belonged to the core products for both Asea and BBC and ABB had, at the start, more than 20 different factories for manufacture of rotating electrical machines, most of them in Europe. Factories for LV induction motors were located in Waasa in Finland, Västerås in Sweden, Odense in Denmark, Saarbrücken in Germany, Decine in France, Sabadell in Spain, Mexico City in Mexico and Melbourne in Australia.

ABB was structured in a number of business areas (BA), which each held the responsibility of development, manufacture and sales of its products. The BA for motors immediately started coordination activities aiming at a coherent product program and rationalized production. An assembly of the design and production managers, from the various factories, constituted a "Technical Forum" for exchange of information and planning product development etc. The BA management made the decisions. Many local managers, who earlier had been very independent, felt in this new situation that they had become distant from the decision process.

6.5.3   The latest motor generations

ABB started with four different series of standard induction motors in the power range covered by Asea's MBT type. There were, in addition, one series from Strömberg and two from BBC. The BBC factories had approximately the same production volume as Asea including Strömberg, but they operated with substantial financial losses. They had difficulties to compete with standard motors and built quite a lot of special designs. The performances of the different motors were fairly equal but BBC's were more difficult to manufacture. It was with this background that the decision was made to develop one new common series of motors, the "M2000". The project started, after completed pre-studies, in the beginning of 1991 [166].

The objectives for the M2000 project were to reduce production costs through structure rationalization and to be able to offer a common series of motors with high performance. Efficiency and noise level were to be at least equal to the best competitor motors. Siemens and Leroy Somer served as benchmarks, in both cases through comparison of published data and testing of samples.

The factories in Odense, Västerås and Waasa shared the responsibility for development. The frame sizes 112 – 250 mm were allocated to Västerås. The dimensioning of the active parts was always performed in parallel by two teams, e.g. sizes 112 and 132 were calculated both in Västerås and in Decine. Each team used existing programs and this created a spontaneous competition, which improved the final result. The project did not result in any radical design changes, but optimization reduced the costs for active material by 5 – 7 percent. Some earlier compromises, imposed by tooling costs, could now be eliminated due to ABB's larger production volumes. Production development accounted for much of the efforts. The factory in Västerås transitioned from casting the stator housing around the stator core to pressing an already wound core into a machined housing. This simplified automation of the entire winding process and it also enabled more rational production through the supply of wound stator cores between the factories. Figure 6.24 shows voltage testing of wound stator cores.

Figure 6.24 High-voltage testing of wound stator cores

The M2000 machines had class F insulation, but the standard versions were rated for class B temperature rises. The traditional safety margin remained and the new series constituted no big step in utilization. However, some large OEM customers, e.g. the compressor manufacturer Atlas Copco, wanted more compact motors to help them reduce the size of their own products. ABB Motors in Västerås therefore made a special version in which the next higher standard power was chosen for each frame size, in some cases even the second higher. This was achieved through increased active length and utilization of class F temperatures. One reason for stressing two steps higher power rating into the Västerås factory's largest frame sizes was to "steal" that power level from the sister factory in Waasa. Both internal and external competition is a major force behind much development. Figure 6.25 shows a sketch of an M2000 motor and such a motor installed in a compressor unit from Atlas Copco, a typical OEM application.

Figure 6.25 a) Sketch of M2000 motor and b) M2000 motor installed in a compressor unit from Atlas Copco

Variable speed operation had become common at the time of the M2000 development and this had, of course, to be taken into account. The access to frequency converters, with much improved switching properties, simplified this requirement. It was not necessary to specify any de-rating. Fairly large airgaps were chosen and the extra losses were so small that they could be contained within the normal margins. This also helped to reduce noise.

The M2000 project required large engineering resources and investments in production lines. The design and development department in Västerås increased its staff from 12 to almost 30 persons during the development period, not all of them for this project. The necessary know-how was available internally within the BA.

The motors developed within the M2000 projects received the type designation M2AA. A revision has been made, and the present series of ABB's LV induction motors has been renamed M3AA. This latest series has only been subject to minor changes, but the new motors have somewhat higher efficiencies. The motor factory in Västerås is now also producing cast iron motors, which it took over after the factory in Saarbrücken was closed. This does not cause any major complications, because all castings, both for aluminum and cast iron motors, are purchased from external foundries. It was mentioned in section 6.3 that the EU Commission had specified three efficiency classes for standard induction motors. ABB is, like some of its competitors, offering motors in accordance with the requirements for the two upper classes. The highest efficiency is obtained through reduced current and flux densities, and such motors therefore become heavier and more expensive. A review of data for ABB standard motors in the range 11 – 55 kW indicates a 2 percent average improvement of efficiency from class "eff 2" to class "eff 1" motors. The weights of the latter have, as an average, increased by 15 percent and the price by 12.5 percent.

Standard induction motors can be seen almost as a commodity, and the price is of great importance. Lower costs have therefore been a major target for development projects since very long ago. The production costs for a modern motor is only a fraction of what it was for a motor with the same output 50 or 100 years ago. The diagram in figure 6.26 shows the development over the years. It is based on a few different sources specified in the reference list. The different sources are not consistent. The 19th century values are based on Asea's total sales of electrical machines, not only induction motors [104]. Helén's index diagram presented in Asea's History 1883-1948 [110] was obviously based on current prices, which has been converted to fixed values in the diagram below. Helen's diagram is very detailed and shows clearly the large price increases during the Word Wars, most pronounced during World War I. The index presented in reference [167] confirms this. Hans Hedström presented, in an article in Asea's Tidning [107], a diagram showing the development of motor prices from 1905 until 1960 in which all short-time variations had been eliminated. This is also the case for the later part of the diagram that is based on recent information [168].

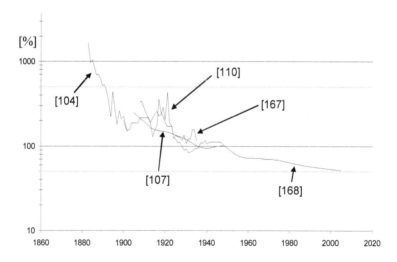

Figure 6.26 Development of price for standard induction motors

155

6.6 Domestic and international competition

Some early Swedish manufacturers of asynchronous motors for industrial applications, in addition to Asea, have been mentioned in section 6.2.4. The largest was Elektromekano in Hälsingborg, presented in section 6.2.5, which developed and manufactured motors under its own trade mark until the beginning of the 1960's. Other manufacturers during the last 50 years have been Hägglunds in Örnsköldsvik, Elmo/Danaher Motion in Flen and BEVI in Blomstermåla. ITT Flygt in Emmaboda is an example of a manufacturer of special asynchronous motors. Hägglunds was, like Elektromekano, acquired by Asea. The other three are still in business as independent motor manufacturers.

6.6.1 Hägglunds

AB Hägglund & Söner, located in Örnsköldsvik in northern Sweden, started as a carpenter factory in 1899 and was, for a long period of time, a true family enterprise. It was managed first by the founder Johan Hägglund (1866-1956) and later by his eight sons until 1967. Production during these years became much diversified going from furniture to coachwork for buses, trams and railcars. Further products were added in the blockade period during World War II, such as wood gas generators, spot-welding machines and even aircrafts. An electrical engineering department was set up in 1940-41, initially to secure competence in electric motors for tramcars. The post-war years were characterized by a shortage in manufacturing capacity for many products; one of them was electric motors. Hägglunds believed in diversification as a way of making the company less vulnerable and therefore started the manufacturing of industrial induction motors in 1948. Other products, added during the 1950's and 60's, were mining equipment, military tanks, ships' cranes and hydraulic motors. The company has been known for taking on new products and for developing them itself.

The motor production in Örnsköldsvik comprised standard induction motors in the power range 0.2 – 200 kW. A brief review of order documents* showed that the bulk of sales was small motors below 10 kW. Very few motors were larger than 100 kW. Hägglunds had its major customers in the domestic market such as large OEM companies manufacturing fans, pumps and other mechanical machinery. Hägglunds also supplied a couple of OEMs with separate stators and rotors for integration in products like submersible pumps. Hägglunds' direct export was very limited. Motors were sold to Norway and to some remote markets such as South Africa and the Far East. Jan Glete wrote in his book "Asea under hundra år" [169] that Hägglunds, like earlier

---

*. Hägglund's order documents stored in ABB's central archive in
    Västerås.

Härnoverken, established electrical motor production mainly in order to serve the local forest based industry. Whatever the original intentions may have been, Hägglunds' market eventually broadened beyond the local industry. Many of Hägglunds' major customers were located in the Stockholm region and south of it.

Hägglunds gained much of the necessary experience and know-how for motor production from manufacture of a number of prototypes, several years before it decided to start commercial production. The company also employed a few experienced electrical engineers from Asea and other companies. One of the Asea engineers, Carl-Gunnar Östberg (1913-1986), later became manager for Hägglunds' motor activities, a position he held as long as Hägglunds manufactured electrical motors. Two more engineers were recruited for electrical calculations of induction motors; one from Elektromekano and one from Motorfabriken Eck. The development and design office for electrical motors had in total 10 – 15 employees. Asea's corresponding department had around 50 – 60 employees, but also a much wider product program. During the start-up phase, Hägglunds had Asea's motor series as a guideline, but it never had any formal cooperation, neither with Asea or any other motor manufacturers.

Hägglunds manufactured, except from some DC traction motors, only LV squirrel-cage induction motors with closed cast iron housings and surface cooling. The production volume became as high as 90 000 motors per year in 1966 – 67. Hägglunds' motors had a good reputation, especially in the heavy process industry, and the main reason was the robust cast iron housings. There is no evidence that Hägglunds had any support from universities concerning induction motors, but it received help from Professor Emil Alm (1878-1963), KTH, for the development of the traction motors.

Hägglunds had undergone a dramatic expansion and faced, as a result of this, big financial difficulties in the mid 1960's. In order to save the company from bankruptcy, the electrical motor production was sold to Asea in 1967, an affair that received massive public criticism. Asea closed the motor factory in Örnsköldsvik in 1968 and moved the production to Västerås and to Denmark. A few years later, at the end of 1971, Asea bought the rest of the large family company and Hägglunds became a subsidiary of Asea [170, 171]. Hägglunds has since then been subject to several restructures and is now split into three independent companies manufacturing military terrain vehicles, ship's cranes and hydraulic motors respectively.

6.6.2 Elmo/Danaher Motion in Flen

The second largest Swedish manufacturer of electric motors, during most of the last 50 years, has been AB Elmo located in the little town of Flen, 150 km southwest of Stockholm. It was founded as a family company in 1948 by an electrical engineer, Per Pettersson (1912-1962). He started with a repair shop but soon began manufacturing of standard induction motors. It was very much a seller's market during the 1950's and the company expanded rapidly. The product program has included standard and customized squirrel cage induction motors up to 30 kW. A specialty for the company is still motors for larger washing machines, for which it has a world-leading position. Figure 6.27 shows a picture of such a motor, which is open-ventilated with end shields that are fitted directly to the core. Induction motors for forklifts also became an important product in recent years. The development of induction motors has been performed internally and in accordance with a cautious strategy, where new types and larger sizes were based on experience from previous ones. Wilhelm Almroth (1913-1998), general manager for Asea's motor division, wrote in a report from a visit to Elmo in June 1965: *"... but there is practically no development work. They try to copy well-known designs as far as possible."* (My translation) [172].

The production has remained "in-house", i.e. there is no outsourcing of manufacture of any major components. Elmo refrained from production of standard motors in the early 1980's, due to low-cost competition from Eastern Europe, and has instead focused on customized products. Neither Elmo nor Asea/ABB has considered each other as severe competitors, probably due to focus on different market segments. There has not been any technical cooperation between these two companies, but some engineers have switched their employment between them. Elmo has, in spite of its limited size and resources, tried to maintain a good competence illustrated by the fact that it, during later years, has had two PhDs in electrical engineering out of a total staff of less than 20 for development and design.

Figure 6.27 Washing machine motor from Elmo

Elmo faced a big challenge in the mid 1980's when it started to develop PM servomotors. The company therefore engaged Professor Vilmos Török as a consultant. Török, who left Hungaryin 1956, had been working for Asea before he was appointed professor of electrical machine theory at KTH. The development was successful and Elmo became supplier of motors for Asea/ABB's industrial robots.

The founder's family controlled Elmo until 1983, when a private business group acquired it. It was declared bankrupt in 1991, but another private investment company took over Elmo and reconstructed it. It was sold to an American company in 1999 and is now a subsidiary of Danaher Motion. Elmo had, around the year 2000, a turnover of 250 MSEK, 260 – 270 employees in Flen and the annual production was approximately 290 000 units. This has been reduced due to transfer of production to low-cost countries and the company has now only a little less than 100 employees left in Flen [173].

6.6.3 BEVI

BEVI AB is a medium sized Swedish company engaged in the manufacturing and sales of electrical machines and related equipment. It was started as an electrical installation and repair firm in 1931 by an electrician, Edvin Peterson (1907-1988). Repair and rewinding of electrical machines and transformers became an important part of the business. In 1971, the company signed an agreement with Polish motor manufacturers, to represent them as a distributor on the Swedish market. Five years later, BEVI acquired a small plant for the manufacture of motors and moved it to Blomstermåla, a small community in south-eastern Sweden, which has been BEVI's place of residence from the very beginning.

BEVI is today a substantial seller of electric motors in Sweden, and it also has sales offices in other Nordic and Baltic countries. The company acts, in case of standard motors, as distributor, while special, usually customized machines are of BEVI's own design and manufacture. These include squirrel-cage and slipring induction motors and generators as well as synchronous machines up to around 1000 kW. Examples of applications are submersible motors, motors for equipment with large vibrations and shock, motors for extra high ambient temperatures, active motor parts for integration in other machinery, and small hydropower generators [174]. The company has a remarkably small department for machine development and design, only a couple of persons, and no established cooperation with external engineering bureaus. The reason why BEVI succeeds is that the technical staff is very experienced and focuses on orders that fit technically and production-wise. This situation presents of course a risk for difficulties if one or two key persons must be replaced. Close contacts were established between BEVI and KTH in 2000, and BEVI cooperates today with both KTH and LTH with focus on manufacturing of PM motor prototypes.

BEVI is today a management owned company and, as such, it is less sensitive for fluctuations in the stock value due to profit variations. Such companies can often accept longer payback time on investments, compared to public companies, as long as the economy is sound. Development of new machines is usually carried out as specific customer projects. Sales of imported motors constitute a larger share of the turnover than BEVI's own manufacture. The company represents a number of foreign manufacturers. According to BEVI's technical director, Jan Folkhammar, it was earlier necessary to specify several details and requirements for motors imported from Eastern Europe, but these manufacturers have now improved so that their standard products can be accepted. Folkhammar also mentioned that some foreign companies have become more severe competitors in the Swedish motor market than ABB.

6.6.4   ITT Flygt

The second largest manufacturer of electrical motors in Sweden is, at present, ITT Flygt with its headquarters and development in the Stockholm area and its main factory in Emmaboda in southern Sweden. ITT Flygt is a world leading company that makes submersible pumps, and that develops and manufactures induction motors for its own pumps. ITT Flygt has no external sales of motors. The first part of the company name indicates that it is a part of the large American corporation ITT (International Telephone & Telegraph); the second part refers to the Swedish engineer Hilding Flygt (1865-1955), who started the company's pump business in the 1920's. He had actually been president of Förenade Elektriska AB in Ludvika (later NFEA, mentioned in section 6.2.4), which was Asea's main domestic competitor until Asea acquired it. Hilding Flygt was an entrepreneur, who managed to expand the business but he had to leave "Förenade Elektriska" in 1909 due to difficulties to finance the expansion.

Flygt started to manufacture submersible pumps in 1947. The electrical motor is an integral part of the submersible unit and the company needed, therefore, to purchase only the active parts of the motor, i.e. the stator core with winding and the rotor. Flygt's main supplier of these motor parts was, for many years, Hägglunds in Örnsköldsvik but, to some extent, also the Austrian company Elin. Hägglunds production of electrical motors was closed in 1968, when it became a member of the Asea Group. It was not in line with Asea's policy to sell non-standard, incomplete motors so Flygt had to find a new solution. The company had already started small scale manufacture of one type of motor a couple of years earlier, so the decision was made to establish a workshop, which could manufacture all motors supplied by Hägglunds. The transfer of know-how was secured through six persons, who moved with the products from Örnsköldsvik to Emmaboda. Flygt had problems with deliveries from the remaining external suppliers and decided therefore, in the mid 1970's, to set up a new motor workshop and manufacture all motors in house [175]. Flygt was acquired by ITT in 1968.

ITT Flygt's motor factory in Emmaboda manufactures around 80 – 100 000 motors annually. The production of cores is outsourced, so the factory is actually a highly advanced winding workshop. The product program comprises squirrel cage induction motors in the range 0.5 – 800 kW, 2 – 16 poles, 50 and 60 Hz. Production development is handled locally in Emmaboda, while product development is the responsibility of the development department in Stockholm [176].

Important development targets are of course low costs and high reliability, but a very significant goal is also to maximize the output for a motor placed in a confined space. Figure 6.25 shows a cross section of a submersible pump with motor. Flygt has managed to maximize the output by shifting as much as possible of the losses from the rotor to the stator, which is then efficiently cooled by the surrounding water. This approach results in a motor with a larger ratio between rotor and stator core diameters compared with normal standard motors. Flygt's motors are thus very specific and the internal motor production is somewhat less exposed to external competition than other manufacturers. That specificity, however, could at the same time create an obstacle for Flygt if the company ever should try to market its motors to other customers.

Figure 6.28 Submersible pump with induction motor from ITT Flygt

Flygt has had a larger freedom in developing motors than manufacturers building standard motors. Therefore, it is interesting to compare the utilization of Flygt's motors with normal industry motors. The present series of motors is, according to information from Rolf Linderborg who is manager for Flygt's motor development, approximately 15 percent smaller in volume than standard motors, without being more stressed thermally. The design of the company's latest series is slimmer than the previous one, which was more similar to standard motors. Efficiency has during the last 10 years become increasingly important while there are no severe requirements on starting torque. The pumps always start without any load.

Around 5 – 10 percent of the pump motors are fed from frequency converters. It has not been necessary to especially consider this in the design. The motors can be used without any down-rating provided they are fed from modern

IGBT converters.

ITT Flygt's development group has acquired its competence mainly through long experience, but it has also employed PhDs with electric machine background directly from universities. The company has had close contacts with KTH, for instance through participation, together with other industries, in KTH's competence centre for PM motor drives. Linderborg emphasized the value the access to the university's theoretical specialists has for a small electrical machine development department. Flygt has also informal exchange of information with ABB, e.g. in matters regarding production development, a situation that is facilitated by the non-competitive relation between the companies.

6.6.5 International competition

The period after World War II was characterized by a huge demand for electro-technical products. Asea had practically no competition. AEG's and Siemens' factories were in ruins. Asea's challenge was to increase the production capacity. It was necessary to extend the factories so Asea built, during the period between 1945 – 49, new workshops for motor manufacture in the western part of Västerås, the so-called "Örjan factory" [177]. It was a sellers' market until the 1950's, when Asea again faced Siemens, AEG and BBC as main competitors, especially the first two for induction motors.

Asea's main markets for standard induction motors during the 1960's were the Nordic countries, to some extent Western Europe, Australia and Mexico, where the company had established local production. Asea's main competitor was Siemens followed by AEG. Strömberg had a strong position, not only in Finland but also in Sweden. Several other companies, the Austrian Elin and the British Crompton Parkinson for instance, were important players in their respective markets. Most manufacturers had started to launch standard motors in accordance with the IEC standards, and the product differentiation between the leading companies became fairly small.

The low-price competition from Eastern Europe, which grew in importance during the late 1970's, has been described in section 6.4.4. Otherwise, Siemens and Strömberg are the companies which are mentioned in interviews as Asea Motors most important competitors during those years. Benchmarking was primarily performed against their motors.

The motor market in Western Europe did not change much during the 1980's. Besides Siemens, and a few other large actors, Asea also faced local competitors in every country. The low-price competition from Eastern European manufacturers was widespread, even if it was somewhat handicapped through inferior quality. The situation has changed drastically since then. Practically all local companies, with too small domestic markets, have disappeared. The market for standard motors is now dominated by a number of large manufacturers, which operate more or less globally. The fall of the socialistic regimes changed much. Many of the former East European countries are now members of the European Union and their motor companies have had opportunities to improve and are now both technically and commercially on par with West European competitors. A few overseas manufacturers have also entered the European scene.

The development in UK is worth a couple of comments. There were around ten domestic motor manufacturers back in the 1980's, but one by one they have since then been acquired by the largest of them, Brook Crompton. This company ended up with a market share around 80 percent, a situation that was unacceptable for the British customers and the market was consequently opened up for more international competition. This can be compared with Asea's ambitions to control the Swedish market between the World Wars (see section 6.2.4). UK was especially vulnerable to competition from abroad for two reasons; it is a large market and there is no language barrier as almost every company has documentation in English. Countries like Sweden and Finland are of course less attractive from both these points of view.

The current main actors in the European market, for standard induction motors, are in alphabetical order: ABB, AEG, ATB (Austria), Brook Crompton, Leroy-Somer, Siemens, TECO (Taiwan), VEM (Germany) and WEG (Brazil). ABB has, together with the U.S. manufacturer Baldor, the largest share of the global market, six percent each, closely followed by Siemens and WEG. The latter has built up its international position partly with export support from the Brazilian government. Neither TECO nor WEG can be considered as low-cost competitors offering cheap, low-performance products. Most customers require reliable motors with good efficiency and it is necessary for the manufacturers to be competitive both with respect to technical performance and commercial conditions.

In the wake of the globalization, motor manufacturers in China, India, Iran and other countries have become competitors to European and American companies, even in export markets. China alone has hundreds of motor factories, but only "one" standard motor design, which has been developed by a governmental institute. This means that the development costs for a motor is practically nil, and that the manufacturers cannot compete with performance but only with price. ABB's motor factory in Shanghai, established in 1996, builds also these Chinese standard motors and the modifications for creating an ABB design have, so far, mainly been cosmetic. This factory exports around half of its production to Europe. The low cost competition is most pronounced for smaller motors for which customers pay less attention to efficiency and other performance factors.

The product differentiation is low in case of standard induction motors, but each one of the leading manufacturers claims, nevertheless, to have a better product. Benchmarking is a commonly used method for comparing a company's own motors with motors from relevant competitors. ABB in Västerås has recently carried out such benchmarking tests against motors from a number of the competitors mentioned above. The results are for natural reasons kept confidential, but the author has had an opportunity to see the measured efficiency values. It can be concluded that, even if there are certain variations, all the large manufacturers build motors that are competitive in this respect. Benchmarking tests and other comparisons give often a signal that a certain motor type needs to be further developed or replaced, they are important for determining the development targets.

It has been mentioned in sections 6.4.3 and 6.5.3 that Asea specified "equal or better than Siemens". It is evident from this study, and that is no surprise, that Siemens has been Asea's, and later ABB's, main competitor from the very beginning until to day. The opposite was not the case until ABB had been established. Siemens, like ABB, has a very comprehensive program of induction motors that meet the two highest efficiency standards [178]. Siemens has factories for production of standard motors in Germany, the Czech Republic, Colombia and now also in China. For these products, there has never been any kind of direct cooperation between Siemens and Asea/ ABB, but both Lage Becker and Sven Sjöberg have witnessed about positive experiences from contacts with their colleagues at Siemens. It is probable that both companies are content with the present state of affairs, and that they instead focus on meeting competition from other parts of the world.

6.7   Speed controlled motors

6.7.1   Why speed control?

Many types of electric motors offer excellent possibilities to vary the speed of the driven equipment without the use of complicated mechanical transmissions. The motors have therefore been used for such purposes, more or less, from the very beginning back in the 19th century. For a long period, such motors were only used for those applications where variable speed was a necessity. This has in later years been extended also to other applications where such drive systems are used mainly due to better life cycle economy.

Among the first examples of variable speed drives are electrical trams. Siemens & Halske demonstrated a small locomotive at an exhibition in Berlin in 1879 and two years later opened the first tramline between Berlin and the suburb Lichterfelde. On the other side of the Atlantic, Edison built a small train at Menlo Park in 1880. The most rapid construction of electrical tram systems then took place in America. The need for speed control in this application is evident. There were also a number of the early industrial applications, in which the processes often required variable speed drives, e.g. mine hoists, rolling mills, paper machines, textile machinery and printing presses. An extensive overview of the development of variable-speed drives was presented by B. L. Jones and J. E. Brown in 1984[179]. The paper also includes a very comprehensive list of references.

Increased energy costs in combination with availability of new technology have later led to the use of variable speed drives in common applications such as fans and pumps. These drives are, from a technical point of view, fairly simple. The normal method for flow control has been throttling, which is not especially energy efficient. Centrifugal fans and pumps follow the same physical laws, which can be used for explaining the difference between throttle control and speed control of the flow. The flow, Q [lit/min], is roughly proportional to the speed, n [rpm],

$$Q \sim n \qquad (6\text{-}1)$$

the pressure head, H [m] is proportional to the square of the speed

$$H \sim n^2 \qquad (6\text{-}2)$$

and the power, P [W], is proportional to the cube of the speed.

$$P \sim n^3 \sim Q H \qquad (6\text{-}3)$$

Figure 6.29 is an example of a pump diagram, in which the two methods for flow control are indicated. It is obvious, both from the diagram and from equation 6.3, that throttling wastes much energy, especially if it is done for longer periods. Speed controlled motors are therefore used more and more for driving pumps, fans, compressors etc. [180].

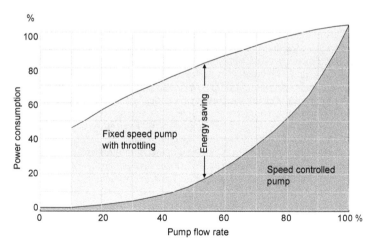

Figure 6.29 Energy savings by using speed controlled pump motor

6.7.2   Speed control of DC motors

A variable speed electrical driveline consists basically of a power supply, some sort of converter, an electrical motor, the driven object and a control unit. Figure 6.30 is an illustration of such a driveline. There are examples of drive systems, which do not fit entirely into this simple description, but those can be left without consideration for the moment.

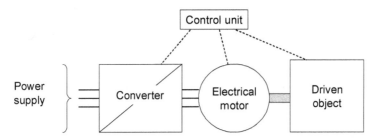

Figure 6.30 Electrical driveline

The first electrical drive systems contained DC motors and one important advantage of such a motor is that it is easy to vary its speed. Equation 4-16 indicates how. This equation can be transformed to the following:

$$n = \frac{kE}{\Phi} \tag{6-4}$$

where n is the speed [rpm], E is the emf or approximate voltage [V], Φ is the flux [Wb] and k is a machine related constant depending on machine dimensions and winding data. This means that the speed can be controlled by the voltage and through the excitation, which creates the flux. The first "converter" was a variable resistance in series with the motor and the "control unit" was an operator, who watched the speed and changed the resistance. This was an acceptable method for starting a driveline, but it was very uneconomical for regular operation due to large losses in the series resistance. Another method was to use a variable resistance in the excitation circuit and thereby control the flux. A reduction of the flux increases the speed but at the same time reduces the torque, which can be a drawback in many cases. The figures 4.28 and 4.29 present schematic DC motor layouts and excitation methods, and figure 6.31 shows typical torque and power characteristics for a separately excited motor.

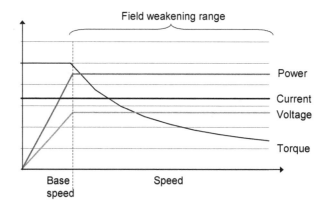

Figure 6.31 Torque and power vs. speed characteristics for DC motor

The American engineer Harry Ward Leonard (1861-1915), who worked for Edison and later GE, invented, in the early 1890's, a drive system including a rotary converter. This system consisted primarily of an AC motor driving a DC generator, which could feed a DC motor, or in some cases several, with a variable voltage. It has become known, all over the world, as the Leonard-system and it provided not only good speed control, but it was also a converter that even facilitated regenerative braking. The drawback was, of course, that it required three machines and was therefore both expensive and voluminous. Asea supplied its first Leonard-system for a paper machine drive in 1907 and used it then also for rolling mills, textile machinery, crane drives etc. Figure 6.32 shows a principle diagram of the Leonard-system [181].

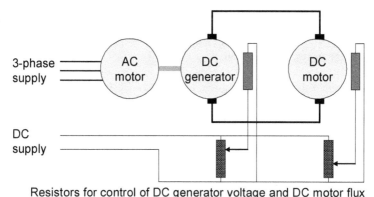

Resistors for control of DC generator voltage and DC motor flux

Figure 6.32 Leonard-system for speed control of DC motor

Stationary converters for DC motor drives were introduced during the 1930's. Mercury-arc rectifiers had been developed since the beginning of the 20th century and early applications were supply of DC grids, e.g. for tram-lines. The American inventor Peter Cooper Hewitt (1861-1921) had found, in 1901, that a mercury-arc discharge between a hot and a cold electrode could be used for rectification. This led to the development of uncontrolled mercury-arc rectifiers, first in evacuated glass bulbs and later also in steel tanks. The famous GE scientist Irving Langmuir (1881-1957) received in 1914 a patent for grid control of this type of rectifiers, which made it possible to control the DC voltage supplied from the rectifier. It was, however, not until the 1920's that this type of equipment was introduced industrially. Langmuir, who had a PhD from the university in Göttingen, Germany, was awarded the Nobel Prize in chemistry in 1932 [182].

Asea built a prototype mercury-arc rectifier in 1928, but its first delivery of DC motor drives, with such converters, was carried out in 1936 for a steel mill in Belgium. The motor was rated 4100 hp. One of the advantages, compared to the Leonard-system, was an improvement of the total efficiency with 4 – 5 percent units. Even the mercury-arc rectifiers were expensive and they were consequently only used for supply of large motors [183].

The next step in the development of DC motor drives came with the introduction of solid-state power electronics, which will be covered in section 6.7.5.

6.7.3 Speed control of AC motors

For a long time a significant drawback for AC motors in comparison with DC motors was that it was more difficult to let them operate with variable speed. This statement does not include single-phase commutator motors or so-called universal motors because they are almost like DC motors in this respect.

The speed, n_s [rpm], of a synchronous motor depends on the supply frequency, f [Hz] and the pole number, p.

$$n_s = \frac{120f}{p} \qquad (6\text{-}5)$$

This means that a variable frequency is required to control the speed of such a motor. The pole number can only be changed in discrete steps. These methods can both be used in case of induction motors, but there is also another possibility, namely slip control. Equation 6-6 gives the induction motor speed, n [rpm], as

$$n = n_s(1 - s) \qquad (6\text{-}6)$$

where n_s is the synchronous speed according to equation 6-5 and s is the slip in per unit, i.e. in relative numbers (1 p.u. = 100 %). The slip is a measure of the speed difference between the rotating flux and the rotor and it is possible to change the slip by means of a variable rotor resistance. This can, in practice, be achieved with an external resistor for motors with slipring rotors. Even the voltage can be used, to some extent, for adjusting the speed of an induction motor; the reason being that the maximum torque is proportional to the square of the voltage, so the intersection point between the motor torque and the load characteristic will move when the voltage changes. The diagram in figure 6.33 illustrates these two methods for induction motor speed control assuming fixed stator frequency.

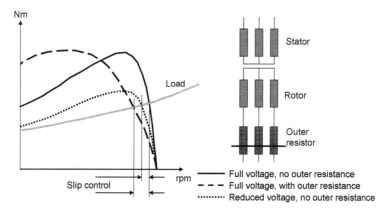

Figure 6.33 Speed control of induction motors through extra rotor resistance and voltage variation respectively

A change of pole number does not make a variable speed drive, but sometimes, a change between two speeds can be a sufficient solution. Arvid Lindström and his Asea colleague Robert Dahlander (1870-1935), later president for Stockholm's power utility, invented and patented in 1897 a wide spread winding arrangement used for switching between two pole numbers with the ratio 1: 2. It has become known globally as the "Dahlander – Lindström pole changing" and the principle is shown in figure 6.34. The method requires that the 3-phase stator winding is made in a special way and the same applies for the rotor winding in the case of a slipring motor. A squirrel cage rotor can be directly used for different pole numbers. The winding in the figure will have eight poles (each "coil" represents one pole pair) if the "B-terminals" are connected to the supply while the "M-terminals are left open, and four poles if the "M-terminals" are connected to the supply and the "B-terminals" are short-circuited. The synchronous speed will thus increase from 750 to 1500 rpm in case of a 50 Hz supply [184].

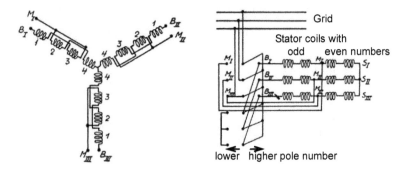

Figure 6.34 Dahlander - Lindström pole changing

Variable speed operation, by means of slip control, as indicated in figure 6.33 was practiced already in the 1890's. The concept was introduced for facilitating starts, which otherwise could be quite problematic due to the weak power supply at that time. The external rotor resistor reduced the start current at the same time as the torque was improved. This start resistance was successively decreased when the motor accelerated. It was possible to use this method even for speed control during continuous operation, but with significant drawbacks. The efficiency suffered from the losses in the external resistance, which had to be dimensioned accordingly and be provided with some sort of cooling. The method was simply very uneconomical, so it is not surprising that different concepts to avoid these extra losses were developed.

The 3-phase commutator motor is an elegant answer to the question, "how do we achieve efficient slip control without a lot of external devices?" It is of interest for this study, because Asea was involved in its invention and has also been manufacturing such motors for a longer period than other manufacturers. A few different solutions were invented in the early 1900's; one of them by Jens Lassen la Cour, but it had a drawback in that it required external equipment such as transformers and controllers. However, in 1910, a young Dutch engineer Hidde K. Schrage (1883-1952), who was working for Asea in Västerås, invented a motor which has become well known over the world as the "Schrage motor". The motor itself is a little complicated and the function is unfamiliar to young electrical machine researchers, so a short description can be motivated. Figure 6.35a presents a winding diagram of such a motor.

The motor has three windings, one in the stator and two in the rotor. It is also provided with both a set of sliprings and a commutator. The primary winding is placed in the rotor and is supplied from the grid via the sliprings. The secondary winding is located in the stator and each phase has both its ends connected to brushes at the commutator. The third winding is a control winding, connected to the commutator in a similar way as in a DC motor.

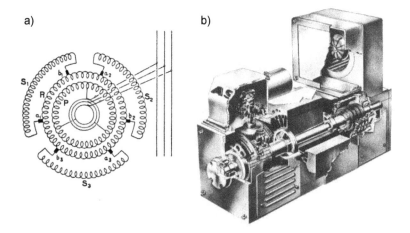

Figure 6.35 a) Principle winding diagram for Schrage motor and
b) sketch of an Asea Schrage motor from the 1970's

The basic idea can best be explained with reference to a normal slipring motor. The emf induced in the secondary winding (usually the rotor winding) is, E_2 [V]

$$E_2 = sE_{20}$$
(6-7)

where E_{20} is the secondary emf at standstill and the frequency of the secondary currents is

$$f_2 = sf$$
(6-8)

E_2 and f_2 can be measured at the sliprings of this type of motor, see figure 6.33. Assuming a delta connected winding connected to a commutator, it would be possible to replace the sliprings with three brushes and still measure the same secondary emf, i.e. E_2. The measured frequency, however, would now be f, and it would thus be possible to transform the secondary power back to the grid. The slip, and hence the speed, could be controlled through a variable transformer. Schrage arranged such a variable transformer inside the motor. He changed the position between the primary and secondary winding, compared with a normal slipring motor, and introduced a control winding placed in the same rotor slots as the primary. It is obvious that the currents in these two windings must have the same frequency. The control winding and the primary winding constitute a transformer, and the secondary winding voltage depends on the position of the brushes at the commutator. Shifting the position of the brushes will change the slip, and thus the speed, of the motor. This shifting is made by a small, geared motor, which drives two geared rings holding the brushes. The motor can operate both under- and over-synchronous [184].

The Schrage motors have been built in sizes up to 200 kW and the speed range is fairly large, in the order of 1:10. The speed response is, of course, not so high due to the time required for shifting the brush position. Typical applications have been textile machinery, printing presses, manoeuvrable bridges, ski lifts etc. The product became important for Asea and the company even had for many years a separate department for development and design of these motors. The manufacturing was transferred from Västerås to Asea's factory in Persan outside Paris in 1968, where the production continued until the 1990´s. A sketch of such a motor developed during the 1970's is shown in figure 6.35b.

Other early examples of slip control of induction motors are the "Krämer system", presented in 1905 by the German inventor Christoph Krämer, and the "Scherbius system" invented in 1907, by the German electrical engineer Arthur Scherbius (1878-1929). Scherbius is otherwise most known for the invention of "Enigma", the cipher machine used by the German forces during World War II. Both Krämer's and Scherbius' systems require two additional electrical machines for feed back of the slip power. The first converts it to mechanical power that is added to the main shaft and the second to electric power regenerated to the grid. Static counterparts, to these electromechanical converters, used for slip control, were introduced in the late 1930's. Attempts to use mercury-arc converters were also made, but without any larger commercial success [182].

6.7.4   Semiconductor components for power electronics

Semiconductors have changed our world. They have long constituted the physical basis for computer and telecommunication development, but also have had a major impact on modern power technology. A semiconductor has conductivity somewhere in between conductors and insulators, but more important, it can be radically changed in a controlled manner. A semiconductor can, under certain conditions, serve either as conductor or insulator, i.e. it can be used as a switch. The first examples used for minor power applications where diodes from copper-oxidule developed in the late 1920's and a few years later from selenium. A diode is a component that is conducting in one direction, the forward direction, and is non-conducting in the reverse direction.

The real break-through came in 1948 with the invention of the transistor, published by the Bell Laboratories scientists John Bardeen (1908-1991), Walter Brattain (1902-1987) and William Shockley (1910-1989). They received the 1956 Nobel Prize in physics for their discoveries. John Bardeen was actually awarded a second Nobel Prize in 1972 for his work on super-conductors. A transistor is a component, which state of conduction can be controlled by a small gate signal allowing it to be used in amplifiers.

The first components, for power applications, were germanium diodes introduced in the beginning of the 1950's. Asea had manufactured selenium rectifiers in its Ludvika factory and took up the development and production of germanium diodes in 1955. The availability of silicon diodes, a few years later, marked a significant step in the power technology. Silicon diodes could withstand higher temperatures as well as higher voltages in the reverse direction. Some of Asea's large competitors found themselves at an advantage from using such components and, as a result, Asea eventually turned its efforts in the same direction. A license for semiconductor diodes was obtained from GE in 1958, which enabled Asea to save time until actual products could be released [185, 186].

After some time, as these new components were being introduced, it became known that GE had invented and patented a controllable silicon based valve, the thyristor. The thyristor is, in principle, a diode in which it is possible to block the component from conducting in the forward direction until it is triggered by a special pulse. This component opened up a lot of new power electronic applications. Asea's cooperation with GE came also to include thyristors. Both diodes and thyristors were designed as so-called bolt components screwed into suitable heat-sinks, see figure 6.36a. A team of Asea engineers, led by Gunnar Mellgren, developed, in 1964, an improved design with double-sided cooling that increased the possible current density on the semi-conducting layer by more than 50 percent. This new design was further developed and was soon known as the "hockey puck" diode or thyristor that is shown in figure 6.36b. The concept was patented in several countries and has later become a general industry standard [187].

a)

b)

Figure 6.36 a) Bolt type and b) disk (puck) type thyristors

The next step, in order to be able to build even more versatile power electronic products, was to develop components which also could turn off the current in forward direction. This was achieved during the 1970's and 80's through the development of the "Gate turn-off thyristor" (GTO) and different "Power transistors". Figure 6.37 illustrates the principal function of the different semiconductor switches mentioned above.

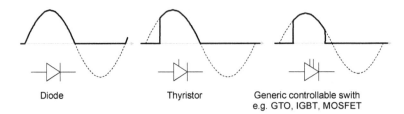

Diode   Thyristor   Generic controllable swith
e.g. GTO, IGBT, MOSFET

Figure 6.37 Symbols and current diagrams illustrating function of
semiconductor switches

The most frequently used power transistors for electrical drives are the MOS-FET (metal-oxide-semiconductor field effect transistor) and the IGBT (insulated gate bipolar transistors). It is outside the scope of this thesis to deal any further with the function and properties of the different semiconductor switches; however, there is a lot of literature available, e.g. the textbook in reference [188], which deals both with the semiconductors and the power electronics, if a more in-depth understanding is desired. Anyhow, a brief comparison between the capabilities of these devices is presented in figure 6.38. This reflects the situation in 2005 and it must be kept in mind that the borderlines change due to ongoing development. There is no manufacture of power semiconductors in Sweden any longer. Japanese companies have been leading the development during the last decades and they also have the largest share of the world market.

Extensive development has been carried out during the last 10 - 15 years on semiconductor switches based on silicon carbide (SiC) instead of silicon. Such devices would allow operation at much higher temperatures, higher frequencies and voltages and still have much lower losses [189]. With so many advantages, why aren't we using these? It is hopefully just a matter of time. The development is difficult and the yield from the production is still low. Some diodes have been available on the market for some time, but, so far, very few controllable devices. Their rated currents are limited to a few amperes. If and when e.g. larger SiC MOSFET or SiC JFET (silicon carbide junction field effect transistor) would be available, it could have a big impact on rotating machine development, because they would offer radically different possibilities for integration. They could even be placed inside the hot machines. Table 6.4 below compares some characteristic properties for Si and SiC devices. Several companies around the world are making big efforts in developing SiC technology, e.g. some large automotive manufacturers, which want to use it in vehicles with electrical drivelines. A recent press release, dated January 2006, reports that a SiC inverter in the 100 kVA size has been demonstrated [190]. It is not expected that SiC converters will be cost-effective below a few tenths of kW, but that will, of course, be very dependant on the applications.

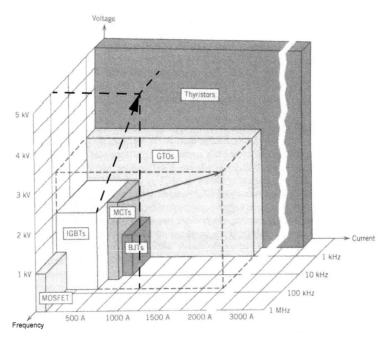

Figure 6.38 Performance of semiconductor switches, original from [204]

Table 6.4   Comparison between Si and SiC devices

Property	Si	SiC
Bandgap [eV]	1.1	3.0
Thermal conductivity [W/cm K]	1.5	3.0 - 5.0
Max temperature [$^\circ$C]	125	600 - 965
Electron mobility [cm^2/Vs]	1400	600
Breakdown electrical field [V/cm]	0.3 x 10^6	4 x 10^6
Potential inverter efficiency* [%]	98.8	99.7

*. Simulation reults, 1 MW inverter

Semiconductor switches are used for conversion of electrical power from one state to another, e.g. from AC to DC or vice versa. A short vocabulary for the most common power electronic converters may be helpful in understanding the following sections.

- Converter   General term, which can include all of the following devices

- Inverter   Converts DC into AC

- Rectifier   Converts AC into DC

- DC/DC converter   Converts DC from one voltage to another, higher or lower

- Frequency converter   Converts AC from one frequency to another

A frequency converter usually consists of a rectifier and an inverter, so the frequency conversion takes place with DC as an intermediate step.

6.7.5   DC drives with power electronics

Asea used its first semiconductor devices, selenium and later germanium diodes, for low voltage applications such as battery chargers and rectifiers for electrolytic processes. These diodes were also used in transductor-controlled rectifiers for small shunt type DC motors [191]. When silicon diodes became available, around 1960, it became possible to implement more powerful drive systems. Interesting examples are of two locomotives with diode rectifiers and series type DC motors that were presented in 1962. The control was, in this case, arranged through a tap changer at the single-phase transformer by means of which the AC voltage, feeding the rectifier, could be varied [192].

The introduction of thyristors in the early 1960's had a thorough impact on DC drive systems. Motors with separate excitation replaced shunt, series and compound excited. The thyristor rectifiers were initially not powerful enough for very large DC motors and were therefore used for small and medium-sized units, up to 300 kW, and for the supply of separate excitation current on large machines, e.g. the generator and motor in a Ward Leonard system [193, 194]. The development of thyristors was fast, and the maximum size of thyrsistor fed motors increased rapidly and reached 2 500 kW around 1965 and 12 000 kW a couple of years later. Examples of important applications are paper machine drive systems and rolling mill drives. The latter required really large motors. A locomotive with thyristor rectifiers and motors with separate excitation was demonstrated in 1965 [195].

Drive systems with thyristor rectifier fed DC motors offered a number of advantages compared with earlier Ward Leonard systems and mercury-arc converter fed motors. Some advantages were better efficiency, faster speed and torque response, smaller foundation and less auxiliary equipment. A disadvantage was lower overload capacity for the thyristor rectifiers depending on their low thermal time constant, i.e. they could easily be overheated. Figure 6.39 displays a comparison of efficiencies for a number of different drive systems published in Asea's Journal 1964 [196].

Efficiency [%]

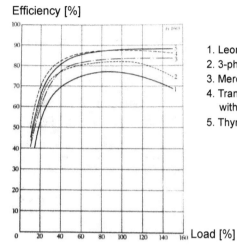

1. Leonard system
2. 3-phace commutator motor
3. Mercury rectifier with DC motor
4. Transductor controlled Si rectifier
   with DC motor
5. Thyristor rectifier with DC motor

*Figure 6.39 Efficiency vs. load for different speed controlled
rolling mill drives*

The use of thyristor converters, for feeding of DC motors, required special attention in the design of these machines. Both the armature current and the excitation current contained harmonics, i.e. AC components, which caused eddy current losses and affected the commutation. Asea's first standard series of large DC motors, which was developed in 1968-69, was therefore built with laminated stator poles and core, compensation winding in the pole plates and armature winding with split and shorted coil pitch. Other targets for the development of this series were product modulization and rationalization of the order specific design work [197].

The end of the DC motor era has been proclaimed over and over during the last two to three decades, but it has managed to survive. It is true that its share of the market for variable speed drives has decreased and there is hardly any manufacture of DC motors in the multi MW range any longer. There is, however, a substantial market for medium-sized motors, which caused ABB, in Västerås, to develop a new generation of such DC motors only a few years ago.

ABB's goal for the project of this new DC motor was to develop a product with significantly higher output, higher torque and higher maximum speed, wider constant power speed range, lower moment of inertia and lower electrical and mechanical stresses compared to existing motors. All these targets were reached, with good margins, according to an article in ABB Review [198]. The most important design improvements introduced were related to the arrangement and cooling of the armature winding, more space for the stator windings and a redesigned commutator. In addition, better materials and refined manufacturing processes contributed to a motor with astonishing performance. A picture of this new DC motor is presented in figure 6.40. This particular motor series covers an output range 25 – 1400 kW.

Figure 6.40 ABB's modern DC motor, type DMI

This new series of DC motors must compete, not only with DC motors from other manufacturers but, which in this thesis is more interesting, even with frequency controlled induction motors. For which applications can it compete and on which merits? Typical applications for these motors are rolling mills, rubber mixers, plastic extruders, container cranes, ski lifts and mine hoists. A comparison of the advantages of different drive systems depends on who makes it. This is what ABB's DC motor department claims to be the merits of their motors in some typical applications:

- Metals (rolling mills, shear etc): Fast acceleration, wide speed range
- Rubber & cement mixers: Overload capability, starting torque
- Cranes & mine hoists: Fast acceleration, compact installations
- Ski lifts & extruders: Starting torque, compact installations
- Test rigs, pulp & papermachinery: Simple and accurate control, wide speed range

A comparison between one of ABB's modern DC motors and a corresponding AC motor is made in table 6.5. The information has been taken from recent ABB catalogues [199, 200]. The choice of motors for the comparison can be criticized because there are several options to chose between, e.g. aluminum or cast iron frames and basic or high-output design for the induction motor. The table contains data for a 4-pole squirrel cage motor with aluminum housing and high-output design. The rated power, nominal torque and speed range are fairly similar for the motors and even dimensions, weight and inertia are not very different. It is somewhat surprising that the DC motor has a utilization factor on par with the induction motor, but it must also be remembered that the former needs inner air-cooling while the latter is totally enclosed. The induction motor has a significantly higher maximum torque provided the frequency converter can supply sufficient current. The induction motor has, however, a disadvantage in case of a large field weakening range, because its maximum torque at higher speeds falls quicker compared to the DC motor. The DC motor is, in this respect, limited by the commutation. Another important difference is the efficiency, which is much higher for the induction motor. On the other hand, the efficiency is higher for a DC power converter compared to a frequency converter. A general, and somewhat trivial, comment is that customers, which are used to DC drive systems, have a tendency to prefer them also for new installations.

Table 6.5   Catalogue data and relative prices for ABB DC and AC motor

Motor type	DC sep. excitation	AC induction
Cooling	Open with built-on fan	Closed, outer
Designation	DMI 180B	M3AA 225SMC
Shaft height [mm]	180	225
Output power [kW]	58	55
Nominal torque [Nm]	363	356
Base speed [rpm]	1535	1480 (at 50 Hz)
Const. power max. speed [rpm]	3694	3600
Max. torque/nominal torque [%]	160	300
Voltage [V]	550 (DC)	400 (AC rms)
Efficiency [%]	85.5	94.6
Weight [kg]	310	265
Length incl. shaft end [mm]	905	803
Inertia [kgm^2]	0.5	0.49
Motor price [%]	100	25
Converter price [%]	30	60

6.7.6 Slip and voltage control of induction motors

Various systems for slip control of induction motors with sliprings have been described in section 6.7.3 above. It was therefore a natural step to implement this type of control by means of thyristor converters. The basic idea is the same as for the Scherbius system, i.e. to feed the "slip-power" back to the line. The slip-power is the sum of the losses in the rotor winding and the external power supplied via the sliprings. The relation between the slip-power, P_{slip}, and the "shaft-power", P_{shaft}, (= output + friction losses) is:

$$\frac{P_{shaft}}{P_{slip}} = \frac{n}{n_s - n} = \frac{1}{s} - 1 \qquad (6\text{-}9)$$

The amount of power fed back to the line will thus change the slip and hence the rotor speed. The system consists basically of a slipring motor, an uncontrolled diode-rectifier, a smoothing inductor, a thyristor inverter and a transformer. The inverter is, in this case, line-commutated, i.e. the transformed line voltage determines when the different thyristors will be conducting. A simple diagram of the system is shown in figure 6.41. This type of system offers a limited speed range but one advantage is that the rectifier and the inverter, also called converters, do not have to be dimensioned for the full motor power, which is usually the case for other systems. The system has a fairly good efficiency. Asea delivered this type of system from the beginning of the 1960's [201].

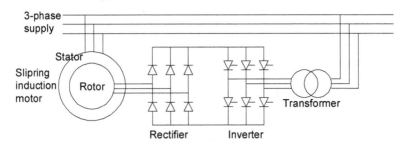

Figure 6.41 Drive system with slip controlled induction motor

Slipring motors are expensive, in comparison with squirrel-cage motors, and it was therefore attractive to develop speed control for such motors also. A simple system is to use a thyristor inverter to vary the motor voltage, more or less in the same way as a light dimmer. The frequency is not changed, it remains the line-frequency, and it is therefore, even in this case, a matter of changing the slip even if it is done through voltage control. The background is that the induction motor torque, at a given speed, is proportional to the square of the voltage. An example on a set of torque curves at different voltages is shown in figure 6.42. It is evident how the speed can be changed through voltage variations for a given load torque. Use of rotors with extra high resistance results in a larger speed range.

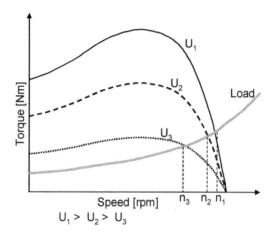

Figure 6.42 Speed control by voltage control

This method of speed control is usually uneconomical due to high rotor losses at lower speeds. It can, however, be justified for smaller motors or such with very intermittent operation, especially if they only operate with high slip for shorter periods. Asea has also built drive systems of this type. An example is a system developed in 1967 with a 100 kW motor for ship cranes [202].

6.7.7 Frequency control of AC motors

Equation 6-5 above tells that the natural way to control the speed of AC motors, synchronous as well as induction, is by means of a variable supply frequency. Thyristor converters made this possible. There had been earlier attempts to develop variable frequency drives, primarily for synchronous machines. A number of "electronic drive systems" with thyratron and mercury-arc converters were demonstrated during the 1930's. Several were technically successful but they could not compete commercially with the established systems. However, these projects contributed with valuable knowledge concerning frequency conversion for synchronous motor drives, which was brushed up when thyristors entered the arena [182].

Frequency control of a synchronous motor was somewhat easier, compared with an induction motor, because in the first case, the inverter could be naturally commutated by the load while the latter required forced commutation.

The reason is that the synchronous machine is separately excited and an emf is induced as soon as this machine rotates, while the induction machine is virtually dead until it is fed with an AC voltage. The synchronous machine emf governs the commutation in the inverter in the same way as if it had been connected to the line. In the case of an induction motor, the control unit has to force the individual thyristors to turn on and off. It is an advantage if the motor can be fed with a sinusoidal current with a low content of harmonics, which reduces losses, torque pulsations and noise. Various methods have been developed to achieve this and the most commonly used is the "pulse-width modulation" (PWM) presented for motor control in 1964 [203].

The function of a PWM controlled three-phase inverter can be briefly explained with the help of figures 6.43 and 6-44. The first shows an inverter consisting of six switches and a diagram indicating how the switches are turned on and off thus converting the input DC voltage into three AC voltages. The second figure contains a diagram, representing one phase, showing how a sinusoidal fundamental can be obtained by variation of the width of the pulses that are created through high frequency switching. A sinusoidal reference signal is used for comparison with a triangular carrier wave to determine when a certain switch shall be turned on and off [204]. The frequency of this triangular wave is called "switching frequency" and is usually in ranges from a few kHz up to 20 kHz for industrial motors. A high switching frequency gives less harmonics and thus lower stray losses in the motor. On the other hand, the so-called switching losses in the inverter increase, so the choice of switching frequency becomes a matter of optimization of the total system.

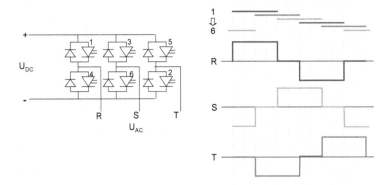

Figure 6.43 Self-commutated inverter with anti-parallel diodes and switching diagram

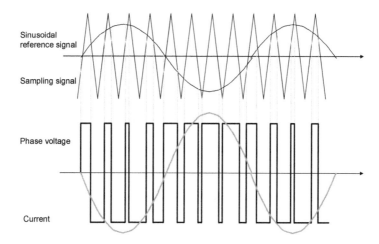

Figure 6.44 Principle of Pulse Width Modulation (PWM)

Asea was late in adopting frequency control for induction motors. The company had instead developed such technology for large synchronous machines in the late 1960's. It was in no way an alternative to speed controlled induction motors, but instead a good solution to very specific problems. The best example is the starting of the motor/generators in the Foyers Pump Storage Plant located at Loch Ness in Scotland. A pump storage unit runs as a turbine – generator set generating power during daytime when the demand is high and as a motor – pump unit, which pumps the water back into a higher reservoir, at night when the need for power is low. Such reversible motor/generators were usually started in no-load motor mode with the help of a special start motor. Frequency control offered a possibility to avoid such extra motors and let the motor/generator self start synchronously. This required no PWM. The inverter had simple forced commutation from stand still to 5 Hz and was then load commutated up to 50 Hz. The two units in Foyers were commissioned in 1974. They were each rated 176.5 MVA, 273 rpm, 18 kV and the power required for starting was 4.3 MW [205, 206].

It is not obvious why Asea did not develop frequency control for induction motors during the 1970's. It was well known that major competitors such as BBC and Siemens were already offering such drive systems. One reason that Asea did not pursue frequency control development was probably that Asea held a very strong position in the field of DC drive systems. The DC drives were, at that time, technically superior and the business was profitable. It seems that the management was not observant enough of the threat from a new technology. The alarm bell finally rang in 1980. The following decision is written in a protocol from a board meeting of the Motor Division on June 2nd that year: *The project to develop a new series of [induction] motors up till 45 kW was approved. The design of the new series shall take into consideration requirements due to frequency control.* (My translation) [207]. Further emphasis on this can be found in a memorandum written by Asea's new CEO Percy Barnevik, in September of the same year, in which he comments on the strategic plan for the Motor Division: ***Strategy for products*** ..... *to maintain such a high level with respect to product performance, quality assurance, technical support and service that Asea continues to compete in the premium price market. The portfolio must also include products sold in small volumes, but which have influence on the demand on the main products. Asea's delay in frequency controlled drive systems must be caught up through internal development of the motors and external purchase of frequency inverters until Asea has introduced its own competitive inverters.* (My translation) [208]. Asea started the development of frequency controlled induction motors and inverters, but didn't manage to fully overcome the delay. The situation changed radically in the summer of 1986 when Asea acquired its Finnish competitor Oy Strömberg.

6.7.8 Development of speed control in Finland

ABB in Finland is nowadays a world leader in terms of electric variable speed drives for industrial applications. Why is it like that and what is the background? An attempt to look into these questions could perhaps contribute with some important information regarding the electrical machine development process. In order to understand what has happened, it is first necessary to know something about the company's historical background.

A Finnish industrialist, Gottfrid Strömberg (1863-1938), started in Helsinki, in 1889, a company for manufacture of DC machines. The company, Oy Strömberg Ab, grew rapidly and became the main electro-technical enterprise in Finland with a wide range of products. Electrical machines were always some of its main products. Strömberg built DC machines as well as asynchronous and synchronous AC machines. The manufacture was first concentrated in Helsinki, but a second factory was established in Waasa, at the Finnish west coast, during World War II, mainly for defence strategic reasons as it was a less exposed location than Helsinki, which suffered from Soviet bombing. The factory in Waasa has since then mainly made low voltage induction motors while the Helsinki factory has manufactured all other machines. Important customers have been the large domestic pulp and paper industry, but also Finnish shipyards building icebreakers with diesel electric propulsion. Both applications required speed controlled drive systems, which for a long time were DC systems. However, Strömberg introduced earlier than most other companies speed controlled AC drive systems.

A family owned company from the beginning, Strömberg soon became publicly owned. The Finnish pulp and paper industries were large shareholders and considered Strömbergs more or less as a service unit providing them with most of the electrical equipment they needed. From time to time, even Asea had some minority share in this Finnish company. Strömberg had a strong position in its domestic market but also in northern Sweden, the latter probably due to the abundant pulp and paper industry in that part of Sweden but also its vicinity to Waasa. Stömberg had also managed to take over many of Hägglunds' customers when this company was acquired by Asea in 1967.

During the mid 1960's, Strömberg had a small development team for electronics, which studied, with great interest, what was published internationally about frequency control, especially with PWM [203]. One of the members of this team was Martti Harmoinen, who later became manager for the company's development department for power electronics. He mentioned, in an interview, that Strömberg started with research and development of AC drive systems in 1969. The reason was that a much larger converter market could be foreseen than for DC drives. There existed "hundred times" more induction motors than DC motors in the world. The development, however, did not receive support from everyone in the company. Many asked, "Why do we spend money on this?" Harmoinen said that it took ten years before at least 50 percent were convinced. This should be kept in mind when Asea's inactivity during this period is criticized.

The first AC drive system, built for driving equipment for changing nuclear fuel rods in the Finnish power plant Lovisa, was delivered in 1974. DC motors were ruled out due to the risk of sparking. The real breakthrough, however, came with the delivery of drive systems for the new Helsinki Metro. These deliveries started in 1976. Finland had since long ago had a considerable bilateral trade with its large neighbour, the USSR. Finland exported industrial products and imported mainly minerals and oil and was therefore subject to political pressure to purchase more advanced products. The trains for the new metro were considered suitable. Both the metro company and of course Strömberg wanted to prevent this and therefore specified a frequency controlled traction drive system, a technology that they knew the Soviet manufacturers did not have. Strömberg took a big risk, even if it was known that BBC had built a diesel electric locomotive with a PWM AC drive. Many problems, first with hardware and later with software, occurred but were eventually overcome.

The oil crises in 1973-74 made more energy efficient technologies economically interesting. Variable speed drives for fans and pumps are examples of this. These are, from a control point of view, rather simple and were very suitable for the introduction of PWM frequency controlled drives. It was thus good timing when Strömberg launched its first series of frequency converters, SAMI A, in 1976. SAMI stands for "Strömbergin Asynkroni Moottori Invertteri." It was later followed by SAMI B and SAMI STAR around 1981 and 1985 respectively. The latter was the first generation with digital control. The performance had now been improved so far that it was also possible to use the AC drives for difficult applications such as paper machine drives [209].

Strömberg managed to reach a technical front-line position in terms of frequency controlled drive systems, but suffered commercially because of that the company was so small. It did not have the necessary resources for wide international marketing. An attempt to overcome this situation in the North American market was made through cooperation with the U.S. company Allen-Bradley, which marketed the SAMI converters during some years in the 1980's.

A radical change took place around Midsummer in 1986 when it was announced that Asea had bought the entire Strömberg company, effective January 1, 1987. The coordination activities were not completed when, one year later, in August of 1987, the merger between Asea and BBC was made public. ABB started its operations January 1, 1988 and the Finnish subsidiary was given a leading position for variable speed drives.

Becoming a part of the big ABB Group gave access to much larger markets and the former Strömberg's sales of frequency converters grew rapidly. The diagram in figure 6.45 shows the increase from 1987 to 2004, which was not only due to an increased market share, because the market itself was also expanding. Anyhow, ABB secured a leading position and could therefore invest heavily in the development of the next generation of converters [209].

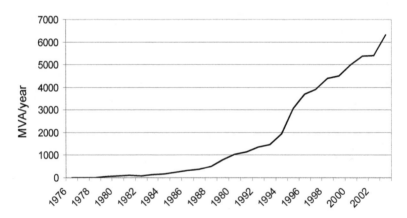

Figure 6.45 Increase of ABB's frequency converter sales

The SAMI converters were based on PWM technology, either open loop frequency control or closed loop flux vector control. The first method was not very accurate and therefore limited to speed control of simple applications such as fans and pumps. The latter had considerably better performance but required the use of an expensive resolver or similar sensor for feedback of rotor speed and angular position. It was against this background that ABB, in Helsinki, started research and development, in 1988, for its next generation of frequency converters. The result was the introduction of "Direct Torque Control" (DTC) in 1993 - 95. This was partly based on the theory on field oriented control of induction machines, first published in 1971-72 by the Siemens researcher Dr. Felix Blaschke [210], and further on the theory of direct self control, published and patented in 1985 by Manfred Depenbrock, professor in Bochum, Germany [211].

ABB, which was the first manufacturer to introduce DTC converters, claimed that the major advantages of DTC were high performance and no necessity for rotor speed and position sensors. The method requires instead a detailed mathematical induction machine model and access to a powerful digital signal processor (DSP). The model calculates continuously, based on measurements of currents and voltages, the actual torque, flux and speed values for comparison with reference values [212]. The torque is calculated as the vector product between stator flux and current as

$$\mathbf{T} = \mathbf{\Psi_s} \times \mathbf{i_s} \qquad (6\text{-}10)$$

where $\qquad \mathbf{T}$ = torque [Nm]
$\qquad\qquad \mathbf{\Psi_s}$ = stator flux [Wb]
$\qquad\qquad \mathbf{i_s}$ = stator current [A]

The stator flux is estimated from the voltage vector and the current according to

$$\mathbf{\Psi_s} = \int (\mathbf{U_s} - \mathbf{R_s i_s}) dt \qquad (6\text{-}11)$$

where $\qquad \mathbf{U_s}$ = stator voltage [V]
$\qquad\qquad \mathbf{R_s}$ = stator resistance [Ω]

DTC is not based on a fixed PWM. The necessary switching is instead determined through on-line calculations. The consequence is a reduced number of switchings and hence losses. It switches only when necessary, which also means that there is no fixed switching frequency. The diagram in figure 6.46 illustrates the principle [213].

ABB has developed several series of DTC type frequency converters covering a range from < 1 kW to 27 MW. The company had, in 2004, a share in the order of 16 percent of the global market, which makes it the world leader in this particular field. Most of the development and production is allocated to ABB in Helsinki [214]. None of it is left in Sweden.

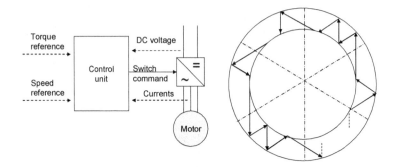

Figure 6.46 Direct Torque Control (DTC) needs no rotor position signal and it calculates necessary switching continuously

ABB Finland's dominant position in the field of frequency converters can be compared with another Finnish company, Nokia, which is the global market leader for mobile phones. Nokia, earlier a manufacturer of rubber boots and cables, but originally a forest industry, started its electronic activities in 1960, i.e. more or less at the same time as Strömberg. It is remarkable that Finnish industry has become so successful in these new, advanced technologies in spite of the small domestic markets. The reason for the success is probably a combination of several factors such as industrial foresight, determination to survive after the trade with the USSR had vanished, consistent support of R&D and higher education, but also a piece of good fortune. Since long, both Finland's electrotechnical education and industry have been strong; one proof is that the electronic- and electrotechnical industry represents around 25 percent of the country's current export [215].

It is, in this context, of interest to note that today, i.e. 2006, more than 50 percent of ABB's total development and production of electrical machines is located in Finland. Strömbergs', and later ABB's in Finland, extensive activities, in the field of frequency controlled drive systems, have had significant impact on the motor development as well. The company's electrical machine engineers have addressed a number of problems caused by these types of drive systems.

For many years, the technical universities and the electrical machine industry in Finland have had a close and fruitful cooperation. Subjects which have received much attention are FEM simulation, high-speed induction motors, and PM machines. The standard is high even from an international perspective, and, since 1990, the electical machine departments at the universities in Helsinki and Lappeenranta have examined 23 and 15 PhD respectively.

6.7.9 Specific requirements and problems

The use of standard induction motors in combination with frequency converters initially caused technical problems and damages. Motors failed due to
over-heating, insulation breakdown and bearing damage. They became nois-
ier, had larger torque ripple and the efficiency was reduced. The common denominator, for most of the problems, was increased current harmonics. A
grid connected motor usually runs with sinusoidal currents, while the currents in converter fed motors contain more or less harmonics depending on
the switching pattern. The diagram in figure 6.47 gives an example of currents in motors with grid connection and supply from a frequency converter
respectively.

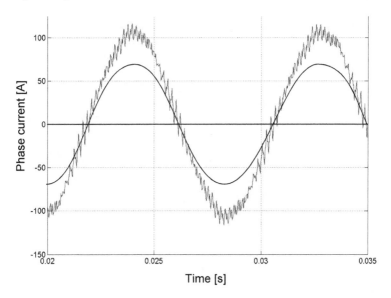

Figure 6.47 Phase currents in grid supplied and inverter supplied motors

Harmonic currents create leakage fluxes, which induce eddy currents both in
the stator and the rotor resulting in significantly increased stray losses and
higher temperatures in the motor. It was therefore necessary to de-rate the
converter fed motors in order to avoid over-heating. Typical reduction of the
rated power was in the order of 10 - 20 percent until new frequency converters, with much better switching performance, were introduced. Now there is
practically no reduction at all [216]. The current harmonics can also increase
noise and torque ripple, which is considered a major drawback. Better converters have of course improved the situation, but even the motor design has
had an important impact. Layout of windings and cores influence the magni-

tude of electro-magnetically created vibrations and the design of the stator house determines the transmission of these vibrations as well as the noise radiation.

In Sweden, the influence of harmonics in inverter-fed induction motors was first studied by Lars Gertmar in his PhD research project which he carried out at Chalmers in the mid 1970's, i.e. several years before Asea started corresponding activities. Gertmar was employed by Asea in 1979 and is now Senior Corporate Scientist in ABB [217].

Initially, variable-speed motors often broke down due to earth faults in the stator windings. The reason was that the inverters produced steep voltage fronts that, through reflection in the motor windings, could create such high voltage spikes that the insulation failed. The problem was not very difficult to identify, but the necessary improvement of the insulation system has required much attention. Motors intended for inverter operation are therefore sometimes provided with reinforced insulation.

Another common problem of frequency converter supplied motors has been bearing failures. Motors fed from modern IGBT inverters could fail within the first months of operation due to erosion of the bearing races caused by bearing currents. The photo in figure 6.48 shows a typical pattern that bearing currents leave in a ball bearing. This kind of failure occurs when the voltage across the insulating oil-film becomes large enough to cause a discharge, which make a small pit in the bearing surface. This process is repeated continuously resulting in increased damage.

Figure 6.48 Ball bearing damaged by bearing currents

Bearing currents are, since long, a well-known phenomenon, especially in large machines, where asymmetries in the magnetic flux could induce low frequency, low voltages (a few volts) but yielding high currents through the bearings. This has not been a problem with small and medium sized standard motors, but it has been necessary to use insulated bearings in larger machines in order to prevent such bearing currents from flowing. Capacitive bearing currents could also arise if a rotor became electrically charged, either from gas friction or capacitive coupling to the stator winding. These phenomena could result in high voltages (> 100 V) but small currents. Such currents, through the bearings, have been avoided through earthing of the rotor via a special brush in contact with the shaft end, thus short-circuiting the bearing oil-films. Standard motors, of common sizes, have not required such protection. These problems were familiar to electrical machine designers, but the failures due to bearing currents, occurring in variable-speed motors caused a lot of confusion in the late 1980's. It could be concluded that it was a matter of high-frequency capacitive currents, but what were the reasons?

ABB, in Finland, started investigations of the new bearing current problems in 1987 and continued for more than a decade; according to their leading motor specialist Tapio Haring certain problems remain even today. It was necessary to adopt a system approach because the high-frequency currents originated from the inverter while the damages occurred in the motor or even in the driven machinery. It is outside the scope of this thesis to explain the results of ABB's work in this field. Some of them can be studied in published documents [218], but a brief example, which illustrates what kind of problems electrical machine specialists have to face nowadays can be helpful.

The problem starts in the inverter. The fundamental frequency components in a 3-phase inverter are balanced but the harmonics are not and, therefore, the voltage of the neutral point is not zero. There is a so-called common mode voltage that contains very high frequencies due to the steep voltage pulses, which are caused by the high switching frequency of the inverter creating a common mode current through the system. The current paths, and the size of the current, are determined by the stray capacitances in various parts of the system, which is illustrated by the circuit diagram in figure 6.49.

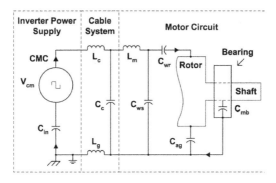

Figure 6.49 Circuit diagram for capacitive common mode currents

The voltage across the bearings depends on capacitive division of the common mode voltage. If it is high enough, discharges through the oil film will occur and constitute a high-frequent bearing current. There are also other ways to create these capacitive bearing currents, but the one described is most common in the case of smaller motors.

The preventive measures against the bearing current problems are use of proper cabling and earthing, insertion of bearing insulation as well as damping of the high-frequency components in the common mode circuit.

There have, of course, been many special motors designed for frequency converter operation, even induction motors, but it is interesting to ask whether manufacturers have developed any separate series of converter fed induction motors as a complement to their series of standard motors. Asea and ABB in Sweden have not done so, except for high speed and integral motors, which will be dealt with in section 6.7.11. Initially, when the volumes were low, it would have been too expensive. Later, the frequency converters have become so good that standard motors can be used without any restrictions.

6.7.10 Interaction between motor and converter development

The most common variable speed drive for industrial applications consists of a frequency converter, a squirrel-cage induction motor and a control unit as shown in figure 6.50. The components have to be carefully selected so that they match each other and this is not only a matter of power and voltage. Many customers prefer to purchase the drive system as one unit from a supplier, which takes full responsibility for function, performance and quality. This often requires some development cooperation or at least sufficient knowledge concerning the other components. Companies like ABB usually have motors, converters and complete drive systems in their product programs. There is coordination of development activities, but it is not too strong, as the components are also sold separately. ABB has, however, introduced tests of complete systems before launching new product series.

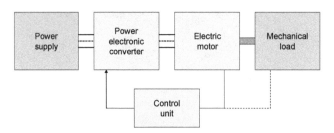

Figure 6.50 Block diagram of drive system

Frequency converters and motors are normally separate units so there is no need for any mechanical integration. Figure 6.51 shows photos of some converters. Voltages are determined by the grid and rated power levels do not require special matching. The cooperation between motor and converter development engineers is instead more focused on control properties, content of harmonics, possible over-voltages, consequences of various failures, optimization of system efficiency etc. Tapio Haring mentioned, as an example, that ABB's converter engineers have developed special filters to meet the motor engineers request for lower harmonics in certain applications.

Figure 6.51 ABB frequency converters for motor drives up to 7.5 kW

One example of interaction between motor and converter specialists concerns the relation between voltage and frequency below base speed, where a constant ratio, V/Hz, is usually assumed, corresponding to a constant flux in the motor. However, it can be advantageous, for many applications, to reduce the voltage and thus the flux which decreases the iron losses enabling the total efficiency to be improved even if a lower voltage means higher currents and copper losses. This kind of optimization cannot be done without sufficient knowledge of both motor and converter. The implementation is then a matter of adjusting an algorithm in the control unit. Ingemar Olofsson verified, in an interview, that this kind of modifications had been carried out for different applications. The diagram in figure 6.52 illustrates the principle.

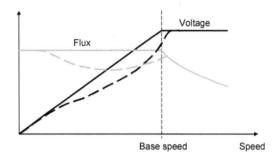

Figure 6.52 Voltage vs. frequency, constant ratio and modified

Most of the university projects carried out in Sweden during the last decades, with focus on induction machines, deal with variable speed applications and the consequences of these. Sweden is no exception in this respect. A necessary condition is knowledge in electrical machine theory, converter technology and control. A couple of doctoral theses from KTH and Chalmers can serve as examples [4, 219].

6.7.11 Development of high-speed and integral motors

The maximum speed of directly grid-connected induction motors is close to 3000 rpm in the case of a 50 Hz supply and 3600 for 60 Hz. Many driven objects require higher speeds which have traditionally been achieved through gearboxes or belt transmissions. The use of frequency converters has opened up possibilities to reach high speeds without any help of such traditional mechanical transmissions. The motor can be directly coupled to the driven machinery as shown in figure 6.53. The investment and maintenance costs, power losses and space for a gearbox are eliminated, but corresponding items have to be considered for the frequency converter. It is not obvious if the high-speed motor itself becomes more or less expensive. It can be smaller than a conventional motor due to a lower torque, but special design of certain components, e.g. the bearings, can increase the costs. A big advantage is that the concept with frequency converter supply implicitly gives an opportunity for energy saving variable speed operation.

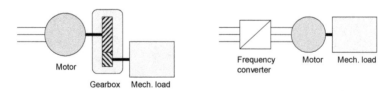

Figure 6.53 Drive system with gearbox and with frequency converter

One common definition of high-speed motors is that they should be able to operate above the synchronous speeds of 3000/3600 rpm for grid-connected 2-pole motors. A better, but less exact, approach is to refer those motors, which are provided with special high-speed design features, to this category. Often 4-pole motors are preferred because of more optimal dimensions, but this means a basic frequency above 100 Hz as soon as the speed gets higher than 3000 rpm. 10 000 rpm corresponds to 333 Hz. High-speed motors must therefore be designed for much higher frequencies than standard motors. Extra thin, low loss lamination has to be applied to reduce eddy current losses in the stator core. The stator winding wire must also be thin for similar reasons. The rotor will be subject to high centrifugal forces, which may lead to the need for special design solutions. It is necessary to choose suitable high-speed bearings and to check if any critical speeds could come close to the operating speed range.

Standard induction motors can be used for high-speed drives, but to a limited extent. In 1988 ABB started, on initiative and leadership from Gunnar Mellgren, a project for the development of high speed drives. Compressor drives were identified as a most promising application and some prototypes were built. The maximum speeds and outputs were not extreme, 10 000 rpm and 200 kW, but nevertheless required motors with modified design; especially the bearings required special attention. Another critical issue was increased stray losses in the rotor winding; a problem which was addressed by Chandur Sadarangani. He invented and patented a new rotor slot shape, shown in figure 6.54, which reduces skin effect phenomena at high frequencies and also increases the slot leakage reactance; both contributing to reduced losses [220, 221]. The high-speed project never resulted in any special series of such motors. The market was not large enough and the standard motors built by ABB in Västerås allow operation up to 4000 – 6000 rpm, depending on the size of these motors, and this was usually sufficient. Sadarangani's invention, however, is used in traction motors designed by Bombardier.

Figure 6.54 Rotor slot with U-shaped bridge for high-speed induction motors

Induction motors can be used for much higher speeds than those discussed above, but that requires quite special designs. Interesting examples of this can be found in Finland, where the technical universities in Helsinki and Lappeenranta have carried out several research projects in this field during the last two decades. The rotors have been made of solid steel in order to withstand excessive centrifugal forces and to provide sufficient stiffness at speeds in the order of 50 000 rpm and higher. Magnetic bearings and airfoil bearings have been used in most cases. The problem with a solid steel rotor, compared with a laminated squirrel-cage rotor, is that its electro-magnetical properties become worse at the same time as it is much improved from a mechanical point of view. Some of the projects have investigated the different possibilities to achieve improvements, e.g. by slotting the solid rotors axially as shown in figure 6.55.

Figure 6.55 Slotted, solid steel rotor for high-speed induction motor

A deeper understanding of this type of high-speed machines can be obtained through some of the PhD dissertations published [222, 223, 224]. These Finnish research projects have resulted in practical industrial activities. A company called Rotatek Finland Oy, which was established in Lappeenranta in 1996, is manufacturing solid rotor high-speed induction motors in varying sizes up to 1000 kW (< 16 000 rpm) and speeds even above 50 000 rpm (< 100 kW) [225]. Compressors, blowers and pumps are typical applications. Another, but power-wise much smaller, application for high-speed motors is so-called spindle drives in machine tools.

An AC variable speed drive system generally consists of a separately located frequency converter and an induction motor connected to each other through a cable. The distance between these two units can often be quite long, especially if they are placed in different rooms. The reason for this is to have the more sensitive electronics in a cleaner area than the motor, but it complicates the installation and contributes to higher costs. These disadvantages could be avoided if the converter and the motor could be integrated into one unit, a solution considered applicable at least for smaller motors. Another advantage is that problems related to EMC could be reduced. Gunnar Mellgren's development team for high-speed drives, mentioned above, realized this possibility and therefore started an "Integral Motor" project in 1988. The idea was to use ABB's standard 4-pole induction motors up to 7.5 kW and provide them with specially designed frequency converters placed at the non-drive end of the motors, see figure 6.56.

The integral motor concept did not complicate the motor design but it created a challenge for the converter designers. The unit had to be totally enclosed as the motor and designed to withstand high temperatures and vibrations. This was achieved and integral motors were introduced to the market in 1995. Such an integral motor was approximately 40 percent longer and weighed about 15 percent more than the corresponding standard motor, and could be installed just as easily as a fixed frequency standard motor. The integral motor also had the same footprint, which opened a possibility to replace existing constant-speed fan and pump motors with variable-speed integral motors. The extra costs would soon be recovered thanks to improved process efficiency [226].

Figure 6.56 ABB's earlier integral motor with frequency converter at the non-driving end of the motor

ABB Motors in Västerås built a small production line for integral motors in the power range of 0.75 – 7.5 kW. The frequency converters were purchased externally from the Danish pump manufacturer Grundfoss, because the concept did not fit in ABB Finland's line production of small converters. The production showed poor profitability because of too small volumes. It was moved from Västerås to Odense, but ABB terminated the manufacture of integral motors in 2001 when the entire Odense factory was closed. ABB has, in spite of this, launched a new series of integral motors in the power range of 0.37 – 2.2 kW. This type of motor has the converter placed at top of the motor. It is now ABB's own converters and these integral motors are built in Italy [227]. There are also other companies which build integral motors using ideas similar to ABB's. For instance, one of the leading American motor manufacturers, Reliance Electric, introduced such a product series in 1997 and has continued to develop it. Siemens is building integral motors up to 4 kW. Leroy-Somer, in France, is another example which also has integral motors in its product portfolio. The Swedish converter manufacturer Emotron is marketing integral motors based on motors from VEM.

The reason why ABB closed the first manufacture of integral motors, in Västerås, was, as mentioned above, too low sales volume. The customers did not find sufficient cost advantages and it is doubtful that integral motors have been a success for any manufacturer. In addition, ABB had chosen a different concept from all others when it was decided to place the inverter at the non-driving end of the motor instead of on top of the motor. ABB's original design may have been better from a strictly technical point of view, but customers usually prefer solutions available from more than one supplier.

6.8   New technical and commercial challenges

6.8.1   Simple …… but complicated

The heading implies a paradox, which actually is relevant for induction motors. Standard motors can be seen as more or less a commodity and there are hundreds of manufacturers all around the world. Many have been established by entrepreneurs with limited technical background. It has been possible to learn how to dimension, design and manufacture these motors almost like a handcraft, i.e. without the deeper theoretical knowledge that development engineers in leading companies usually have. The smaller manufacturers have learned from the larger ones and in many cases become quite successful. A couple of reasons for their success can be that induction motor development depends so much on production development and that small manufacturers are often more flexible with respect to special customer requirements. Nonetheless, the induction motor has long been an established product that has not experienced any radical development steps. You would therefore expect that theoretical predictions of their performance nowadays could be made with exact precision, but that is not the case.

Accurate theoretical analyses of induction motors and other types of electrical machines are difficult, even with the help of powerful numerical simulation tools. The analyses are influenced by non-linearities and many random effects, which are practically impossible to model, and those simulations that can be performed sometimes require several hours of computer time.

A good example of an almost impossible analysis is the calculation of stray losses in an inverter fed induction motor. This problem is so difficult, even for a line-connected, fixed frequency motor, that the IEC standard still specifies that the stray losses shall be considered as 0.5 percent of the rated output [228]. The other losses, i.e. copper, iron and friction losses can be predicted analytically with a fairly high accuracy, but not the stray losses. This can be very critical as the stray losses, in an inverter fed motor, often can be in the order of 20 – 30 percent of the total losses. The losses are usually determined through tests and the results constitute an empirical base for estimation of stray losses in future machines. A new IEC standard is under preparation which will specify tests for determination of induction motor losses.

Why is a correct theoretical analysis so difficult? The stray losses consist mainly of eddy current losses due to leakage and harmonic fluxes in the stator core, rotor surface and to some extent also in structural components such as the end-shields. Losses due to eddy currents and internal circulating currents in the windings are also referred to the stray losses. Stray losses in the stator teeth can be used as an example. They are, per definition, the extra losses not included in the no-load iron losses and they consist mainly of eddy current losses caused by leakage flux and harmonics. The exact distribution of the leakage flux is hard to predict as the permeability of the iron depends on local, non-linear saturation effects, residual stresses in the sheets influenced by the sharpness of the punch tool, etc. There is also a temperature influence as the resistivity of the sheet varies with the temperature. An accurate simulation, which could take the harmonics into account, would be very time consuming and still affected by production related random effects. This is the background on why stray losses, even in quite recent scientific documents, are assessed just as an estimated percentage of the total losses [228]. Determination of stray losses is not the only problem of this character, but it will serve as a sufficient and important example of this situation. Electrical machine designers will have to live with these difficulties even in the future.

Professor Sadarangani has expressed an opinion that significant improvements in performance of the induction motors can only be reached if there will be a major engineering effort, which would include both development of more advanced simulation tools and extensive prototype testing. There is a potential for improvements, but it is doubtful whether any manufacturer will consider the necessary efforts profitable.

6.8.2 Customization vs. standardization

A company like ABB sells slightly more than 2/3 of its LV motors to OEM customers, i.e. companies which integrate the motors in their own products such as compressors, machine tools etc. The rest are sold to end users via internal or external distributors. Standard motors are normally sufficient for the latter category, while many OEMs look for customized solutions. Examples of these can be special shaft ends, flanges or feet, water cooled stator houses, customer specific color but even supply of semi-finished motors, usually only the active parts. Customization has mainly impact on the mechanical design, while the cores and windings remain the same as for the standard motors. It is especially important to utilize the expensive punching tools for the laminated stator and rotor cores. The stator windings are, of course, made for several different voltages.

The request for customized solutions has increased in recent years. The main reason is that an OEM can save more money in his final product than the extra cost he has to pay for a customized motor. He can also have other advantages such as more compact design, less noise, and easier installation to mention a few. Strong competition, as well as rational and flexible production, contributes to low price increases for customized motors built for big OEMs, which order large quantities. A consequence of this trend is increased long-term cooperation between motor manufacturers and their major customers. This can be further enhanced by extensive technical backup that gives an added value to key customers, which get access to the motor supplier's unique know-how for choosing optimal drive solutions. The motor for Atlas Copco's compressor unit, shown in figure 6.25b, is an example of a customized motor.

6.8.3 Globalization

The word "Globalization" can probably be interpreted in different ways depending on the actual subject. In the case of trade of goods and services, globalization stands for an increase of international commercial activities and reduced trade barriers. More and more companies see the world as their market, but also as their source for supplies. Transports are less of a problem; they have become both quick and inexpensive. We can, for instance, at our Swedish breakfast table find orange juice from Florida, yoghurt from France, sausages from Germany, cheese from Holland, coffee from Brazil and fruit from South Africa and New Zeeland.

Globalization today influences almost every kind of economic activity. In relationship to industrial products, it contains some key elements such as increased international trade, outsourcing of production and transfer of technology. What has this globalization meant for the induction motors, especially from a Swedish point of view? It is obvious, from earlier sections in this chapter, that Asea, and later ABB, has operated as a multinational company in this field. Motors have been exported from its factories in Sweden since the end of the 19th century and the company has had motor production, in foreign countries, for almost the same time. There has also always been an import to Sweden of motors from other manufacturers. So, what is the difference; why is globalization seen as a fairly new phenomenon?

The fact that large companies, like ABB, sell their motors in various countries all over the world is nothing new. The focus may shift from European and North American markets to China and India, but this does not change anything in principle. Foreign production is more important in this respect. ABB closed induction motor factories in Decine, France in 1993, in Saarbrücken, Germany in 1998 and sold the Mexican factory to WEG in 1999. Most of the production of cast iron motors was transferred from Saarbrücken to Västerås. During this period, ABB has also started to manufacture motors in China and in India. This has, so far, not influenced the development work because both these factories are producing motors according to existing, older designs. The responsibility for future development of motors in frame sizes 160 – 250 is still allocated to the Västerås factory. The Spanish unit handles the smaller sizes and Waasa in Finland the larger ones. Siemens, in comparison, has moved most of its European LV motor production to the Czech Republic, but the development has remained in Germany.

"Transfer of technology" usually refers to immaterial rights and know-how. Such transfer can take place between persons, organizations and even countries. In the case of electrical machines, it is most common that experienced manufacturers in industrialized countries share parts of their technology with companies in newly industrialized and developing countries. The technology that is transferred can vary from know-how concerning use and maintenance, to a license to manufacture according to existing drawings, further via engineering tools and documentation that enables product design, to extensive knowledge necessary for further independent development.

Transfer of technology is sometimes requested from governments in certain countries. It is difficult for Western companies to refuse such requests, because there are always competitors willing to accept these requirements. In other cases, the motive can be a matter of cost reduction. Engineers in countries like China and India cost, according to ABB, only 1/3 and 1/5 compared with Swedish personnel.

Outsourcing is a trend strongly connected with globalization. The automotive industry has been a pioneer in this field implementing outsourcing in the 1980's. An important prerequisite was "just in time deliveries" from reliable suppliers. Many other industries have followed, among them the electrotechnical companies. The basic idea is to purchase from suppliers everything they can do better or cheaper and instead focus on those activities, which require the company's core competence. Manufacturers have always used subsuppliers, sometimes located in the vicinity, but the extent has increased and their location is now often in low-cost countries in Eastern Europe or Asia. As an example, it is interesting to see how ABB's motor production in Västerås has changed in the last 5 – 10 years from being very integrated to becoming more dependent on external suppliers. Very few components were outsourced in the mid 1990's; only fans, fan covers, terminal boxes, besides purchased components such as bearings and sensors. Ten years later all stator housings and end-shields are supplied from foreign foundries and an external Swedish manufacturer delivers all shafts. The motives may have been to get access to a more efficient production facility, to use low cost labor, or to avoid new investments in production capacity or machine tools.

Outsourcing is not just a matter of manufacturing; it can also include certain development in which the specialized competence of a supplier can contribute to a better and more cost effective product. This has, according to information from ABB Motors, not been the case so far. As mentioned above, the company is outsourcing shafts, stator housings, end-shields etc. but everything has been designed in-house.

An ultimate form of outsourcing is "Branding", i.e. when a company buys complete products from a sub-supplier but puts his own nameplate on them. The products have, in such cases, been developed by the sub-supplier, possibly according to a specification from the former. One example of branding can be found within ABB, which gets one of its smallest frame sizes from a Romanian company. The exterior design is in line with ABB's other motors while the inner parts are the same as the manufacturer use for his own products. Another example is BEVI (see section 6.6.3). This company belongs to the largest in Sweden for sales of induction motors after ABB and most of these motors are imported from foreign manufacturers; often provided with BEVI's nameplate.

New actors have entered the market in recent years. The largest shares of the Swedish market for induction motors, around 1990, had European manufacturers. The situation has changed so that in 2005, import from overseas manufacturers such as the Brazilian company WEG, is substantial. Figure 6.57 shows the companies which had the largest market shares in Sweden during 2005. It is obvious that manufacturers in newly industrialized countries have reached a technical level that allows them to compete with their products also in countries like Sweden. The extra cost for long transportation is compensated for by lower labor costs for production.

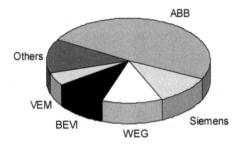

Figure 6.57 Market shares for suppliers of LV induction motors in the Swedish market 2005

Much of the information above was obtained in interviews with Sven Sjöberg, former senior vice-president for market and technical development of LV motors in ABB's business area for automation products, and from Bo Malmros, electrical development engineer. Sjöberg underlined that factors such as logistic system, availability and delivery reliability are very important within the EU. The customers care less about where the manufacture takes place. The name of the manufacturer is also of limited importance. The bonus of a well-known trademark can be estimated at five percent for industrial products like induction motors. This can be compared with 10 - 20 percent for many kinds of consumer products, e.g. domestic appliances. The tendency is, however, that the acceptance of such price differences is decreasing.

6.8.4 Competing technologies

The induction motor has been dominant for industrial applications for more than a century. Is there any real threat to this situation from competing technologies? If so, which are the options?

Old machine types such as DC motors and electrically excited synchronous motors can be disregarded as competitors. They have since long been marginalized by the standard type induction motor for most industrial applications. More viable future alternatives are PM and SR motors.

So far, the majority of induction motors are used for fixed frequency drives, i.e. they are used without any converters. Power electronic converters are still expensive, compared with the motors, and will remain so even though the ratio is decreasing. This means that fixed frequency motors will continue to constitute a major part of industrial drives also in the future. This excludes the use of SR motors, which require inverters to be able to function. PM motors, on the other hand, can operate directly connected to the line, provided that they can be started. Therefore, PM motors have been developed with some sort of cage winding in the rotor, which enables them to start asynchronously and then be automatically pulled into synchronism when it is close to nominal speed [229].The PM motor has better efficiency than an induction motor, but it is also much more expensive. It can consequently only compete in constant speed applications in which the capitalized value of the losses will outweigh the higher price for the motor. Another condition is that the requested starting torque is moderate. A logical conclusion is that the induction motors will continue, almost completely, to be the "one and only" motor type for fixed frequency drives.

The situation becomes more complicated in the case of variable speed drives. The relative share of induction motors used in combination with frequency converters has increased; figures of 25 – 30 percent have been mentioned. This is a significant portion and both PM and SR motors are viable alternatives.

PM motors are well established for many applications, e.g. as servomotors, and have also entered into other areas. The merits they have, compared with induction motors, are higher specific torque and power, better efficiency and high power factor independent of pole number. A high pole number makes the PM motor suitable for low speed / high torque drives, thus eliminating the need for expensive reduction gears. An example is shown in figure 6.56. A better power factor often opens a possibility to use somewhat smaller converters for PM motors. The disadvantages are higher costs and the fact that they are more temperature sensitive. Another drawback for PM motors has been the need for rotor position sensors, often expensive resolvers, which become especially significant in applications requiring operation with high accuracy at very low speeds. However, many R&D projects have focused on sensorless PM drives so the two motor types will probably be on par in this respect in the future, even for more demanding applications [230]. Larger PM machines have, so far, mainly been used as wind power generators, lift motors, high-torque industrial motors and for electrical drivelines in vehicles. ABB has introduced a series of low-speed high-torque PM motors for paper machine drives capable of sensorless operation based on ABB's DTC control [231, 232]. Even in the future, they have their best chance in similar applications, but can probably also gain some share of the total variable speed drive market if the requirements on high efficiency is further enhanced.

Figure 6.58 High torque ABB PM motor for paper machine drive

The SR motor (Switched Reluctance motor) is, as its name indicates, a result of the access to power electronic switching devices. The development started already in the 1970's, but there has not been any real break through for these motors. They have suffered from a reputation of being noisy and having large torque ripple. The performance has been improved and SR motors are presently used for applications such as rotating heat exchangers, dosage pumps and even in vehicles. The motor itself is very simple and should not cost more to manufacture than a corresponding induction motor provided the production volumes are of the same order. The absence of a rotor winding should give the SR motor a somewhat higher efficiency. It is basically a simple and robust machine but the current control is complicated, and some specialists claim that it can be difficult to reach the required performance without certain complex measures. Sensorless control requires, for this type of motor, accurate models and advanced algorithms. In addition, the SR motor needs small airgaps, which requires high precision in the production. SR motors will, therefore, probably remain a niche product, even if it can become a competitor to induction motors for some variable speed applications, provided the level of noise and torque ripple can be accepted [233].

An interesting question is; will these competing technologies influence the development of the induction motors? Probably not, because standard motors for fixed frequency drives will remain dominant in the foreseeable future. Manufacturers of induction motors have to develop their products primarily to meet competition from each other. Competition from other motor types is mainly a matter of variable speed drives, and it is likely that it will enhance the development of high efficiency induction motors, thus focusing on their weakest point in comparison with PM and SR motors.

6.8.5 Competence and resources

The competence and resources for development of induction motors have, from a Swedish perspective, been strongly linked to Asea's, and later ABB's, ability. There have been, and still are, a few other manufacturers but they are, in comparison, niche companies. ABB, in Västerås, maintains a position as a leading motor manufacturer, even if the development and manufacture nowadays are shared with sister factories in foreign countries, primarily Finland and Spain.

Asea's M-division had, in the mid 1970's, a department for design and development employing approximately 50 persons. ABB Motors corresponding department had around 25 persons in the year 2000. It is difficult to compare these numbers because the product range decreased after Asea's acquisition of Strömberg and the merger with BBC, but the production volume remained due to the access to a larger market. The use of more efficient, computer based, engineering tools also has a big impact on the size of the staff. One thing is however obvious, the number of graduated electrical engineers has decreased from roughly ten to five, a fact that has made the organization more sensitive. It can be questioned whether it is below the "critical mass", but on the other hand, the department is now part of a large multi-national business area which helps. The situation is also somewhat improved taking into account that the Corporate Research unit had only a couple of electrical machine specialists in the 70's, but it has close to ten such specialists today.

Cooperation between the Swedish electrical machine industry and the technical universities has grown during the last two decades; a fact that is particularly true for ABB's motor factory in Västerås. Fredrik Gustavson, who for a long period had a leading position at KTH's department for electrical machines, mentioned that KTH's major contacts with the electrical manufacturers, from the 1950's until the 70's, were primarily for testing of various machines and other equipment. The contacts with Asea were, in this respect, very limited, and "Asea looked upon itself as self-sufficient." The technical universities constituted primarily a recruiting ground for Asea. This can be illustrated by the fact that Asea employed eight out of nine PhD with electrical machine background, examined from Chalmers between 1974 and 1983.Only one of them had had a project supported by Asea.

KTH started with more extensive electrical machine research activities first in the late 1980's. Chalmers was, before then, in this particular field, much ahead of KTH. Professor Svante von Zweigbergk (1914-2001), who came from Finland, had initiated electrical machine research at Chalmers already in the 1950's. These activities included some industrial cooperation, but practically nothing with Asea. The present situation is very different and there are many examples of R&D cooperation between the technical universities and the electrical machine industry, not least ABB. The departments for electrical machines and drive systems at KTH, LTH and Chalmers can, therefore, be considered as an important complement to the manufacturers own theoretical resources. The specific production technology, however, has to be secured by the manufacturers themselves.

6.9 Conclusions

The study concerning the development of induction motors, presented in this chapter, covers a long time span, and it involves many actors, not only Asea/ABB, even if this company has received most attention. The technical development was, as expected, most intensive at the beginning; no radical steps have been made since the standards for LV motors were introduced. However, the increased use of frequency converter fed induction motors for variable speed drives has resulted in new challenges for the motor designers. The study has shown that production development is very important for this kind of standard products, which are very little differentiated from one manufacturer to another. The business strategies have primarily been based on market considerations. The most important conclusions are summarized below.

The historical phases:

- The standard LV induction motor has long been an extremely well established product; by far the most common electrical motor for industrial drive systems. It has been manufactured globally by many hundreds of companies, both large and small, and has almost become a commodity.

- The standard LV induction motor has been strictly standardized since more than 50 years, a fact which has minimized product differentiation. It has been easy to copy, thereby enabling manufacturers with limited R&D resources to become successful.

- Asea/ABB established itself as a leading motor manufacturer more than a century ago and has managed to maintain this position. This was typical also for other frontline companies as for instance BBC,

GE, Siemens and Westinghouse; at least until the last decades when the electro industry started to be restructured.

- Asea/ABB has achieved and maintained its position as a leading motor manufacturer for basically four reasons:
  - It developed early a comprehensive line of motors and has continued to do so.
  - It has consistently developed and invested in very rational production facilities.
  - It has successfully strived for dominating positions in certain markets, e.g. the Swedish one.
  - Its motor business has been strengthened by being part of a powerful company manufacturing many other electrical products which has had a wide customer basis.

- Production (or process) development has long been the most important factor for the reduction of product costs.

- The market need for product variants as open ventilated motors and slipring motors decreased successively and Asea decided to exclude them from the product program in the late 1970's.

- Requests for improved efficiency, but even reduced noise and improved starting capability have been the major technical driving factors for the product development since the 1970's.

- The use of induction motors in inverter-fed variable speed drives began in the 1970's, but Asea was in this respect somewhat behind its major competitors. This changed, however, when ABB was established.

- Asea's specialists represented, for a long period, the state-of-art competence in electric motor technology in Sweden, and the cooperation with the technical universities was very limited. Such cooperation has become much more common during later years.

It is possible to identify a number of phases in the development of the induction motors as illustrated by the diagram in figure 6.59.

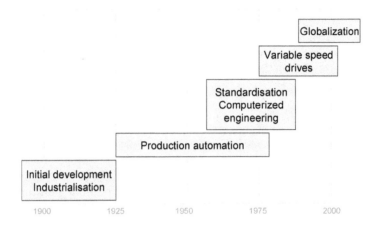

Figure 6.59 Development phases for standard induction motors

The current situation:

- The standard LV induction motor is still dominant for industrial drive systems. An increasing share of them is used for variable speed drives.

- Globalization is an ongoing process that is changing the structure of the electric motor industry. Manufacturers from countries like Brazil, China, India and Taiwan export motors even to Europe.

- Outsourcing is to a large extent replacing in-house manufacturing, especially of components which don't represent the motor manufacturer's core competence.

- Most of the major manufacturers build motors of similar quality and performance. Sales are instead promoted through better technical support, quick and reliable deliveries as well as a well-known trademark.

- Important development targets are reduced costs and improved efficiency. Standard motors have to meet certain stipulated efficiency levels.

- Modern frequency converters cause such low harmonics that the induction motors developed for fixed frequency operation can be used without special restrictions.

- Special motors, e.g. for higher speeds, and motors customized for OEMs are offered by many manufacturers.

- ABB has maintained its position and is today a world leading manufacturer of LV induction motors, but its Swedish factory produce only around 15 percent of ABB's total volume of such motors with a little more than 10 percent of the personnel.

- ABB Motors factory in Västerås handles development of standard products in close cooperation with the sister factories in other countries. It is dependent on support from ABB Corporate Research for special studies and it cooperates with academy and other Swedish manufacturers in KTH's competence centre for advanced motor drive systems.

The future:

- The standard LV induction motor has long been a mature product and no drastic changes are expected for the future; no real threats can be identified.

- The induction motor will remain dominant for industrial applications, primarily due to its superior properties and cost effectiveness for fixed frequency drives.

- There will be no changes of topology and main components. There is too much invested in present designs and production facilities.

- No changes of the current standards are foreseen.

- Ongoing rationalization and globalization will result in larger and more powerful OEM customers, which probably will require more customization, especially mechanical adaptation.

- The share of variable speed drives will continue to increase, but more slowly.

- The induction motors are somewhat challenged by PM and SR motors in case of variable speed drives, but it will remain dominant, of cost reasons, also for such applications

- Requirements for improved efficiency will be even more emphasized, due to increased energy costs and further legal stipulations.

- There is a potential for performance improvements but such improvements require development of better tools for analyses and simulations. Such development is expensive and it is questionable whether it will payoff.

- Better but more expensive materials, e.g. electrical steel with lower losses, will be available and used more frequently. Even squirrel-cage windings from cast copper will be used if high loss evaluations are introduced.

- The globalization can be expected to continue and comprise manufacturers from new low-cost countries. A question mark is, however, whether the presently low transport costs will increase and affect the current trend.

- The Swedish manufacturers, mainly ABB, survive due to excellent production facilities which enable competitive prices as well as quick and reliable deliveries. Competent technical support to major customers is also a decisive factor.

- The development of standard motors will remain a task for the motor manufacturers, but close cooperation with universities will continue for analyses of different phenomena and development of better algorithms to be included in future engineering tools.

The competitive situation for the Swedish electrical motor manufacturers is severe, but obviously not impossible to handle. The existing competences as well as all other resources required are large and represent a huge investment. It is therefore reason to believe, that it will remain strong also in the foreseeable future.

7   Large turbogenerators

Large steam turbine-driven generators rated at a few hundred MW and higher constitute, in many respects, a big engineering challenge. The Swedish manufacturer Asea was faced with this challenge in the late sixties, when the company started to develop such generators for nuclear power plants. It was necessary to choose new concepts which, in many cases, later led to difficult teething problems before the generators could be delivered and operate satisfactory; a process, which took around a decade to be completed. This is an excellent example on how an electrical machine manufacturer decided to take a big step and how the company faced different problems. It is also a very interesting example on advanced technology and interaction between various technical disciplines. In addition, it illustrates very well the development of large machines carried out in connection with customer orders and the risks involved in this.

7.1   The start position

7.1.1   Hydropower generators

Asea had, in the early sixties, a position as one of the world's leading manufacturers of hydropower generators. An important reason for this was the large domestic market for such generators. The harnessing of the abundant energy from the Swedish waterfalls for electricity production started already in the late 19th century and the construction of new hydropower plants continued then for 70 – 80 years. During this long period, there was a stable growth in generator size and several Asea generators have been milestones also in the international development; some were the largest in the world at the time they were put into operation. Early examples were four 10.5 MVA generators for Svaelgfos in 1907, six 18.9 MVA generators for Rjukanfos II in 1915 and two 22 MVA generators for Glomfjord in 1919. A picture of the latter is shown in figure 7.1. None of these power plants were Swedish, but were actually Norwegian. Norway was a very large, almost domestic market for Asea's hydropower generators during the first couple of decades of the last century, before national manufacturing was established [235, 236]. A contributing factor, to the orders from Norway, was also that the Stockholms Enskilda Bank, which had financially supported Asea, had large interests in Norsk Hydro, one of the major Norwegian customers. Asea also managed early to become an important supplier of generators to some remote countries, notably Canada. The leadership in generator size shifted during those early years between Westinghouse, GE, BBC and Asea [237].

Figure 7.1 Stator for Glomfjord generator in Asea's workshop, 1919

Most of the old hydropower generators had horizontal shafts and pedestal bearings. They were air-cooled and self ventilated, either completely open or closed, in which case the ambient air passed through ducts to and from the machine. The stator frames were made of cast iron and the stator cores from paper insulated steel sheet segments. The three-phase windings were normally two or three layers concentric windings with micanite slot insulation. Rotor hub and spider, as well as the rotor rim with pole cores, were usually made from steel castings. The allowable winding temperature rise was up to 65 K ($^\circ$C). Rated load efficiency as high as 97.6 percent was reported for Rjukanfos II [237].

The technical development of the hydropower generators continued successively and component wise. 50 years later, the following points characterized the design of a typical generator. The shaft was vertical with upper and lower guide bearings and a thrust bearing placed in brackets above or below the rotor as shown in figure 7.2. It had a closed system for air-cooling with water-cooled heat exchangers placed around the welded stator frame. The stator core lamination was now insulated by a thin varnish. The stator winding was a lap type diamond coil winding with insulation of epoxy impregnated glass-backed mica tape. The rotor spider had also become a welded construction. The rotor rim was built up from large segments of rolled steel and the poles were now laminated. The winding temperature rise could be as high as 100 – 110 K and the efficiency close to 99.0 percent.

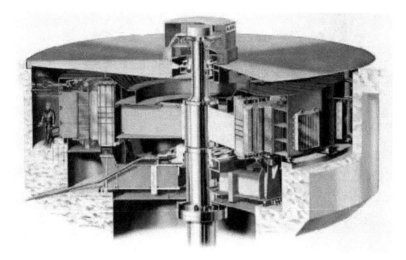

Figure 7.2 Hydropower generator with vertical shaft and thrust bearing below the rotor

Asea maintained its position as a leading manufacturer of hydropower generators. Three 105 MVA generators supplied in 1951 to Vattenfall for Harsprånget in northern Sweden were the largest outside the USA at that time [236] and the three 150 MVA generators to the same customer for Stornorrfors, in 1958, was again a world record in output [238].

Asea had also managed to become a supplier to a number of internationally prestigious hydropower projects. A good example was 22 large generators to four power plants owned by Snowy Mountains Hydro Electric Authority in Australia [239]. A photo from the machine hall at one of the plants is shown in figure 7.3. There is no wonder that the company had a lot of self-confidence as generator manufacturer. Table 7.1 lists a number of large generators, which the company exported from 1957 until 1971, the most interesting period in respect to the start of development of large turbogenerators.

*Figure 7.3 Asea generators in Murray 1 hydropower plant,
Snowy Mountains, Australia*

Table 7.1   Large hydropower generators exported by Asea 1957 - 1971

Year of delivery	Plant	Country	No. of units	MVA	rpm
1957	Tumut 1	Australia	4	84.7	375
1960	Tumut 2	Australia	4	75.7	428
1965	Murray 1	Australia	10	102.7	500
1965	Paulo Afonso	Brazil	3	80	200
1967	Mal Paso	Mexico	4	218	128
1967	Mossy Rock	USA	2	167	128
1967	Murray 2	Australia	4	145.2	333
1968	Estreito	Brazil	4	160	112
1968	Kainji	Nigeria	4	85	115
1971	Estreito	Brazil	2	174	112
1971	Castaic	USA	6	250	257

The largest hydropower generators Asea had built until the mid sixties were
those for Stornorrfors in Sweden. The diagram in figure 7.4 indicates that
there had been a relatively steady growth over the years, which improved the
security in development of this type of large products. Each new generator

would benefit from the experiences from tests and operation of already manufactured machines.

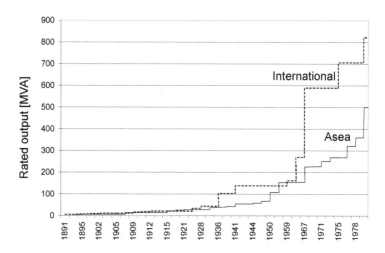

Figure 7.4 Development of hydropower generator size

Hydropower generators are more or less tailor-made, designed according to individual specifications for each plant. Waterfalls are always different. Output, speed and other parameters vary from case to case. The generators were usually sold directly to the end-customers, i.e. the power companies. The water turbines were, in a corresponding way, acquired directly from the turbine manufacturers. As a consequence, Asea was used to quote and deliver customer specific hydropower generators to many customers, in combination with turbines from different manufacturers, and in international competition with other large electro-technical companies. Therefore, in order to be competitive, Asea had to have a commercially and technically very competent sales department for these generators. Many of the sale engineers had first worked in the electrical design department for a few years. The design engineers were used to work in close contact with their colleagues in the sales department and they became very customer oriented.

7.1.2   Steam turbines and turbogenerators

The situation for Asea as manufacturer of turbogenerators was very different, even if the company had already delivered its first directly steam turbine-driven generator in 1903 [235]. Initially, these generators were delivered together with turbines from AB de Laval Ångturbin, a company that also made their own turbogenerators. This company had been founded by a famous

Swedish industrialist, Dr. Gustaf de Laval, who invented the action[*] or impulse type axial flow steam turbine in 1882. It was the first useful steam turbine in the world. Factories for building de Laval turbines were then established, not only in Sweden, but also in England and the USA around the previous turn of century. In spite of a close, but financially somewhat complicated, relation between Asea and de Laval from 1896 until 1902, Asea never became the main supplier of turbogenerators to the de Laval factories [238, 240].

Contemporary with Gustaf de Laval was the British inventor Charles A. Parson (1854-1931). In 1884 he invented the reaction[*] type axial flow steam turbine, which had better efficiency than de Laval's turbine, but was also more bulky. The Parson-turbine soon became a widely used turbine type for both marine and stationary applications. Several companies, e.g. BBC and Westinghouse, obtained license from Parsons for these turbines [48, 241]. An improved axial flow impulse turbine was patented in 1896 by the American engineer Charles G. Curtis (1860-1953), who sold the patent rights to GE one year later. Based on this, GE later obtained the position as the largest turbine manufacturer in the world. The Curtis-turbine had a somewhat inferior efficiency but was smaller and cheaper. As the various patents expired, turbine manufacturers combined the best features of each type, so the modern axial flow steam turbine could be considered as a hybrid of the original concepts [242].

Steam turbines were already used for driving dynamos during the late 1880's, but the directly coupled turbogenerator with cylindrical rotor was first introduced after the turn of century. It had been invented by Charles E. L. Brown, who was one of the founders of BBC. The invention was made in 1901 and the design was protected by a number of patents [243]. The photo in figure 7.5 shows such a rotor from 1902, now at display at Deutsches Museum in Munich. The rotor had a number of radial slots machined axially along the rotor. The excitation winding was placed in these slots and secured by dovetail wedges and retaining rings over the end-turns. Note that the rotor body was made up of several pieces and that cooling-air passed axially along the shaft and then radially between the solid rotor core sections.

Some years later, in 1908, two other Swedes, the brothers Birger and Fredrik Ljungström (1872-1948 and 1875-1964) invented and developed another type of steam turbine, the so-called double rotation, radial flow turbine. They established a company, AB Ljungströms Ångturbin (ALÅ), which licensed the patent rights for the Nordic countries in 1913, to a new company, Svenska Turbinfabriks AB Ljungström (STAL), located in Finspång, a small Swedish

---

*. The difference between an action and reaction turbine is that the steam in an action turbine expands and receives high speed before it enters the turbine wheel, while in a reaction turbine, this takes place when the steam passes through the turbine wheel.

town with a long industrial tradition. ALÅ also licensed its turbine to several other companies including even GE. For some time Asea's board of directors, led by Sigfrid Edström, had realized the necessity of a close cooperation with a steam turbine company. All the large competitors had started manufacturing of steam turbines. It was concluded, after certain studies, that the "Ljungström-turbine" was the best available option. The majority of the shares in STAL were therefore acquired by Asea in 1916 in something that could be considered as an hostile takeover [238, 244, 245].

Figure 7.5 BBC 2-pole rotor from 1902 (435 kW, 3000 rpm) and the inventor Charles E. L. Brown

The STAL-turbine was a reaction type turbine, in which the steam expands in radial direction from the steam inlet through two counter rotating disks as shown in figure 7.6 a. Each disk was directly coupled to a generator rotor, so this concept implied that two identical generators shared the turbine power. For somewhat larger outputs, axial flow turbines were combined with the radial turbine as figure 7.6 b shows.

The advantages with the STAL-turbine were that they were very compact and thus cost-effective. The efficiency was relatively high due to low steam leakage and the foundation became much simpler and smaller than for axial turbines because all forces were contained inside the turbine/generator unit itself. The main disadvantages were that each unit needed two generators and the radial turbine was limited in size due to centrifugal forces and steam flow restrictions [246].

This type of turbine proved to be very suitable for industrial backpressure applications, in which the outlet steam had a certain pressure and was used for heating various processes, e.g. in pulp and paper mills. STAL manufactured large numbers of such units for installation all over the world.

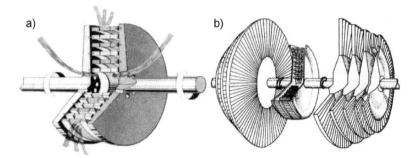

Figure 7.6 a) Principle of counter-rotating radial flow steam turbine
b) Radial flow high-pressure turbine combined with axial flow
low-pressure turbines turbines

Usually, the industrial turbines were rated below 50 MW, consequently the two generators less than 25 MW each, corresponding to approximately 30 MVA [247]. The STAL-turbines could also be used for pure electric power production. In such cases, the outlet steam was condensed in separate condensers, so that the total pressure-drop, down to vacuum, could be utilized. Figure 7.7 shows a picture of a complete backpressure unit with a counter-rotating radial flow turbine driving two identical turbogenerators.

Figure 7.7 Counter-rotating radial flow turbine with generators

Asea became an important manufacturer of smaller turbogenerators. The development in size is illustrated as follows. A two-pole 4.35 MVA, 3000 rpm generator was delivered in 1911 and an 8.75 MVA in 1917. Figure 7.8 shows a turbine/generator unit delivered to a Swedish customer in 1913. Four-pole generators rated 25 MVA, 1500 rpm were supplied to Vattenfall's thermal power plant in Västerås, Sweden in 1932 and a couple of 36 MVA generators to the same power plant in 1949 [235].

Figure 7.8 Axial steam turbine and Asea turbogenerator rated 1500 hp
(1100 kW), 3000 rpm delivered to the city of Borås in 1913

The delivery in 1965 of four 76.4 MVA, 3000 rpm generators for Vattenfall's thermal power plant in Stenungsund at the Swedish west coast was a milestone. The turbine plant, which was a 240 MW condensing unit, had the following concept. In principle, it consisted of two radial/axial turbines. One of the radial turbines was the high-pressure unit and the other the intermediate-pressure unit. The four axial turbines were connected in parallel and together constituted the low-pressure unit. Figure 7.9 illustrates the layout of this unit. The Stenungsund-generators were extra interesting, because they were the largest turbogenerators with which Asea had real experience when the rapid development of the much larger generators for the nuclear power plants started. Technical problems, which occurred in the Stenungsund-turbine, also had a big impact on subsequent strategic decisions as will be seen later.

In order to avoid confusion, it should be noted that STAL and de Laval Ångturbin merged in 1959 and formed Stal-Laval Turbin AB with almost all its activities concentrated to Finspång.

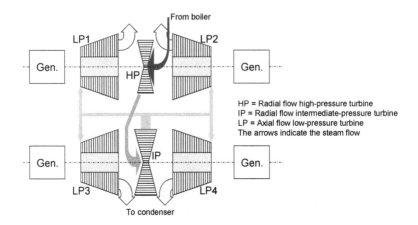

Figure 7.9 Concept of Stal-Laval's 240 MW turbine-generator unit in Stenungsund

The diagram in figure 7.10 shows how the size of Asea's turbogenerators had developed until the mid 1960's. The corresponding international development has been included for comparison [248]. It is evident from the diagrams in figures 7.4 and 7.10 that Asea, at that time, had a pronounced profile as manufacturer of hydropower generators.

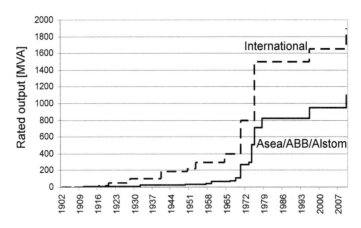

Figure 7.10 Development of turbogenerator size

Turbogenerators were almost always delivered together with the steam turbines. All the large electrical companies built complete units consisting of both turbines and generators. The integration between the turbine and the generator was more complicated in the case of thermal power plants, compared with hydropower units, therefore steam turbines and turbogenerators were always sold to the final customers as complete units. There was hardly any separate market for turbogenerators and Asea had, in reality, only one customer for these generators, its own daughter company Stal-Laval. Stal-Laval was the main supplier and Asea didn't need a highly qualified sales force as they did for the hydropower generators. There was only a small department for making quotations and handling orders. The design engineers had less market contacts than their colleagues, dealing with hydropower generators, and they didn't face competition in the same way. Asea had, at the end of 1965, 24 employees in its turbogenerator design office and 56 in the design offices for large salient pole machines [249].

7.1.3   Asea's turbogenerator designs

The rotor is usually the bottleneck in a turbogenerator, especially in two-pole machines, and the design of such rotors is very crucial. Asea had, since long, used two quite different rotor concepts, parallel slot rotors and radial slot rotors. Most common were the parallel slot rotors, which were used for all generators up to approximately 17 MVA. This type of rotor consisted of a cylindrical steel core in which parallel slots had been machined. Copper coils and insulation material were wound into these slots and kept in place by special brass wedges. Separate shaft ends were then bolted to the rotor core [250]. A sketch of slots in a parallel slot rotor is shown in figure 7.11 together with a photo from the winding process. This type of rotor had a cost-effective design but with clear limitations. The cooling was poor because the rotor losses had to be dissipated through the rotor surface. The diameter had to be limited due to the centrifugal side forces created on the teeth. The length was restricted because it was not considered acceptable to operate rotors with bolted shaft ends at overcritical speeds.

Figure 7.11 Section of rotor slot and winding of parallel slot rotor

Asea was almost the only turbogenerator manufacturer in the world using parallel slot rotors. Meidensha in Japan built them on licence from Asea and the Hungarian company Ganz Electric had its own, somewhat different concept [251].

Like other turbogenerator manufacturers, Asea used radial slot rotors for larger sizes. The rotor core and the shaft ends were forged in one piece. The rotor winding was placed in radial slots, machined along the rotor core, but with the coil ends protruding at each end of the core. Special retaining rings were shrunk over the coil ends in order to carry the centrifugal forces. The trapezoidal teeth allowed space for cooling ducts, which of course improved the cooling of the rotor winding; see figure 7.12. All the drawbacks related to the parallel slot rotor were more or less eliminated, but at the cost of a more expensive product [250].

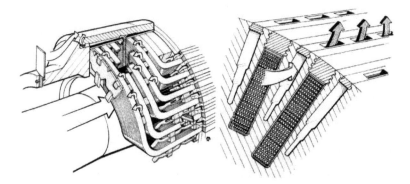

Figure 7.12 End region and slot portion of radial slot rotor

From the concept point of view, the stators were rather similar for the parallel and radial slot generators. The stator cores were built from packages of punched steel segments insulated by heat resistant varnish. The packages were axially separated by radial cooling ducts. Diamond lap type windings with one or two coil sides per slot had become the most frequently used stator winding. The insulation system had gone from micanite via asphalt-compound to very modern epoxy impregnated systems. Asea had developed two such systems. In both systems the insulation consisted of epoxy-impregnated glass-backed mica. The first system, Micapact, was based on vacuum-pressure impregnation and curing of individual coil sides and it was used for windings with rated voltages above 14 kV. The other system, Micarex, used between 3 – 14 kV, utilized pre-impregnated tape and the insulation was cured through heating of the completed winding [252, 253, 254].

The stator housing consisted of the stator frame, which carried the core, and the end-shields. Coolers were usually placed somewhere in the stator frame.

Oil lubricated sleeve bearings and sealings were placed in the end-shields. The drive-end end-shield was generally shared with the steam turbine. Gas turbine-driven generators were built as separate units. Figure 7.13 shows such an air-cooled Asea turbogenerator from the late 1960's.

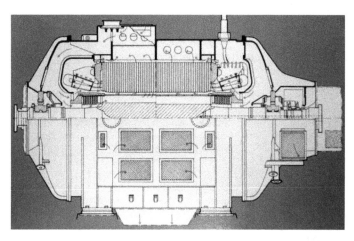

Figure 7.13 Axial section of Asea air-cooled generator with radial slot rotor

As explained in section 4.7, cooling is always very essential for all types of electrical machines and it became more critical when the machine size increased. Therefore, Asea had introduced hydrogen as the cooling medium in turbogenerators above 50 MVA. The inner parts were in principle similar to those in the air-cooled generators with radial slot rotors, but they were contained in a tight pressure vessel. Asea's first hydrogen cooled machine was a 75 MVA synchronous condensator supplied to Vattenfall in 1949 for installation in Hallsberg transformer station [235]. The first hydrogen-cooled turbogenerator was rated 50 MVA and it was delivered to a gas turbine power plant in Västervik, also with Vattenfall as customer. This was the largest gas turbine unit in the world when it was commissioned in October 1959 [255].

7.1.4   The international situation

Most industrialized countries had been much more dependent than Sweden on fossil fuels for electric power production. Oil and coal fired power plants generated most of the worlds electricity and the power plants had become bigger and bigger. To have fewer, but larger units in each plant was cost-effective and therefore, there had been a pressure on development of very large steam turbines and turbogenerators. Asea had not been subject to this, to the full extent, and was clearly behind its important competitors in this field.

The development of directly steam turbine-driven generators had started around the previous turn of century and the cylindrical rotor concept with radial slots became successively more common during the first decade of the 20th century. Both 2- and 4-pole generators were developed. The 4-pole generators grew more rapidly in output, depending on lower stress levels and easier construction, compared with the 2-pole machines. However, after some time the focus shifted towards the 2-pole machines as they became smaller. Both turbines and generators were then easier to handle during manufacturing, transport, and installation, and they were also less expensive. Hydrogen cooling had been introduced as early as 1928 in the USA for synchronous condensers and in 1937 for turbogenerators. The use of direct water-cooling of stator windings was reported from Great Britain in 1956 [13]. Much earlier attempts were made in the USA to use liquid cooled turbogenerators, e.g. with oil passing through the stator core, but in such a case with the winding indirectly cooled [256].

What is characteristic for the design of large turbogenerators, i.e. those rated above 200 – 300 MW? It is most interesting to look at the 2-pole generators because they are, in almost every respect, more difficult than the 4-pole generators. As already mentioned, the rotor is the most critical component determining the size of the machine. The rotor diameter is limited by the centrifugal stresses and the strength of the available materials for rotor body, retaining rings and slot wedges. The active length is limited by rotor dynamic properties. If a rotor becomes too slim, it would be very difficult to avoid severe resonance vibrations induced at or near critical speeds. Furthermore, the total flux has to pass the rotor centre and it is important to avoid that this region becomes magnetically saturated. So, in the case of a 2-pole machine, both geometric dimensions and magnetic flux are limited by the rotor.

The only remaining parameter for determining the output of the generator is, according to equation 4-19, the linear current loading in the stator. Therefore, large 2-pole generators have stator slots, which are both deep and wide and carrying high currents. Cooling becomes very essential, but also fixation and support of the coils, as there would be high electro-dynamic forces both inside the slots and in the end regions. Generator manufacturers have to pay a lot of attention to the design of the stator windings, especially the end-winding bracing which has to withstand both steady state vibrations and high transient forces, e.g. at short circuits, but at the same time has to allow movement due to thermal expansion. Transport limitations are also a factor with high impact on the design of the stators for these large turbogenerators. Many different solutions have been developed and introduced for solving the problems mentioned, even if the fundamental prerequisites are universal.

The importance of cooling has already been addressed in chapter 4. Air-cooling was clearly insufficient for large turbogenerators and the introduction of more effective cooling has been the main road towards higher ratings. The chosen cooling concept has also had a big impact on the generator design in many other respects and it became a factor, which differentiated the manufacturers from each other. The three main parts requiring cooling were the stator core, the stator winding and the rotor winding. There were also other parts that needed cooling but those can be neglected for the moment. The windings were most difficult to cool because of the higher specific losses than in the core and also because they were encapsulated by insulation material constituting a thermal barrier. It has been mentioned that hydrogen was introduced as a more efficient cooling medium than air. However, the heat dissipation through the winding insulation remained very limiting. The next step was to blow the hydrogen directly through the windings. Directly hydrogen cooled stator windings were developed in which hydrogen passed axially through rectangular channels from one end of the stator winding to the other. This method had the disadvantage of a high pressure-drop in the long stator bars even if the gas ducts were rather large. A powerful compressor was needed for blowing the hydrogen through the stator winding. Therefore, direct liquid cooling soon became dominant for stator winding cooling. There were cases reported where oil was used as coolant but normally it has been de-ionised water. All manufacturers used fairly similar concepts, i.e. Roebel bars with integrated hollow copper conductors as shown in figure 7.14. The rotor winding, stator core and other parts remained hydrogen cooled. A Roebel bar is a conductor, divided into many parallel strands, which have been fully transposed in the active part of the coil side, thus enabling the strands to be inter-connected at both ends of the coil side. This concept was invented and patented, in 1912, by the German BBC engineer Ludwig Roebel (1878-1934), as a method to reduce large eddy current losses in stator windings.

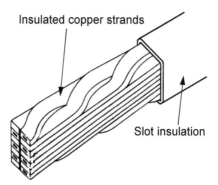

Insulated copper strands

Slot insulation

Figure 7.14 Directly water-cooled Roebel bar for stator winding

Some different concepts were developed for direct hydrogen cooling of the rotor windings and, in several respects, these concepts characterized the whole generator design. Hence it is important to briefly look at them and compare their advantages and disadvantages. Two basic concepts could be identified, axial cooling and gap-pickup cooling, but both of them can then divided in a few variants [257, 258].

Figure 7.15 shows the principles of axial cooling. The rotor conductors are made from hollow copper and the hydrogen enters into these conductors through special openings at both end-winding regions. The hydrogen is then discharged into the airgap through radial holes in the conductors and the slot wedges, located in the axially central part of the rotor. Many manufacturers used this system even if it had some obvious disadvantages. In very large machines, separate high-pressure blowers could be required to force the hydrogen through the long cooling channels. The temperature difference between gas inlet and outlet zones is fairly large. Advantages are that the rotor slot section is used in an efficient way and that the temperature of the inlet gas is low.

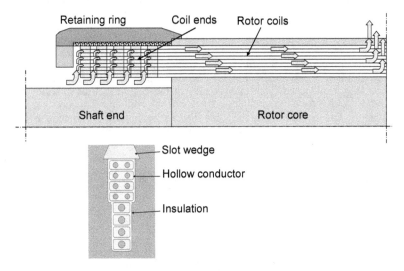

Figure 7.15 Direct axial hydrogen-cooling of rotor winding

A variant of the axial cooling is the so-called sub-slot system, which also could be referred to as radial cooling. The principle is shown in figure 7.16. The rotor is provided with sub-slots placed radially inside the main slots. The gas enters the sub-slots at both ends of the rotor and is then discharged through radial holes in the conductors and slot wedges along the entire active length of the rotor. The main advantage is that the entire winding is provided with cold gas resulting in a more uniform temperature along the rotor winding. Disadvantages are that the sub-slots take up valuable rotor space from the magnetic flux and that the inner cooling surfaces are smaller than for the axial cooling system. The problem with high pressure-drop for very long rotors remains.

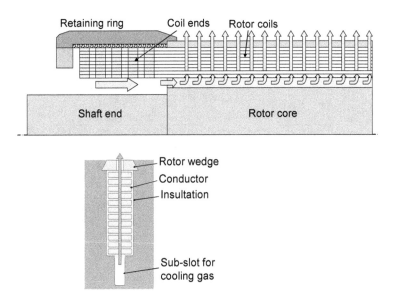

Figure 7.16 Direct radial hydrogen-cooling of winding in rotor provided with sub-slots

A very different approach to provide the rotor winding with hydrogen for cooling is the gap-pickup principle. In this case, the slot wedges are provided with inlet and outlet holes so that gas can be taken from the airgap and forced through channels in the conductors before it is discharged back to the airgap. Fig 7.17 shows two different concepts for gap-pickup cooling, the diagonal flow system developed by GE and the cross-flow design invented by Ganz Electric.

In the first case, the airgap is axially divided in a number of inlet and outlet zones for avoiding mixing of ingoing and outgoing gas. This makes the stator rather complicated. The second solution mixes the gas, but this could be acceptable due to low temperature rise when the gas passes the winding and because the airgap is efficiently ventilated. The advantages of gap-pickup cooling are independence of the machine length, uniform temperature rise and low pressure-drop and hence it is not necessary to have high-pressure blowers. The main disadvantage is that the hydrogen has already been preheated in other parts of the machine before it reaches the airgap.

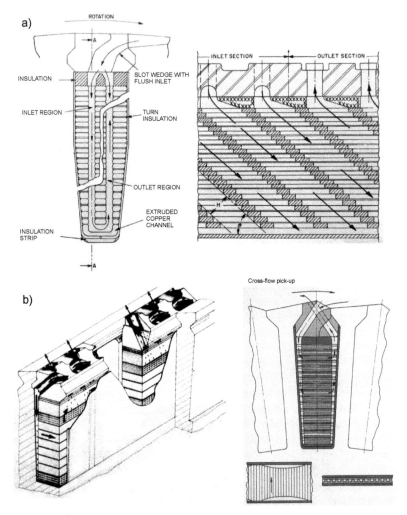

Figure 7.17 Gap-pickup cooling with a) diagonal flow and b) cross-flow

Common for all these rotor cooling concepts was a fairly complicated manufacturing process, but also the necessity to build the generators as pressure vessels with efficient sealings, and to provide them with external systems for hydrogen supply and control.

Manufacturing of large turbogenerators was an important and also prestigious industrial activity during the sixties and seventies. In those years, there were around 20 companies, most of them in Europe, which more or less independently developed such large generators. Table 7.2 lists important manufacturers, which were active around 1970. The table also contains some available information on the technologies used.

Table 7.2   Manufacturers of large turbogenerators around 1970

Company	Country	Licenses	Rotor cooling
ACEC	Belgium	Westinghouse	Axial
AEI	UK		
Alsthom	France		Axial + gap-pickup
Ansaldo	Italy		
Asea	Sweden		Water
BBC	Switzerland/Germany		Axial + water
CEM	France	Owner BBC	Axial
Electrosila	USSR		Gap-pickup + water
Electrojashmasch	USSR		
Elin	Austria	KWU	Axial
English Electric	UK		Axial + subslot
Ganz	Hungary		Gap-pickup
Jeumont-Schneider	France	Westinghouse	Axial
KWU (Siemens + AEG)	Germany		Axial + water
Parsons	UK		Axial + subslot
Skoda	Czechoslovakia		Axial
TIBB	Italy	Owner BBC	Axial
GE	USA		Subslot + gap-pickup
Westinghouse	USA		Axial
Hitachi	Japan		
Mitsubishi	Japan		
Toshiba	Japan		

Direct water-cooling of stator windings represented state of the art in the late 1950's and had received general acceptance as a very efficient solution. Therefore, it was a natural question whether it would be advantageous to use this method also for cooling rotor windings. It was easy to figure out that theoretically water-cooling was superior to all other cooling methods investigated, but the practical problems in providing rotating parts with cooling water caused hesitation. Nevertheless, several manufacturers started studies, including experiments, regarding water-cooled rotor windings. GE reports it had such a program from 1957 until 1963, but abandoned it in order to focus on development of gap-pickup cooling [258]. Other companies pursued the water-cooling much further, notably BBC and Siemens (later KWU), but also the USSR manufacturer Electrosila, even if directly hydrogen-cooled rotors remained their main concept. The first turbogenerator with water-cooled rotor was installed for regular operation in USSR in 1959 [13]. BBC started to develop both salient pole machines and turbogenerators with water-cooled rotors in the early 1960's and presented, for instance, a theoretical study of such a turbogenerator rated 1333 MVA, 3000 rpm at the 1966 Cigrè session. The report showed that a water-cooled machine would be smaller than a hydrogen-cooled with the same rating, but having somewhat lower efficiency [259]. One year later, Siemens/KWU received an order for its first 2-pole generator of this kind, rated 400 MVA, 3000 rpm, for a power plant in Kiel, Germany [260].

Much of the international interest during the 1960's was thus focused on cooling systems for large turbogenerators. A quick review of the reports on turbogenerators presented at the bi-annual Cigré-sessions 1964 to 1972 reveals that the generator industry considered this as the most important topic. Stresses on turbogenerators under abnormal operation, e.g. electrically unbalanced conditions, also received much attention and so did even excitation systems for very large generators [261].

In those days, the generator industry expected the machine sizes to continue to grow. An evidence of this can be obtained from a discussion at the Cigré meeting in Paris in August 1968. GE's chief engineer for turbogenerators, C. H. Holley, gave the following comment: *"At the present time in the U.S. there are under consideration a substantial number of two-pole, 3600 rpm generators rated approximately 1000 MVA. There are also a significant number of four-pole, 1800 rpm generators rated about 1300 MVA. Several studies of future requirements have indicated that within 10 years from now, two-pole, 3600 rpm generators about 50 percent larger than those now under construction will be required. Four-pole, 1800 rpm, approximately 100 percent larger than now under manufacture, will be needed."* [262] Five years later, GE's world famous specialist on power systems and synchronous machines, Dr. Charles Concordia (1908-2003), presented a paper to the Royal Society in London, in which he projected power systems and generators to grow more than 10 times within the next 50 years [263].

7.1.5 Asea's start position in short

During the 1960's, Asea held a position as one of the world leaders in development and manufacture of hydropower generators. In the case of turbogenerators, the situation was different. Asea had a long tradition of building such generators as a sub-supplier to its own daughter company Stal-Laval. The generator sizes were limited by the turbine concept used, so that indirect air- and hydrogen-cooling were sufficient. Many other generator manufacturers had developed and built much larger turbogenerators using directly water-cooled stator windings and different concepts for direct hydrogen-cooling of the rotor windings and other parts. Some interest had also been shown in directly water-cooled rotor windings.

7.2 The nuclear power programme

7.2.1 Initial phase

The electric power supply in Sweden was almost entirely based on hydropower until the mid 1960's [264]. The power consumption had shown a steady increase, since the end of the Second World War, and it was projected to continue to rise. Many waterfalls had been harnessed and it had become evident that the remaining hydropower resources would be insufficient for the future demand of electricity. In addition, environmentalists had begun campaigning for the preservation of the remaining rivers. It was hence necessary to start developing other power sources. Sweden didn't have any fossil fuels and it was therefore natural that the Swedish government initiated research aiming at the development of nuclear power. The first of such activities were started in the late 1940's [265].

A special company, AB Atomenergi, was established for the development of nuclear power technology. This company, the government, Asea and Vattenfall became then, for several years, the main actors in a political play about the coming nuclear power. There were many intricate tours before things settled, but that interesting story is outside the scope of this thesis. It is, however, described in detail in other published works such as Sigfrid Leijonhuvud's "(parantes?" published 1994 for ABB ATOM's 25 year anniversary and Jan Glete's "Asea under hundra år" written for Asea's 100 year anniversary in 1983. Several detailed references to these books are included in the reference list.

After two research reactors called R1 and R2 (R1 was actually located at KTH in Stockholm), Atomenergi planned two larger units R3 and R4. Asea had, together with Vattenfall, in parallel, discussed two units called Adam and Eva. These plans were more or less forced together by the government into R3/Adam and R4/Eva. The first one was implemented in a combined heat and power plant located at Ågesta, a suburb south of Stockholm. It was rated 10 MWe and 55 MWt and it was in operation from 1964 until 1974 [265]. The reactor was heavy water moderated and had natural uranium as fuel. The government had given preference to this concept because of Sweden's own uranium resources, thus avoiding dependence on the import of enriched uranium. R4/Eva later became Marviken, a nuclear power plant located at the Swedish east coast, close to Norrköping and owned by Vattenfall. The rated output was, after some changes, set at 200 MWe. Marviken was also a heavy water moderated reactor but it was provided with superheating. This concept was eventually found to be unstable, so the reactor was never put into service. The power plant was instead converted to an oil-fuelled condensing plant [265]. The two indirectly hydrogen-cooled turbogenerators were rated at 111.5 MVA each and Asea delivered them in 1968. The plant was commissioned as late as 1974 and since then has only been used occasionally as a stand-by unit. The Marviken-generators came therefore too late and were in fact also too small for an improvement of Asea's knowledge base for the following larger units.

7.2.2 Oskarshamn 1

In parallel with the above-mentioned projects, some private and municipal Swedish power companies formed a consortium for building a nuclear power plant, preferably with a light water reactor. The American companies GE and Westinghouse had developed such reactors, but with different concepts. GE had chosen the boiling water reactor (BWR) and Westinghouse the pressurized water reactor (PWR). Asea had, in spite of the involvement in the work with Ågesta and Marviken, successively come to the conclusion that the light water reactor technology was more attractive.

The consortium of power companies was transformed, in July 1965, into a company called Oskarshamnsverkets Kraftgrupp AB (OKG), which at the same time placed a turnkey order with Asea for a 400 MWe nuclear power plant with a BWR reactor. It was later increased to 440 MWe. Many alternatives had been studied before this decision was reached, and Asea had even negotiated a licence agreement with GE, but this was never signed. Asea decided instead to develop its own large light water reactor, which must be considered as a very brave and visionary step. Asea's president and CEO, Curt Nicolin, had a key role in this decision, and he managed to create the necessary confidence in Asea's possibilities to carry out this huge project. Asea happened to be the only company in the world that developed light water reactors without licence from GE or Westinghouse [265].

The steam turbine was ordered from Stal-Laval. It was a double rotation radial/axial turbine of Stal-Laval's traditional type, but much larger than earlier units. The two generators were rated 271 MVA each, by far the largest turbogenerators Asea had received order for at that time. It was decided to use the hydrogen-cooled design for these generators, but with directly water-cooled stator windings. The rotors were of the radial slot type with indirect hydrogen-cooling of the windings. The dimensions were, of course, much larger than previous turbogenerators, especially the length. They had a rotor-diameter of 1050 mm and the active length was 4800 mm [247]. A sketch of the entire turbine/generator unit is shown in figure 7.18.

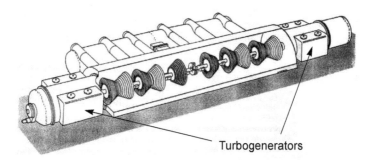

Turbogenerators

Figure 7.18 Turbine/generator unit in Oskarshamn 1

OKG's largest shareholder, Sydkraft AB, was Sweden's second largest power company, and it signed, more or less simultaneously with the order for Oskarshamn, a contract with BBC for the supply of three 340 MW, 3000 rpm turbine/generator units for what would become Sweden's largest fossil fuelled power plant, Karlshamn. Stal-Laval had offered similar type of units as for Stenungsund but Sydkraft was doubtful about the radial turbine technology and preferred axial turbines. The timing of publishing the orders for Karlshamn and Oskarshamn were coordinated in order to avoid too much criticism for placing the first of these orders abroad.

Asea's turbogenerator design office started the order design on the Oskarshamn-generators in 1966 and the manufacturing began a year later. A couple of problems occurred, which could clearly be attributed to the size increase. One was related to the water-cooled stator winding. The winding consisted of a number of half-coils, so-called Roebel bars, in which some of the rectangular copper strands had been replaced by small stainless steel tubes. Several of the steel tubes were accidentally clogged by epoxy resin, used for impregnation of the insulation material. The resin, which had cured, could be removed by means of formic acid and a long, laborious work. An important lesson was that it is not enough to document the product itself, but it is equally important to specify the manufacturing process in detail. The introduction of new concepts requires a very close cooperation between design engineers, production engineers and the workshop.

The next problem was rotor balancing. The rotors were slimmer than earlier ones which increased the requirements on accurate balancing. Initially, it was impossible to balance these rotors. Investigations showed that the reason was thermal instabilities in the rotor forgings. The manufacturer's equipment for cooling the forgings had not been in proper order and the properties of the forgings became asymmetric. New forgings were ordered and the rotors were manufactured a second time. It was still difficult to balance them and obtain low rotor vibrations, which could depend on insufficient damping in the bearings. A change from cylindrical sleeve bearings to pad type bearings was discussed but not implemented. The generators were finally shipped to Oskarshamn and have performed well in combination with the turbine. The bending modes and damping factors were obviously more favourable in the complete unit than for a separate rotor.

The turbine had also its share of problems, mainly related to the radial system, and the commissioning of the plant was delayed 18 months compared to the original plan. Oskarshamn 1 was put into commercial operation in 1972. Therefore, the experiences gained had hardly any influence on the decisions taken for the development of the really large turbogenerators [266].

7.2.3 More nuclear power plants

Vattenfall had, during a few years, looked into the possibilities to build a nuclear power plant with a light water reactor, even while the organization was occupied with the construction of Marviken. The private power company OKG's decision to build Oskarshamn 1, pushed the government and Vattenfall to go ahead and plan a large nuclear power plant at Ringhals, located south of Gothenburg at the Swedish west coast. Asea and Westinghouse were invited to prepare bids during 1967-68. During this preparation phase, a decision was made but not announced, that Vattenfall should buy two units at the same time, though with different concepts and suppliers. Orders were placed in July 1968. Each unit should consist of one reactor and two parallel turbine/generator sets. Asea received an order for a 750 MWe BWR for Ringhals 1, while the turbines and generators should be supplied by English Electric. The American company, Westinghouse, received the order for an 800 MWe PWR for Ringhals 2. Stal-Laval was chosen as supplier of the two turbines with generators from Asea. These generators were rated 504 MVA each. The same time these contracts were signed, it was also announced to establish Asea-Atom as a new company, jointly owned by Asea and the Swedish state [267]. Figure 7.19 shows a schematic layout of a nuclear power block like that in Ringhals 2. The main difference between a PWR and a BWR plant is that the latter has no steam generator. The steam from a BWR goes directly to the turbine.

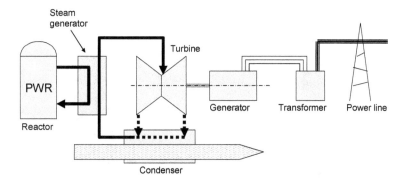

Figure 7.19 Schematic layout of PWR nuclear power block

Less than a year later, in May and June of 1969, the private power companies were ready for the next steps. OKG had decided to build a second unit, Oskarshamn 2, but with the power increased up to 600 MWe. The BWR should be delivered by Asea-Atom and the turbine/generator unit by BBC. The turbine order was, however, taken over by Stal-Laval while the generator remained to be supplied by BBC. Sydkraft, which was the leading partner in OKG, placed an order, only a few weeks later, for a unit more or less identical except for the generator, which should be made by Asea. Gunnar Tedestål, who was Sydkraft's leading generator specialist, mentioned in an interview that the motivation for Sydkraft to choose a new generator from Asea was a substantial discount on the generator price. Sydkraft's power plant was located at Barsebäck, a little north of Malmö in southern Sweden. The contract for Barsebäck 1 also included an option for a second unit, Barsebäck 2. Asea had thus, in addition to the Ringhals generators, an order for a 710 MVA turbogenerator, an enormous challenge taking into account that experiences from operation were still limited to turbogenerators below 75MVA. The option for Barsebäck 2 was turned into an order in June 1972 [267].

Vattenfall had chosen a different strategy compared with OKG and Sydkraft. They continued to order both BWRs and PWRs, they wanted larger outputs and two parallel turbines per reactor. Therefore, Vattenfalls's next purchase resulted in a substantial increase of Asea's order stock for large turbogenerators. Vattenfall decided in November 1971 to order a 900 MWe PWR for Ringhals 3 from Westinghouse, and two turbines from Stal-Laval with Asea generators. At the same time they ordered a 900 MWe BWR from Asea-Atom for Forsmark, with two turbines from Stal-Laval and generators from Asea. In addition, the contracts included options for Ringhals 4 and Forsmark 2. The power plant in Forsmark, shown in figure 7.20, is located at the east coast north of Stockholm, and it was jointly owned by Vattenfall and some private companies. Asea had in one day received firm orders for four 577 MVA generators and options for another four [267]. These options were turned into firm orders in 1973 for Ringhals 4 and in October 1974 for Forsmark 4.

Figure 7.20 Forsmark nuclear power plant

Even Finland needed more electricity and had also decided on using nuclear power. The state owned power company, Imatran Voima Oy (IVO), built the first nuclear plant at Lovisaa with two units, both delivered from USSR. The private Finnish power company, Teollisuuden Voima Oy (TVO), planned to build the next plant. During 1972, TVO and Asea Atom started negotiations concerning turnkey supply of a 2 x 650 MWe BWR power plant located in Olkiluoto at the Finnish west coast. The contract for TVO 1 was signed in June 1973 and for TVO 2 in September 1974 [267]. Stal-Laval was chosen as the turbine supplier including generators from Asea. The generators should be rated 825 MVA each, thus the largest ever designed and built by Asea.

Some years later, in 1976, contracts were also signed for two 1150 MWe units, Forsmark 3 and Oskarshamn 3. The reactors were ordered from Asea-Atom and the 1500 rpm turbines from Stal-Laval, but in this case with BBC as the supplier of the 4-pole generators [267]. It was considered that Asea had more than enough to do with all the 2-pole generators, so generators of Swedish design were never seriously discussed for these last units [268].

7.2.4 Summary of nuclear plants

Almost half of the electricity produced in Sweden and Finland has since long been generated in a few nuclear power plants built during the seventies. The Asea Group was in most cases the main supplier, not only for the nuclear reactors but also for steam turbines and generators. An overview is given in the map in figure 7.21.

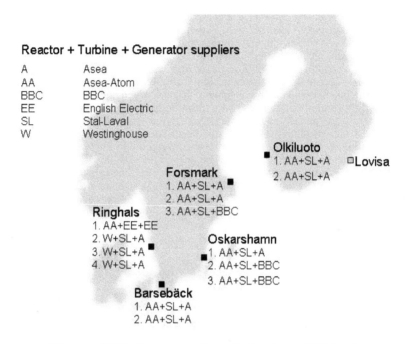

Reactor + Turbine + Generator suppliers

A   Asea
AA   Asea-Atom
BBC   BBC
EE   English Electric
SL   Stal-Laval
W   Westinghouse

Olkiluoto
1. AA+SL+A   □Lovisa
2. AA+SL+A

Forsmark
1. AA+SL+A
2. AA+SL+A
3. AA+SL+BBC

Ringhals
1. AA+EE+EE
2. W+SL+A
3. W+SL+A
4. W+SL+A

Oskarshamn
1. AA+SL+A
2. AA+SL+BBC
3. AA+SL+BBC

Barsebäck
1. AA+SL+A
2. AA+SL+A

Figure 7.21 Nuclear power plants in Sweden and Finland

7.3   The strategic decisions

7.3.1   No generator licence

Stal-Laval's old turbine concept had reached the end of the road. The combined radial-axial flow turbines could not handle the large steam flow from really big nuclear reactors, in addition, the experience from the thermal power plant in Stenungsund was discouraging. The intermediate-pressure unit in Stenungsund 3 (see figure 7.9) was the largest and heaviest radial turbine built. It was impossible to accurately predict the critical speeds with the calculation tools available at that time, and it was very difficult to balance the turbine rotors. Excessive vibrations caused the counter-rotating disks to touch each other resulting in a total break down of the rotating parts of this turbine. The same radial turbine had been chosen as high-pressure unit in Oskarshamn 1, but the need for even larger units was urgent. According to

Carsten Olesen, chief engineer for Stal-Laval's large steam turbines, it was not only a matter of getting the large steam flow through the radial turbine, but also cost disadvantages when the radial system became only a minor part of the whole unit. Therefore, the company had started a project to design its own axial turbine. The order from Vattenfall for Ringhals 2 in July 1968 was, in principle, based on this new design, but the matter was not finally settled. In view of the problems in Stenungsund, Vattenfall's technical director, Ingvar Wivstad (1924-1999), more or less required that Stal-Laval should acquire a licence on an existing design. Thus, in addition to the design study mentioned, Stal-Laval's management discussed and investigated different possibilities to obtain the necessary technology through a licence agreement with some experienced manufacturer of large axial steam turbines. Stal-Laval's vice president for sales, Åke von Sydow and Carsten Olesen have described, in interviews, that many companies, both in Europe and America, were contacted but the final decision was either BBC or English Electric. Finally, a licence agreement was signed with BBC in April 1969 for steam turbines larger than 200 MW [269]. All the nuclear units mentioned above, starting with Ringhals 2 and Oskarshamn 2, were consequently provided with turbines designed in accordance with this licence. A photo of one of the turbines in Forsmark is shown as figure 7.22.

Figure 7.22 1150 MW, 1500 rpm steam turbine in Forsmark nuclear power plant. The high-pressure turbine is seen in the front and the three low-pressure units behind

A very important question is now: "Why didn't Asea also take a licence for the corresponding generators?" The increase in size was the same. The generators also required new design concepts. The company had no experience with really large turbogenerators. Looking from the outside, it seemed like the prerequisites were more or less the same for the generators as for the steam turbines.

The main arguments for a licence are access to proven technology, substantially reduced development costs, and often certain market allocations. The most important arguments against a licence are the loss of freedom to market the products worldwide, and the costs incurred by the licence fees. The market issue had no real relevance in this case, as this type of generator was only sold in combination with the turbines. Thus, the potential market was in practice defined by the licence agreement for the turbines. It is also doubtful, or even unlikely, that the costs for development versus the costs for licence fees were subject to any deeper analysis. So the question remains: "Why did Asea decide to develop its own generator design without any licence ...... or was this never a matter of discussion?"

It has been more than 35 years since the decisions were made and several key persons are not any longer available for interviews. Neither is the existing documentation very explicit on this point. However, some information has been obtained, which supports the following explanation.

As mentioned above, Vattenfall placed the order for Ringhals 2 in July 1968 and Stal-Laval signed the licence agreement in April 1969. Asea had therefore started to work on its own generator design before the turbine licence was ready. Asea had a reputation as a successful supplier of generators, let be mainly for hydropower, but was confident that it was also capable of developing large turbogenerators. Electrical machines of all kinds were core business for the company and acquiring licences had never been part of the strategy. Therefore, according to well-informed sources, the alternative to take a licence also for the generators was never investigated or seriously considered. This is based on interviews with Torsten Lindström (1921-2005), who then was Asea's executive vice president for technology as well as Jan Liljeblad and Carl Rönnevig, both deeply involved in the management of turbogenerator development.

Looking at Asea's history, it is evident that the company had a long tradition of developing the necessary technology in-house. Not only products, which could be prototype tested but also large, complicated systems. Good examples are the first ever 380 kV transmission system commissioned in 1952, the first HVDC transmission in the world put into operation 1954, and the development of nuclear reactors already mentioned above [270]. There are not many signs of hesitation preceding these huge projects, something that can partly be attributed to Ragnar Liljeblad (1885-1967), legendary technical manager of Asea from 1918 until 1944 and then consulting technical director until 1963. So the question should perhaps be re-phrased: "Why did Stal-Laval take a licence on the turbines?" One answer has already been given: "The customer required it!" A contributing fact is also that Stal-Laval had always been technically very independent from the mother-company. But it is worth noting that Stal-Laval started the licence cooperation with BBC at a minimum scope. From the beginning the condensers were not part of the licence, in spite of that they were built by Stal-Laval's own subsidiary Stal-Laval Apparat AB.

Was there never any discussion concerning licence for the generators? Yes there was. Interviews with several of the persons mentioned above have confirmed that representatives of Stal-Laval had expressed a different opinion from Asea, claiming that the risks in developing a new generator design were too big and Asea lacked experience from really large turbogenerators. This opinion was also supported by BBC. Åke von Sydow, who was deeply involved in the licence negotiations from Stal-Laval's side, reports an informal comment from BBC's marketing director: "Why don't you take a licence also for the generators. The turbines are fairly straight forward, but the generators are really an art!" This opinion can be argued, both are very difficult machines, but the comment is still interesting. The interviews have revealed that there were obviously some animated discussions between Asea and Stal-Laval, and Vattenfall might have also expressed support for Stal-Laval's point of view. Even if it is difficult to prove, it is very likely that prestige had a significant influence on both sides. However, Asea's position was reluctantly accepted. Whether or not this decision was due to confidence in Asea's executive management, or due to its powerful position, remains an open question.

7.3.2   Directly water-cooled rotors

The single most important technical decision for the development of the large turbogenerators was to use direct water-cooling not only in the stator windings but also in the rotors. This was different from what other manufacturers used to do and it had a profound impact on the entire concept and the course of events that followed. It is therefore very important to analyse and understand the background for this decision, which implied that all the large turbogenerators, from Ringhals 2 and on, were made in this way.

The importance of efficient cooling has been dealt with in chapter 4, Technical overview, where it also was presented that water is a superior coolant compared with other available alternatives. Direct water-cooling results in lighter and more compact machines, which also are potentially more cost effective. The temperatures are lower than in gas-cooled machines, a fact, which usually decrease temperature gradients and stresses due to differential thermal expansion, something that could improve the lifetime of the machines. The drawback is that it is complicated to have the water circulating directly through the windings and other active parts of a machine. This cooling principle is therefore only used when necessary, mainly for very large machines. As mentioned earlier, in case of large turbogenerators, the common solution was to have the stator winding directly water-cooled while the stator core, the rotor winding and other parts were cooled by hydrogen. Asea decided to avoid the hydrogen and apply direct water-cooling even in the rotor.

Asea's first use of direct water-cooling in turbogenerators was for the stator windings in the Oskarshamn 1 generators as mentioned above. However, complete direct water-cooling was introduced earlier in a couple of other synchronous machines. The initial attempt was made during 1963 – 65 within a special project. It had actually started in 1962 as a design study made within Asea's nuclear department on initiative of Kristian Dahl-Madsen, a young and creative Danish engineer working with concept studies for nuclear reactors. The reason for this study was the estimation that future nuclear plants might need turbogenerators in the 1000 MW class. The first alternative studied was exceptional. It had a super conducting excitation winding in the stator and a liquid cooled 3-phase armature winding in the rotor! It was foreseen so-called airgap windings, i.e. such a winding is placed in the airgap instead of in slots [271, 272] The idea of a super conducting winding was soon abandoned, but an experimental machine was then developed and built with oil-cooled windings. The excitation winding was still in the stator and the rotating armature winding was an airgap winding. According to Dahl-Madsen, a trial and error philosophy was behind the project at this stage. "Let's build, test and see what happens!" Some tests were performed and the experimental machine was also shown to Vattenfall. The photo in figure 7.23 is from such an occasion.

Figure 7.23 Demonstration of liquid-cooled experimental machine in presence of J. Liljeblad, C-G. Stensson, K. Dahl-Madsen, G. von Geijer (Vattenfall) and T. Strömberg

Also, the rotating armature winding was deleted, primarily due to expected complications with sliprings for both high voltages and high currents. In the concept finally chosen, both stator and rotor windings were to be water-cooled and the first remained an airgap winding, i.e. the stator core had no slots and teeth. The idea was to use as much as possible of the circumference for conductors with high current density, and to combine this with a very high airgap flux density, in the order of 1.5 T. The result would then be a very compact machine with high specific output. The excitation winding in the rotor would have conductors of hollow copper, while the cooling channels in the stator winding would be stainless steel tubes. A sketch of the machine concept is shown in figure 7.24 [273].

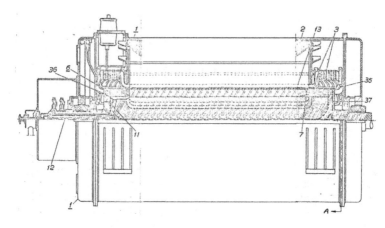

Figure 7.24 Sketch of synchronous machine with concentric water-cooled airgap stator winding (from Swedish patent 315654)

Asea tried to interest Vattenfall in this very advanced machine concept and a discussion between Asea's sales director Curt Mileikowsky (1923-2005), who had supported the project from the very beginning, and Vattenfall's technical director Gottschalk von Geijer (1900-1981) resulted in an order of a 125 MVA synchronous condenser for Vattenfall's Hamra transformer station. The order was placed in November 1964 and commercial operation was scheduled for September 1st 1966. The following lines from Asea's technical director's annual report for 1964 is worth of note: *"The order on "Hamra G3" was received 9.11.64. The delivery time is very short, even for a machine of normal design. If it is taken into account that G3 represents the largest step from conventional turbo-machine design during the last 50 years, it seems as almost a wonder if we manage to get the machine ready for workshop assembly in December 1965."* (My translation) [274]. Asea was obviously prepared to take risks, however, an emergency exit existed. The contract had a clause stating that it was possible to switch to delivery of a capacitor bank instead of the synchronous condenser in case of problems. Such problems occurred during manufacturing and the project was finally abandoned in October 1965, basically due to financial considerations as the manufacturing costs became far too high. The main reason was that the stator winding was much too difficult to manufacture. The conductors were fully exposed to the alternating main flux and hence had to be divided in very small strands to avoid excessive eddy current losses. The cross section of the stator coils became too complicated as can be seen in figure 7.25. It was however concluded that the concept was technically possible, but it had no chance of becoming cost-effective [275].

Figure 7.25 Stator coil cross section for Hamra G3

The same year, 1964, as the contract for the "Hamra project" was signed, Vattenfall also ordered the largest hydropower generator Asea had designed so far. It was a 225 MVA, 200 rpm salient pole synchronous generator for the Seitevare power plant in northern Sweden. Asea proposed to develop a wa-ter-cooled generator for this plant and it was agreed to build it with direct wa-ter-cooling in both stator and rotor. The main reason for this decision was the possibility to build a more compact and also more cost-effective generator [276]. The "Seitevare project" required development of many new compo-nents and solutions including manufacturing processes, and many of these became a know-how base for subsequent projects. In focus for the develop-ment were the water-cooling of three main components, namely the stator winding, the stator core, and the rotor winding. Specialists from Asea's Cen-tral Laboratory recommended stainless steel tubes as water channels inside the conductors and in the rest of the system, because welded joints in this ma-terial were considered safer than soldered copper joints. Test rigs were set up for long term testing of stator and rotor coils, especially the water connec-tions. The outline of the Seitevare generator is shown in figure 7.26.

The development of the Seitevare generator was quite independent from the "Hamra-project", but the latter had contributed to awareness and interest in more advanced cooling methods. The importance of close relations between customer and supplier for full-scale introduction of new technology is obvi-ous. Vattenfall took the risk and ordered both Hamra G3 and Seitevare during the same year. The main reason was most probably the long tradition of de-velopment cooperation that existed between Asea and Vattenfall, but inter-views have also pointed at a desire to be in the frontline, applying new technical solutions [270].

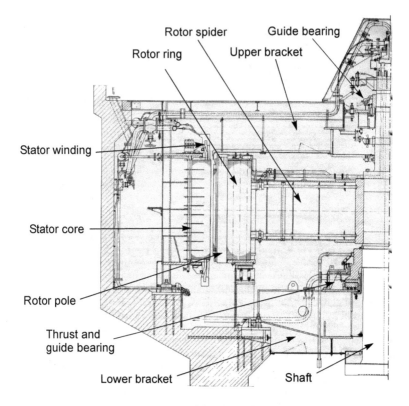

Figure 7.26 Outline drawing of the directly water-cooled Seitevare generator

The manufacturing of the Seitevare generator was not free from problems, but the organization learned successively to deal with the new components and more strict requirements. The generator was commissioned in October 1967, i.e. long before the development of the turbogenerators for Ringhals 2 was started. The full load tests showed that temperatures and temperature gradients were considerably lower than in conventional generators. However, the stator end windings were unfortunately damaged at a rated voltage 3-phase short-circuit test and a re-winding became necessary. This accident emphasised the importance of correct end winding support for large machines [277, 278].

Hamra G3 was a separate activity even if the intention was to develop concepts for large turbogenerators. A different project, carried out in 1965-66, was a water-cooled turbogenerator test rotor. It was not a full size rotor, but several relevant tests concerning heat transfer, risks for corrosion and thermal unbalance etc. were performed [279]. The results were promising and Jan Liljeblad, chief engineer and manager for development and design of DC and large AC machines, wrote in a memo to the executive management: *"Since the tests of the water-cooled test-rotor is more or less completed and the design principle used has met the expectations, there is reasons to re-investigate the possibilities and the consequences of implementing this concept on the generators ordered for Oskarshamn."* (My translation) [280]. As described earlier, the generators for Oskarshamn 1 were not made in accordance with this recommendation, but in a much more conventional way. Only the stator windings were water-cooled. However, the interest in fully water-cooled turbogenerators remained, and the following short quote from a meeting in December 1966 with the board for the DC and Large AC Machine Division (O-division) is interesting: *"All efforts shall be made to receive an order, as soon as possible, for a water-cooled turbogenerator, preferably for a peak power plant or a large gas turbine."* (My translation) [281]. The second part of the sentence indicates that the risk to go directly for a nuclear power application was considered too great. Present at this meeting were from the executive management, Curt Nicolin, Curt Mileikowsky, Asea's technical director Halvard Liander (1902-1990) and from the O-division Eric Sjökvist (1919-1996), who was general manager for the division, and Jan Liljeblad. The intentions of the management seem to have been clear.

Figure 7.27 Curt Nicolin, Eric Sjökvist, Jan Liljeblad, Carl Rönnevig

Asea's next completely water-cooled machine was, in spite of what has been said above, also a salient pole machine. A 345 MVA, 900 rpm synchronous condenser ordered by American Electric Power Company (AEP) for a transformer station in Dumont, Indiana. AEP's inquiry specified first a much smaller, hydrogen-cooled condenser, but after a number of proposals, AEP asked Asea, "what is the largest condensor you can make?" The reason for this was purely economical [282]. With Seitevare as a base, Asea decided to respond to AEP's request by offering a fully water-cooled condenser. Transport restrictions would have limited the output of a hydrogen-cooled machine to approximately 250 MVA. AEP ordered the water-cooled condenser and the development started in 1967. It was a very advanced concept with two water-cooled rotor windings, both the excitation winding and the damper winding. The damper winding served as a squirrel-cage winding for asynchronous start of the machine [283]. The Dumont condenser became an even more important reference for the water-cooled turbogenerators than the hydropower generator in Seitevare. The organization had learnt from earlier mistakes and the manufacturing was much more efficient and trouble free. For instance, special "clean rooms" had been arranged in the workshop for sensitive operations. Carl Rönnevig, a Norwegian engineer deeply involved in the development of the first water-cooled machines, said during an interview that the manufacturing and testing of the Dumont condenser ran so smooth, that the awareness of the difficulties in developing water-cooled turbogenerators was veiled. The commissioning in Dumont took place in July 1971. It is no big surprise that Asea decided on directly water-cooled rotors for the large turbogenerators, because the organization had for long considered this alternative and had already become familiar with the concept through the salient pole machines. Figure 7.28 shows assembly of the Dumont synchronous condenser for testing in the workshop.

Figure 7.28 Assembly of 345 MVAr synchronous condenser for test

Asea was not the only manufacturer working with directly water-cooled rotors. Even if all the others used directly hydrogen-cooled turbogenerator rotors, some of them also developed and built a few with water-cooled rotors. Manufacturers already mentioned in section 7.1.4 were BBC, KWU (Siemens + AEG) and Electrosila in former USSR. During the period of interest, the second half of the 1960's, BBC and KWU built one 2-pole generator each, and Electrosila also built a few 2-pole generators with this type of cooling. BBC delivered a 330 MVA generator to the Danish power plant Skaerbeak in 1971, and KWU its 400 MVA generator for Kiel roughly at the same time. Later, during the 1970's, both BBC and KWU also built some very large 4-pole generators with water-cooled rotors. In case of the USSR manufacturers, information is not easily available, but Electrosila's first turbogenerator with water-cooled rotor was rated only 60 MW. The next size was 200 MW, built in 1964, and a 500 MW machine was put in service in 1968 [284]. At the Cigrè session in Paris in August, 1970, one of BBC's leading experts, Mr. R. Noser, presented BBC's view on water-cooling with the following contribution to the discussions: *"The authors of 11-06 (from GE) reach the conclusion that rotors cooled by liquid could only be justified for rated outputs which are very high at some time in the future. This opinion is not shared by all manufacturers and neither by the company to which I belong. ... ... On the basis of all the experience acquired, we are confident that water-cooling of turboalternators will in the future replace, over a considerable area, hydrogen-cooling such as is practised today."* [285].

The fact that both BBC and KWU were developing generators with water-cooled rotors must certainly have increased the confidence within Asea that this was the right path to follow. Asea's president Curt Nicolin had visited BBC's generator factory in January 1965 and wrote in an internal report that he had seen that all turbogenerators above a certain size were water-cooled in both stators and rotors [286]. This was not true, because it was only the Skaerbaek generator and a few larger 4-pole generators, which later were built so, but that is not the point; he could have seen some prototype and misinterpreted the information. What was important was that Asea's management had been strengthened in its belief in direct water-cooling.

The electric power consumption in Sweden as well as many other countries increased year after year during the 1960's at a rate of 5 – 10 % annually. Prognoses made for Sweden indicated a need for around 20 large nuclear reactors towards the end of the 1980's [287]. The generator size had also grown and there were no reasons to believe that it would stop growing. Turbogenerators in the 1000 – 2000 MVA size were anticipated [288]. For such large generators direct water-cooling of both stator and rotor was considered a necessity, at least within Asea. Therefore, Asea was of the opinion that by developing generators with water-cooled rotors, the intermediate step with directly hydrogen-cooled rotors could be omitted. This was probably the most relevant and also most important reason for the decision.

It is an interesting question whether Asea investigated and compared several different concepts before the decision was made. Interviews with persons engaged in this process give the fairly clear answer that there were no such studies performed, at least no extensive studies of directly hydrogen-cooled rotors. It seems surprising, but it is probably the truth. Some comparisons had been made for Oskarshamn 1, but that was a comparison between indirect hydrogen-cooling and water-cooling. Table 7.3 was found in a message from Jan Liljeblad to Curt Nicolin written in August 1966 with a proposal to build the Oskarshamn generators with water-cooled rotors [289]. Liljeblad later mentioned in his annual report for 1966 that a pre-study of directly hydrogen-cooled rotors had been initiated, but there was no evidence that it was ever completed [290].

Table 7.3   Comparison of hydrogen- and water-cooled generators for Oskarshamn 1

Alternative	Hydrogen	Water
Stator winding cooling	Water	Water
Rotor winding cooling	Indirect hydrogen	Water
Hydrogen pressure [bar]	4	-
Rotor current [A]	1120	4100
Rotor diameter [m]	1.05	1.05
Active length [m]	4.8	3.0
Losses [MW]	1.9	2.4
Cost (for 2 generators) [MSEK]	5.3	4.25
Loss evaluation (700 SEK/kW) [MSEK] (for 2 generators)	2.66	3.36
Comparative cost [MSEK]	7.96	7.61

However, some sort of comparison was made also for Ringhals 2 in which a 4-pole generator with directly hydrogen-cooled rotor was studied in parallel to the 2-pole generator with water-cooled rotor, according to an interview with Sven Nilsson, electrical design engineer. Such a study had very little relevance for the type of rotor cooling to apply. Vice versa would have been more logical, because a 4-pole rotor is easier to cool by water than a 2-pole machine.

Managers at both executive and operative levels did not question the direct water-cooling. On the contrary, it was almost a policy to prioritize concepts, which would put Asea in the technical forefront. It had been witnessed that a few key persons from the old turbogenerator design office had expressed a strong preference for hydrogen cooling, but it is not clear at which point in time. As a conclusion, the following reasons for choosing directly water-cooled rotors have been identified:

- Water is the most efficient cooling medium resulting in more compact and, for larger units, more cost effective machines.

- Water-cooling was also applicable for very large generators expected in the future when hydrogen-cooling would be insufficient.

- The company had started to use water-cooled rotors for salient pole machines, so this technology was already familiar to the organization and several synergies could be expected.

- Hydrogen-cooling is not a realistic option for hydropower generators, so by choosing direct water-cooling, it would be enough to develop one technology.

- A few other leading manufacturers were also developing generators with this type of cooling.

- It was possible to avoid costly development of an intermediate step with direct hydrogen cooling.

- It was an advantage to avoid hydrogen due to the explosion risk, especially in nuclear plants with sophisticated ventilation systems.

- The stator housing did not have to be a pressure vessel with hydrogen sealings around shaft ends, terminals etc.

- No external hydrogen system was required.

- It was a state-of-art concept, which emphasized Asea's high-tech profile and this was preferred by the management.

7.3.3   Resource and competence build-up

The development of much larger and more advanced electrical machines than in the past constitutes a challenge for any organization. It is a matter of both competence and capacity. Important questions are: to which extent did Asea realize this, and how did they act? Was this ever a strategic question?

Asea had, more or less, the same organization for development and design of electrical equipment from the early 1920's; a central design department divided in offices for various products, supported by a common laboratory for special tasks [291]. A big reorganization was then made during the sixties when a number of divisions were formed. One of them, the "Division for Large AC and DC Machines" (the O-division) was established from January 1, 1965. Each division had full responsibility for development, design and manufacturing. The O-division had a department for development and design, OK, headed by the chief engineer Jan Liljeblad, and this department was divided into a number of offices as the diagram in figure 7.29 below shows.

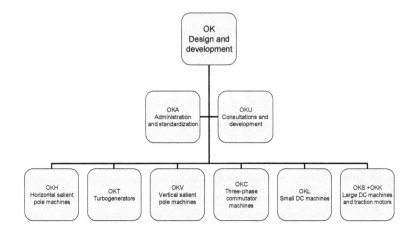

Figure 7.29 Organisation diagram for dept OK beginning of 1965

The organization remained fairly intact until the beginning of 1970 although the number of employees varied. Development of the directly water-cooled machines was handled by the design offices, which had responsibility for the corresponding conventional machines. The hydropower generator for Seitevare was designed by the office "OKV", the synchronous condensor for Dumont by "OKH" and the turbogenerators for Oskarshamn 1 by "OKT". This indicates that the introduction of the new cooling technology was not considered so exceptional that the machine development couldn't be handled more or less in the traditional way. However, a special development office "OKU" had been set up in January 1964 with the main responsibility of developing components for water-cooling assigned to this office. The OKU manager, Erik Agerman, initially recruited a handful of development engineers, mainly from the Central Laboratory. It should be observed that OKU was also engaged in other projects, which had nothing to do with the water-cooled machines. Only a couple of engineers worked in the first years with the direct water-cooling. Then in 1967, OKU was given the responsibility of designing the complete rotor for the Dumont synchronous condenser, and further in 1968 the development of the Ringhals generators. The dedicated crew was accordingly increased to around 15 - 20 members. Table 7.4 shows the number of persons working in OK's officies for "large AC machines" during 1963 – 76, according to the annual reports for OK.

Table 7.4   Number of people employed within OK during 1965 - 70

Year	University graduated	Technical college	Administr. staff	Total
1963	24	82	5	111
1964	25	85	6	116
1965	29	77	2	108
1966	29	81	2	112
1967	27	72	3	102
1968	19	59	4	82
1969	18	66	3	87
1970	19	67	2	88
1971	19	66	4	89
1972	18	64	5	87
1973	27	67	5	99
1974	27	68	3	98
1975	33	67	4	104
1976	30	62	5	97

The simplest way to understand the build-up of competence and capacity is to look at the staffing of the individual projects, including cooperation with other units. The "Hamra project" was the first and was in many respects very different. Kristian Dahl Madsen, who was very entrepreneurial, had presented his advanced, futuristic machine concept for Asea's executive management. The project had started as a study within the nuclear power department but was transferred to the generator department in 1963, where it was organized as a special development project. A year later, it became a separate unit, reporting directly to Asea's technical director Halvard Liander. It is not obvious why this decision was made, but it is quite clear that Dahl Madsen had a close communication with the executive management. As already mentioned, in 1964 the project became focused on the synchronous condenser ordered by Vattenfall for Hamra [274]. The development team grew from six persons in the beginning of 1963 to 18 persons by 1965. The initial group came, together with Dahl Madsen, from the nuclear power department. The connections with the existing generator design offices were very limited. Only one electrical design specialist, Richard Sivertsen, came from the turbogenerator office. All the others had no electrical machine background. The project was, as already described, not successful and the manufacturing was terminated, but the question remains open whether this depended on lack of collaboration with the more experienced electrical machine departments.

Only two people from the project team were transferred to those design of-fices, which had just started with new, directly water-cooled machines. The importance of the Hamra-project for subsequent development was more as a catalyst and less as a source of know-how.

The hydropower generator for Seitevare was developed by the Design Office for Vertical AC Machines, OKV, with the assistance from the development office OKU. The idea of making this generator directly water-cooled was probably initiated by Carl Rönnevig, manager of electric design within OKV. Vattenfall had ordered the generator in 1964, but the decision to make it wa-ter-cooled in both stator and rotor was not made until 1965. A stator winding test rig and a rotor test rig with two salient poles helped to convince the cus-tomer. The development of this new generator concept did not greatly affect the number of development and design engineers employed. The personnel in OKV + OKU increased during 1965 and 1966 from 53 to 59, i.e. nothing exceptional. The attitude of the organization was confirmed by Rönnevig who said: "The project was never considered as something very complicat-ed!" Asea's executive management was not closely engaged in this project. Seitevare was nevertheless important for later projects even if a hydropower generator is very different from a turbogenerator. Design and handling of components for the stainless steel cooling circuits gave especially valuable experience. It also helped building confidence in the technology.

In the case of the Oskarshamn 1 generators, it was only the stator winding that would be directly water-cooled. The remaining parts were hydrogen cooled in the same way as the earlier, smaller turbogenerators. The develop-ment was therefore handled within the existing Turbogenerator Design Of-fice, OKT. No efforts were made to increase the staff, it was very much "business as usual". However, OKU handled the development of the compo-nents necessary for the stator winding cooling, thus obtaining some synergy with the Seitevare work. As mentioned earlier in section 7.2.2, some prob-lems occurred during manufacturing and testing of the Oskarshamn genera-tors, but too late to have an impact on the design. The most severe problems were related to rotor dynamics and had nothing to do with the water-cooling. It was instead an example of the classical problem of extrapolating an exist-ing design too far.

The next project was the big synchronous condenser for Dumont. This order from AEP was received in the autumn of 1967, after the Seitevare generator had already been delivered, and therefore, OKU had the ability to take on this new machine with some experience. The Design Office for Horizontal AC Machines, OKH was responsible for this project, but the development was shared so that OKH made the stator and OKU the rotor design. Carl Rönnevig, who had been deeply engaged in the Seitevare project, had succeeded Erik Agerman as manager for the development office, which of course helped with the internal technology transfer. A handful of persons came from the earlier projects to this new. The Dumont project was consid-ered as a good preparation for designing completely water-cooled turbogenerators. It is somewhat surprising to see that the total staff for the involved offices decreased from 59 to 50 persons during 1968 when the main part of the design was made. Nevertheless, the design turned out to be successful and the manufacture went fairly well.

It is possible to conclude from both documentation and interviews that no external expertise was consulted for any of these three projects: Seitevare, Oskarshamn and Dumont. The O-division relied on its own engineers, supplemented with some support from specialists of Asea's Central Laboratories. It has been expressed by many persons that "you could get answers on almost all kinds of technical questions by clever use of Asea's internal telephone directory." The only external sources of know how were some of the suppliers, e.g. Sandvik concerning the stainless steel tubes. Technical literature was also important. Several key persons have mentioned BBC's and Siemens' journals, the German Elektrotechniche Zeitung (ETZ) and Cigrè reports; a clear German influence. There was no cooperation with universities or consultants except for discussion of the balancing problems and the bearings for the Oskarshamn generators.

The order for the two turbogenerators for Ringhals was placed in 1968. The final customer was Vattenfall, which once again accepted a radically new machine design from Asea. A clear evidence that the "development pair" described in Mats Fridlund's thesis "The mutual development" [270] still existed even if it was informal. First Hamra G3, then Seitevare and now Ringhals, all of them were advanced prototype machines and the first one not even successful. The Design Office for Turbogenerators, OKT, had the main responsibility for these new generators of much larger design than before. A special development group was established within OKT for designing these generators in close cooperation with the development office OKU. Furthermore, Carl Rönnevig was appointed as new manager of OKT in the autumn of 1968, a decision taken in order to maintain the connection to the previous projects. In contradiction to the earlier projects, in which the existing engineering staff made most of the design, only a few experienced turbogenerator designers were involved in this new project. The crew of 21 engineers, working directly with the Ringhals project in late 1969, was composed as follows: 11 came from the salient pole machine projects, especially the Dumont synchronous condenser, 5 came from the turbogenerator design office; two of them were electrical design engineers, 5 were newly employed, more or less directly from school.

The total number of people in the design department decreased during 1968 from 102 to 82 (13 due to a re-organization), indicating that the management underestimated the work to be done. One reason could have been that the Dumont project was too successful, which made it easy to underestimate the extent of the turbogenerator challenge at this point in time.

The brief overview of the initial projects is based on interviews and internal company documentation. The conclusion is that the question of resources and competence was never a strategic matter. These more advanced machines were handled in the usual way, i.e. most of the development was made on customer orders by the people already working in the design offices. Expertise for answering special questions was available in Asea's Central Laboratory. This approach was confirmed in an interview with Jan Liljeblad who added: "I am somewhat embarrassed when I think about the lack of structured development the O-division had at that time. No tradition existed for systematic planning of development projects for very large machines." The big decision was to build completely water-cooled machines. The matter of competence and resources was a matter to be solved along the road.

7.3.4   A prototype generator in the backyard

In 1969, Stal-Laval and Asea received an order for a 250 MW steam turbine and a 294 MVA generator from AB Aroskraft in Västerås, Sweden. The unit would be installed in a fossil fueled, combined heat and power plant. Siemens had delivered the first unit of this size to the same power plant. The delivery time was short, less than three years. The generator would be delivered in February 1972, approximately nine months before the first Ringhals 2 generator. Västerås was Asea's hometown and it would be, of course, an advantage to have the first turbogenerator with direct water-cooling close to the factory. It was also smaller, which would turn it into a very suitable prototype. Therefore, Asea started to negotiate with Aroskraft and succeeded in getting acceptance for delivery of a fully water-cooled generator. Arguments supporting this proposal were that Vattenfall had already ordered this concept for Ringhals 2, and the short length of the water-cooled generator would decrease the costs for building and foundation. This decision increased the working load on the design department quite a bit, especially in view of the shorter delivery time for this new order. The negotiations about a prototype were probably simplified by the fact that after his retirement, Tage Strömberg (1905-1970), former manager of Asea's large AC machine department, had become a local politician and acted as a representative for the city of Västerås in Aroskraft's board. As a parenthesis in this historical study, it is interesting to note that Tage Strömberg was the first ever PhD graduated from Chalmers. He presented his dissertation on "fractional-pitch wave windings", particularly for hydropower generators, in 1943 [292].

The diagram in figure 7.30 below shows how the development projects of the different water-cooled machines were distributed over time.

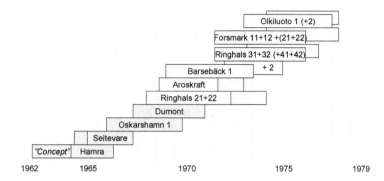

Figure 7.30 Timing of the different projects with large water-cooled synchronous machines

7.4 The challenges

The company faced a number of tough challenges with the development and the manufacturing of the new, large turbogenerators. The most obvious challenges were technical. An almost 10-fold extrapolation in size resulted, of necessity, in larger stresses, new phenomena, and new concepts. The next types of challenges were economical. It was difficult to predict the costs both for development and manufacturing of products, which were so different from the traditional ones. The risks for underestimation of the costs were high. Finally, another big challenge was to build quickly the competence and resources required for the development and the manufacturing.

7.4.1 Technical key problems

Which technical problems did Asea foresee when the company started to develop the completely water-cooled turbogenerators? Was any systematical risk analysis performed? What kind of special studies and experiments were carried out? What did the people engaged know about problems encountered by other manufacturers of large turbogenerators?

The easiest question to answer concerns risk analysis. The answer is simply no. There was no comprehensive analysis made. Today, this may seem irresponsible, but it is not very surprising. The methods for performing such analyses, e.g. FMEA[*], were not widely spread at that time, so the organiza-

---

[*]. Failure Mode Effect Analysis, introduced in the aircraft industry in the 1960's and in some automotive companies in the 1970's. [293]

---

tion was not prepared for such an activity. People just addressed items that they felt could be critical. Interviews with some people engaged during the initial development have given examples on such items.

Several of the identified questions were related to problems experienced on other machines as, of course, can be expected. For instance, there were several theoretical studies performed concerning the type of bearings to use because of the difficulties with balancing and testing of the Oskarshamn 1 generators. The risk for thermal unbalance due to uneven distribution of the cooling-water in the rotor windings had been investigated by means of the test rotor in 1966 [279], but was also discussed in light of the first Oskarshamn 1 rotors being scrapped due to thermal instability.

The support of the stator end-windings was also a matter of concern. This was due to the fact that the stator winding in Seitevare, as mentioned earlier, had been severely damaged at an instantaneous short-circuit test in the power plant. It was clear that the large turbogenerators would be subjected to much higher forces and the design had to take this into account. It was also a matter that received quite a lot of attention internationally, e.g. in Cigrè's study committee for large machines. [294]

Heating of the end regions, both in stator and rotor, due to eddy currents caused by leakage flux, was a well known but difficult problem. This phenomenon became more important for large machines with low pole numbers and it also received wide spread attention, which Asea's engineers learned from published reports. [295]

Of course, the risk for water leakages was also a matter of concern. The focus was primarily on erosion due to the water flow and galvanic corrosion in the vicinity of joints. The retaining rings were another item, which was identified as critical. In an interview, Rönnevig underlined that they were prepared for unforeseen problems, which only could be detected through tests. This is a big risk, which a company has to face, when large, advanced products are developed on customer orders.

7.4.2 Cost targets

What kind of cost targets did Asea have for the development of the large turbogenerators? Were there any market prices which could be used as reference? It has been mentioned earlier that there was hardly any market for separate large turbogenerators. They were generally sold together with the turbines and represented only a minor portion of the total cost for a complete turbine/generator plant. Therefore, it was not easy to find a correct sales price for these products, especially as they were much larger than previous generators. The company had experience from the manufacturing of a lot of other machines and that formed a base for how to calculate manufacturing costs. The principle for setting prices for the new generators was simply to estimate the costs based on existing information and rules and then add applicable overheads and a profit margin. One cost target was thus to keep the final manufacturing costs within the pre-calculated ones. This was the most important target. Others were related to the development costs, which partly were financed directly by the customer orders and partly by the expense budget.

It was obviously very difficult to estimate in advance all the costs of these large and complicated machines and their development. In hindsight, they were considerably underestimated. Some results are presented later in section 7.7.3.

7.4.3 Lack of know-how

It is not evident to what extent Asea realized that there was a lack of know-how when the development of the large turbogenerators commenced. Acquirement of a license was never a serious option, so it was obviously considered possible to overcome any deficiencies. There was already an experienced crew of engineers in the design department and it was only considered necessary to build-up new competence concerning the details required for water-cooling. This task was given to the development office OKU and a few key persons were transferred there from the Central Laboratories. All recruitments were made internally within Asea.

However, Asea's president, Curt Nicolin, must have been aware that know-how could be missing. He decided to find an external advisor and was recommended by a former Stal-Laval employee, Richard Söderberg (1895-1979) since long a professor in America, to contact professor Herbert Woodson at MIT. Nicolin wrote to Woodson in October 1968: *"...... I think we have in ASEA a reasonably good technical team, but it might be said that our experience in large turbo alternators really is rather scarce and that we consequently do not have sufficiently wide background of experience supporting our development work. Our men are feeling this and would very much estimate the opportunity to consult on various plans and ideas of theirs ..."* [296]. Professor Woodson was engaged as consultant, but it turned out to have very little impact on the development.

Another attempt to acquire knowledge was a three weeks study tour to North-America in 1968. Ove Tjernström, who was one of the key persons behind the water-cooled design, had replaced Rönnevig as manager for the development office. Tjernström mentioned, in an interview, about this study tour, which even included a visit to GE in Schenectady. He considered, however, that the result of the tour was meagre; the Asea engineers had to continue according to their own ideas.

The lack of know-how was further enhanced by the fact that a few key persons left the O-division at the end of 1969 and beginning of 1970. The chief engineer, Jan Liljeblad, moved from his position and was replaced, half a year later, as head of the development and design department by Bertil I. Larsson. Both Carl Rönnevig and Richard Sivertsen returned to new jobs in Norway and were replaced by Sture Eriksson, author of this thesis, as manager for the turbogenerator office, and Karl-Erik Sjöström as responsible for electrical design. Rönnevig claimed, in the interview, that he left Asea mainly due to insufficient resource allocation. Bertil Larsson, whose background was marketing, had been recruited directly by Curt Nicolin. He expressed in an interview: "I certainly didn't get the job on technical merits, which indicates that the executive management was more concerned about possible organizational problems than technical ones." The author of this thesis does not want to make comments referring to his own personal role, but it is necessary to make an exception, because it underlines that the technical difficulties must have been underestimated. My personal comment is: "Instead of giving the job as manager of the turbogenerator development to a 32 years old, still inexperienced, electrical engineer, it would have required someone with the experience I had when I left the department 12 years later."

7.5 The initial design

7.5.1 Main data and performance

The generator for Aroskraft and the two generators for Ringhals, called Ringhals 21 and 22 respectively, were very similar in design except for the difference in length, depending on the much higher output for the Ringhals generators. The generator for Barsebäck, which followed immediately after those for Aroskraft and Ringhals, was designed according to the same principles, but it had both larger diameter and length, and there were also some minor conceptual modifications. However, all these generators, including the second Barsebäck generator, can be treated as an homogenous group representing the initial design concept.

The ratings and a few other key data for these first generators are given in table 7.5.

Table 7.5   Main data for Aroskraft, Ringhals 2 and Barsebäck

Unit	Aroskraft	Ringhals 2	Barsebäck
Rated output [MVA]	294	506.5	710
Power factor	0.85	0.85	0.85
Active power [MW]	250	430	600
Rated voltage [kV]	17.5	19.6	17.5
Rated current [kA]	9.7	15.0	23.4

Table 7.5   Main data for Aroskraft, Ringhals 2 and Barsebäck

Unit	Aroskraft	Ringhals 2	Barsebäck
Rotor diameter [mm]	1150	1150	1250
Active length [mm]	2600	4150	4900
Utilization factor [kVA/rpm m^3]	20.7	22.3	22.4
Losses [kW]	3370	6157	6850
Efficiency [%]	98.67	98.59	98.87

7.5.2   Stator design

The stator core was built up from punched segments of thin, silicon alloyed steel sheet. The segments were glued together after the stacking had been finished. The reason for this was to improve the heat transfer in the axial direction. Cooling segments of aluminum were placed at equidistant intervals in the back of the core as shown in figure 7.31. Pressure fingers of stainless steel, interconnected to form complete pressure rings, transferred the necessary axial pressure on the core. This was important for avoiding axial vibrations, especially in the stator teeth. Axial leakage flux from the rotor and from the stator end-windings could cause overheating in the stator core end regions. Spiral wound leakage flux rings made from a thin electro steel sheet were therefore placed at both ends axially outside the pressure rings as figure 7.31 shows. The intention was that these laminated rings should protect the core by absorbing the leakage flux.

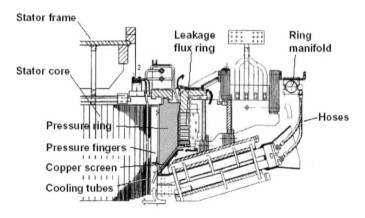

Figure 7.31 Stator end region in first generation GTD generators

In principle, the stator windings in these large generators were traditional 3-phase lap type windings with two coil sides per slot. The windings were high voltage windings and the coil sides had to have adequate insulation. Each coil side, which was made up from a number of parallel individually insulated copper strands, had a main insulation of glass backed mica tape, vacuum pressure impregnated with epoxy resin. It was a high-class insulation system with a sensitive manufacturing process, but this system already existed and it was also used for other large generators and didn't require special development for these turbogenerators.

The coil sides were so called Roebel bars, each consisting of a large number of small, rectangular copper strands, fully transposed in the straight portion of the coil side, as previously shown in figure 7.14. Some of the strands had been replaced by stainless steel tubes, thus providing channels for the cooling water. The reasons for dividing the conductors in smaller strands and transposing them is to minimize extra losses due to eddy currents and circulating currents. Figure 7.32 below shows a cross section of a stator slot in which both coil sides are double Roebel bars.

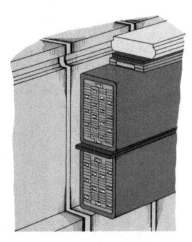

Figure 7.32 Cross section of stator slot with water-cooled Roebel bars

A number of theoretical studies concerning the losses in double Roebel bars were performed by Dr. Erik Morath (1914-1977) before the design was decided [297]. It was also necessary to perform different manufacturing experiments related to both the assembly of the bars and the cooling tubes. The results were good from the technical point of view, but the coil sides became very expensive. A popular comparison was that the cost for each coil side was roughly the same as for a small passenger car.

The stator windings in large turbogenerators are subjected to large electro-dynamic forces both at normal operation and at certain transient conditions. The slot portion tends to vibrate with double rated frequency, i.e. 100 Hz and it is important to avoid these vibrations, which would wear the insulation. Asea had experienced such problems with smaller turbogenerators when the epoxy impregnated system had replaced the old asphalt-compound impregnated windings. Therefore, the problem was given proper attention from the beginning and both theoretical and experimental investigations were made as a base for the final design [298]. The coil sides should be permanently kept under pre-tension in the slots both in radial and tangential direction. This required suitable wedges combined with springs and/or dampers for compensation of long-term dependent changes, as well as thermal movements.

The stator end-windings are large in two-pole generators and their design is critical. The coil-ends forms a "basket" at each end outside the active portion of the machine. There are electro-dynamic forces acting both in between the individual coil-ends and between the total end-winding assembly and the adjacent stator core. Figure 7.33 shows a photo of such an end-winding. At normal operation, it is important to avoid resonance with multiples of the rated frequency for various modes of vibrations. It is also necessary to support the end-windings so that no damages are caused by the large forces, which would occur at an instantaneous short-circuit or other severe transient conditions. However, the end-winding support must, at the same time be designed in order to allow for differential thermal expansion between the winding and the core.

Figure 7.33 Stator end winding in 577 MVA turbogenerator and sketch of end winding basket

Problems with end-windings on large turbogenerators had been experienced by many other manufacturers receiving wide attention. Several reports had been published, e.g. [294], so Asea's engineers were already aware of the importance of this issue at the start of the design work. Bertil I. Larsson reminded the author of an illustrative example of stator end problems. In the mid 1970's, a large number of turbogenerators in British power plants were out of service due to such problems, that it actually led to a hearing in the House of Parliament. Those generators had been built by different British manufacturers.

The stator housing consists mainly of the stator frame, end-covers and end-shields. This was, in some respects, easier to design than corresponding hydrogen cooled turbogenerators, for which the stator housing must be a completely tight pressure vessel. The stator core was stacked directly inside the stator frame. One decision to be made was whether the core should be flexibly mounted in the frame in order to prevent core vibrations to be transmitted to the concrete foundation. The decision was to make the radial height of the back of the core large enough to limit the vibration amplitude to an acceptable level for a fixed mounted core.

The end-covers and the end-shields had different designs for this first group of generators. The Aroskraft and Ringhals generators had end-shield mounted bearings requiring heavy and rigid end-shields and end-covers. The Barsebäck generator had pedestal bearings and the heavy end-shields could be omitted and replaced by light end-covers.

A terminal box on a small motor is fairly simple. It becomes much more difficult for large generators with phase currents in the order of 20 000 A. The stator winding terminals were arranged in a special aluminum cubicle attached to the stator housing. Each phase conductor had to be kept in place by insulating supports and was provided with current transformers for control and protection systems. One problem, which was significant for many details in these large machines, was to combine strength and flexibility. In this case, the conductors had to be firmly supported due to the large electro-dynamic forces, but the connections to the winding and to the busbar had to be very flexible and allow thermal movements. Figure 7.34 shows the Barsebäck generator including the terminals underneath the generator.

The rotor water-inlet device, the slipring shaft bearing, the sliprings with brush gears, and the slipring-end main bearing can be seen in the left part of the sketch. All water connections to the rotor coils are arranged at the slipring-end of the rotor. The stator core is cooled by means of cooling segments inserted in the core as shown in the sketch.

Figure 7.34 710 MVA, 3000 rpm directly water-cooled turbogenerator

7.5.3 Rotor design

The rotor body is a key part of the magnetic circuit. The main flux shall pass directly through it, therefore, the magnetic properties of the material are essential. It is also subjected to large centrifugal forces so it must be made from high-strength steel. The rotor body, including the shaft ends, is made in one piece forged by some qualified steel mill. As an example, the rotor forging for the Barsebäck generator had a weight of 65 tons and a length of roughly 9 meters. The yield strength was 730 MPa and the ultimate strength 830 MPa [299]. Each forging had to undergo extensive quality control before delivery from the steel mill, e.g. measurement of magnetization curve, control of mechanical properties on a number of test pieces, and ultrasonic control of the entire forging. Asea sharpened the quality requirements after the experience with the earlier Oskarshamn rotor forgings.

Each rotor forging was then machined in a number of different operations, such as turning, milling, drilling etc. Radial slots for the winding were milled along the active part of the rotor as shown in figure 7.35 a. In order to avoid double frequent vertical oscillation of a slim two-pole rotor, it is necessary to design it with approximately the same bending stiffness in both direct and quadrature directions. This can be achieved either by milling narrow transversal slots, or by fairly large axial slots in the poles. In the latter case, which Asea used, the slots are filled with magnetic steel blocks.

*Figure 7.35 a) Machining of rotor body for an 825 MVA generator.
b) Cross-section of rotor slot with directly water-cooled coil*

The rotor winding was made up from a number of concentric coils connected in series. Each coil contained a certain number of turns. Figure 7.36 a shows insertion of coils in a 2-pole rotor. The conductor consisted of two parallel copper bars with groves for containing the cooling tube, see figure 7.35 b. The insulation between the turns consisted of thin strips of insulating material and the slot insulation consisted of pre-fabricated L-shaped pieces for side insulation and flat pieces for top and bottom insulation.

Figure 7.36 a) Insertion of rotor coils. b) Retaining ring shrinking

The coils were subjected to a press operation to prevent later radial displacement due to centrifugal forces. Slot wedges made from special brass were placed at the top of each slot. Insulating blocks were placed between the core and the end-windings and also in between the different coil-ends. One reason, for these blocks, was to support the coil-ends against the forces created by the large current, but also to avoid successive axial movement, mainly due to differential thermal forces.

A difficult problem for turbogenerators has always been the risk for overheating the rotor surface and the slot wedges through eddy currents caused by flux harmonics, and by negative sequence currents due to asymmetric loading. The flux harmonics were not considered as the main problem because the airgap was very large in these generators, 100 – 115 mm. The negative sequence current was of more concern. An asymmetric three-phase system can be divided into one positive and one negative sequence system [300]. The mmf caused by the positive sequence currents rotates synchronously with the rotor, while the mmf from the negative sequence currents is counter rotating and hence induces double frequent eddy currents in the rotor. This problem had received wide international attention as can be seen in some Cigrè reports, e.g. [301]. Requirements were specified for both continuous negative sequence current ($I_2 = 0.08$ p.u.) and transient capability ($i_2{}^2 t = 4$ s) [302]. One way, to meet these requirements, was to provide the rotor with a damper winding. Therefore, Asea decided to form such a winding by inserting conductors in small slots, located in the top of the teeth. Short-circuit rings were then arranged at both ends by special copper pieces placed under the retaining rings.

Large centrifugal forces act upon the end-windings, which therefore need an additional mechanical support. For this purpose one retaining ring is shrunk onto each end of the rotor surrounding the end-windings, as shown in figure 7.36 b. The retaining rings are made from very ductile, high-strength, non-magnetic steel. The rings, which are forged, could have a weight in the order of two tons each. The extremely high strength is achieved by cold stretching of the rings. The reason why they should be non-magnetic is that they should not contribute to the increase of leakage fluxes in the end regions. The data presented in table 7.6 was specified for the original retaining rings for Barsebäck [303].

A cylinder of insulation material was placed between the end-windings and the retaining ring. A supporting ring was shrunk into the outer end of the retaining ring to prevent it from becoming elliptical due to the non-circular load from the end-windings. Special locks were arranged at the rotor body, close to the shrink fit, with the purpose to prevent the retaining ring from sliding tangentially, for instance in case of a short-circuit.

Table 7.6   Retaining ring material specification

Type of material:	Austenitic maganese steel
Composition:	Mn 16 - 20 %, Cr 3.5 - 6.0 %, Ni 2.0 %, C 0.4 - 0.6 %, Si 0.4 - 0.9 %
Tensile yield strength:	> 1130 [MPa]
Ultimate tensile strength:	> 1210 [MPa]
Elongation:	> 17 [%]
Reduction of area:	> 28 [%]
Impact strength:	> 30 [J]

The rotor winding terminals were placed in holes in the shaft-end at the non-driving end of the rotor and were connected to the sliprings, one for each polarity. The excitation current was in the order of 6000 - 7000 amperes DC, which required a large number of carbon brushes at each slipring. It was foreseen that it could be difficult to get all these parallel brushes to divide the current equally and they were therefore provided with series resistors. A slipring test rig was built in order to investigate this problem.

7.5.4   Cooling

Asea's original idea was to cool practically everything with the cooling water. This philosophy was implemented on the Aroskraft generator as well as Ringhals 21 and 22. In the case of Barsebäck, some modifications were introduced. The water-cooling was certainly very efficient, but it had to reach the right places. Eddy current losses could occur in many components that didn't have good contact with the larger "heat sinks" and thus had poor heat transfer possibilities. In order to cool all these different components, the generators had to be equipped with a very advanced and extensive "plumbing system", actually too advanced and too extensive. Gas cooling, even if less efficient, is more forgiving. It is usually more global in the sense that it reaches more easily different components e.g. in the end regions. That is why there were some changes introduced in the Barsebäck generator.

The decision to use stainless steel tubes was not difficult. It had actually been made long before for the Seitevare generator and Asea found no reason to choose anything else. The only technical disadvantage of steel tubes compared with hollow copper is the extra thermal resistance the steel tubes represent, although there are several advantages. A higher water flow speed could be used without risk for erosion, resulting in a smaller cross section required for the water channels, even including the area of the tube walls. A system with only stainless steel, plus insulation material, constitutes a lower risk for galvanic corrosion compared with a mixed copper – stainless steel system. Furthermore, as mentioned earlier, welded stainless steel joints were considered much safer than soldered copper joints.

In this type of cooling system, the cooling-water passes through both stator and rotor coils having different electrical potential. It is therefore necessary that the water connections between these coils and the rest of the system are insulators and that the water itself is also an insulator. This means that normal tap water is useless. De-ionized water, which has a very low conductivity, is required. The high-voltage stator coils were supplied with cooling-water through long, reinforced Teflon hoses provided with stainless steel connectors at each end. They could withstand a pressure test of approximately 3000 kPa (30 bar). The requirements on the insulators for connection to the rotor coils were quite different. The voltage was much lower but it was DC, which could increase the risk for galvanic phenomena. The pressure was much higher due to centrifugal forces. The test pressure for the rotor components was in the order of 17 500 kPa (175 bar). A couple of different concepts were developed and tested. The design used for the first generators is shown in figure 7.37.

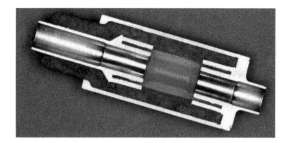

Figure 7.37 Insulating sealing for rotor cooling circuit

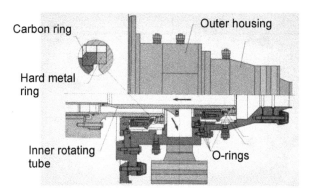

Figure 7.38 Rotor water inlet device

An obvious problem for a generator with a directly water-cooled rotor is how to get the water into and out of the rotor rotating at 3000 rpm. This was done by means of a water-inlet device containing two water-lubricated, axially spring loaded sealings. The rotating parts were steel rings, and the stationary parts were carbon rings. One sealing separated water inlet and outlet and the other the outlet-water from the ambient air, see figure 7.38. The device worked, but with a slight leakage, which the system had to make up for with fresh water. The water flowed, to and from the rotor winding, through two concentric tubes placed in the center of the shaft end with the inlet water flowing in the inner tube. Radial tubes connected the concentric inlet and outlet tubes with circular water chambers shrunk on the shaft end. The water flowed, from here, in many small parallel tubes, through the insulators and into the rotor coils, and returned back after a few turns in a corresponding way [304].

It was mentioned above that almost everything had to be water-cooled in the first generators. The water-cooled components in the stator were the winding, the core, the pressure fingers, the leakage flux rings (flux shields) and the terminal conductors. The corresponding rotor components were the excitation winding, the rotor surface with damper winding, the short-circuit rings and the sliprings. A water flow diagram can be seen in figure 7.39.

Some modifications were introduced for the Barsebäck generators. It was decided that it was better to cool some parts in the end regions with air, so the rotor was provided with small fan blades at the supporting rings for the retaining rings (see figure 7.58), and with heat exchangers, air/water, located in the stator end covers. The approach became thus somewhat less fundamentalistic.

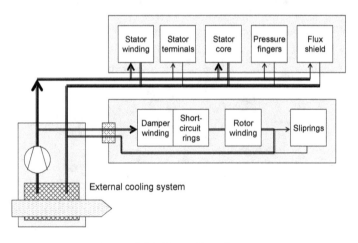

Figure 7.39 Water flow diagram

Each of these generators had to be equipped with an external cooling system containing circulation pumps, heat exchanger for cooling the primary cooling-water with raw water, de-ionizer, filters, make-up water supply, pressure gauges, flow meters, valves etc. These systems became quite large, which is no surprise, as they had to dissipate roughly 5 MW of losses. The requirements on reliability and redundancy contributed as well to their complexity.

7.5.5 Sensors

These large generators were naturally provided with a number of different sensors for monitoring and protection. Voltage and current transformers were installed in the terminal cubicle.

Temperature sensors in the stator winding, resistance elements type Pt 100, were placed in each slot. In more conventional machines there are usually not more than two per phase. The reason for one in each slot was that these sensors were used not only to determine the temperature rise of the winding, but also for checking that all the individual coils had sufficient cooling water flow. Temperature sensors were also inserted at certain spots in the stator core.

Vibration sensors were placed at the bearings for monitoring lateral rotor vibrations. The piezo-electric accelerometers recorded the vibrations in both horizontal and vertical direction. Very important sensors for these completely water-cooled generators were the leakage detectors. There were two types, humidity sensors and water detectors placed at the bottom of the stator frame. Water flow meters, pressure gauges and conductivity meters were installed in the external cooling system.

7.6 The process

7.6.1 Calculation and design

The calculation process determines the dimensions of the machine and its performance. It comprises electrical, thermal and mechanical calculations. The first two focus very much on the active parts while the latter deals with the entire machine. These calculations are fundamental parts of the entire development process. How were they performed for these new turbogenerators, much larger than those with which the company had experience? Was it necessary to introduce new methods?

The electrical calculation was the first step, keeping in mind that the entire process was iterative. The basic step, in which the magnetic circuit and the windings were outlined, was carried out in the traditional way. The algorithms and the rules had been established over several decades and had been

used by electrical design engineers for manual calculations of numerous turbogenerators. Asea acquired its first computer[*] for technical calculations in
1958 and development of calculation programs for some products, turbogenerators was one, started directly thereafter. In principle, it was an automation
of the manual procedure and the main advantage was a rationalization of the
electrical calculation work. The engineers suddenly had the possibility to calculate and compare more alternatives. The theoretical approach was the
same. This program, which had been developed by Richard Sivertsen, was
still in use 10 years later although somewhat improved. It was thus also used
for the electrical calculations of the water-cooled turbogenerators. Interviews
with Richard Sivertsen and Sven Nilsson, who had made most of these calculations, reveal that the electrical calculations where never considered as a
big problem, except for some very specific items such as eddy current losses
in the stator winding and in the end regions of the stator core.

Many of these specific items were handed over, more or less as mathematical
problems, to Erik Morath, who made theoretical analyses and wrote technical
memoranda as answers. He was a prominent expert on using methods such
as Laplace transformations, and conformal transformations etc. Some examples on such investigations are: "Induced teeth currents in a long turbo rotor
without damper winding due to negative sequence load currents" [306],
"Strand resistances and reactances for calculation of circulating currents between strands in slot conductors" [297], "Contraction factors for shallow machine slots" [307] and "The electrical resistance of rectangular conductors
with reduced cross section" [308]. It was then up to the electrical design engineers to interpret and apply Morath's results. All the electrical calculations
were at this time purely analytical.

Also the thermal calculations were usually handled by the electrical design
engineers. The analytical methods were completed by extensive empirical
data. However, in the case of these new generators with a different cooling
concept, mechanical engineers from the development office specializing in
thermal analyses were engaged. In this case, it was also a matter of analytical
calculations using thermal resistance grids by means of which the temperatures in different parts of the generators were determined [309]. Despite this,
the thermal calculations for a liquid cooled machine were considered simpler
than for a gas-cooled machine and tests later confirmed that these calculations had been fairly accurate.

The mechanical calculations were more extensive for these large generators.
Examples are: rotor stress analysis due to centrifugal forces, determination of
lateral and torsional critical speeds, dimensioning of bearings, calculation of
stresses due to differential thermal expansion, analysis of forces and stresses
in the stator due to steady state electro-dynamic forces as well as transients,

---

[*]. The computer was a Swedish Facit EDB with 10 kB CPU memory and
40 kB drum memory and it filled a large office hall. [305]

270

especially short-circuit torques [310]. Except for fairly trivial mechanical calculations, it was common practice that university graduated specialists, belonging to the development office, performed these calculations. Most of the mechanical design engineers in the different product offices had a less theoretical background. All the calculations were originally analytical, but Folke Pettersson, manager for the mechanical analysis team, mentioned in an interview that the numerical Finite Element Method (FEM) was soon used, e.g. for analysis of stresses in rotor slot wedges and teeth. Experiences soon proved that the mechanical problems were more difficult than initially expected, and consequently the amount of calculations grew significantly during the development period.

The entire development process consisted of several types of activities and it was to large extent iterative. A central activity is the mechanical design in which the conceptual layout as well as all components are designed in detail. Most of this work was carried out at the drawing boards. It was documented through hundreds or thousands of drawings, purchasing specifications, inspection instructions etc. An example is the main assembly drawing in figure 7.40 showing the generator for Ringhals 2.

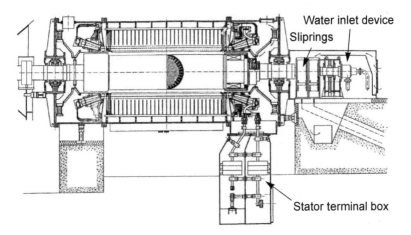

Figure 7.40 Assembly drawing for 506 MVA, 3000 rpm turbogenerator for Ringhals 2

The development of these large generators was mainly carried out on specific customer orders and was usually referred to as "design work". Only minor parts, primarily test rigs, were financed over the annual budget for development. The order specific design work became quite extensive and did amount to 27 027 man-hours for Aroskraft and 32 668 for Ringhals 2. This was significantly more than planned. In case of Aroskraft it was estimated 12 500 man-hours and correspondingly 20 000 for Ringhals[311]. This large discrepancy clearly indicates that the company had underestimated the difficulties in designing these generators.

The Turbogenerator Design Office was, in September of 1971, divided in four sections, one for electrical design and the others for mechanical design. Six out of totally 37 persons were electrical design engineers. Three electrical and 15 mechanical engineers worked with the GTD generators. In addition, there were a few mechanical engineers in the Development Office dealing with stress analysis and thermal calculations. This shows that roughly 85 percent of the development work was actually made by mechanical engineers. An interesting observation is that four of the five managers had a background as electrical engineers, a situation which, in retrospect, can be questioned [312].

The design work was based on input from the electrical calculations, while the mechanical calculations were performed later after concept drawings had been made. The process was iterative, but it was often felt that the electrical engineers had an advantage because the mechanical engineers found it difficult to dispute the former's arguments. It was easy to close a technical discussion by claiming, "the transient reactance will become too large." It is not easy to find specific examples on serious mistakes due to this, but certainly less good solutions due to misunderstandings or lack of communication between various specialists.

Figure 7.41 Internal functions engaged in development of large turbogenerators

The large, water-cooled generators constituted also a big challenge for the production department. These machines were in many respects difficult to manufacture and frequent consultations with the production specialists were necessary before the design could be finalized. The chart in figure 7.41 shows the central role of the "mechanical design function" in the development process.

7.6.2 Experiments

It has been mentioned that Asea's policy was to develop large products and systems on customer orders. A request from the development team during 1969 to build a full-size test rotor was turned down due to the high costs for such an experiment. Full-scale prototypes were built only in the case of smaller machines. However, a number of test rigs were made as part of the development for the water-cooled generators. The most important are listed below.

- Long term test rig for 20 kV, water-cooled stator coils at the Central Laboratory. This started already in 1965 for verification of the design of the coils for Seitevare, but had big importance for subsequent machines [276].

- Water-cooled test rotor with 800 mm diameter and 1560 mm active length for the study of connections, erosion, risk for thermal unbalance etc. These tests were performed during 1965-66 [279].

- Slipring test rig with brush gear and resistors for control of current distribution between parallel brushes.

- Test rig for study of stator coil vibrations and evaluation of methods for fixation of the coils in the slots. These tests were completed in 1971 [298]. The test was arranged as a resonance circuit by means of a large capacitor bank.

- Comparative tests of alternative types of insulating sealings for rotor cooling tubes.

- There were a large number of smaller experiments, most of them focused on the possibilities to perform various manufacturing operations, e.g. welding tests.

7.6.3 Development and production costs

Most development engineers do not consider cost calculations as the most glamorous part of their work, but it is nevertheless of utmost importance. Asea's O-division had a separate office for pre- and post-calculation of production costs for the various products. However, development costs financed from the expense budget were calculated by the different departments concerned.

Asea used the so-called "self cost principle" for calculation of the cost for a certain product. This method was actually first developed during the 1920's by Asea's technical director Ragnar Liljeblad. It takes into account the costs for material plus direct labor while most other costs are added as overheads. Equation 7-1 below represents the full manufacturing costs for a product. The sale price for a certain number of these products is then calculated according to equation 7-2.

$$C_{manuf} = (1 + P_{type}) \, [(1 + P_{mat}) \, M + A \, (1 + P_A)] \qquad (7\text{-}1)$$

$$S_{order} = (1 + P_{profit} + P_{sale}) \, (n \, C_{manuf} + C_{eng} + C_{tool}) \qquad (7\text{-}2)$$

where

C_{manuf} = manufacturing costs
P_{type} = product type bound overhead covering general development, quality costs, administration
P_{mat} = overhead on material covering costs for purchasing and storage
M = price paid to suppliers for purchased material and components
A = direct labor cost
P_A = overhead on labor covering costs including depreciation on machine tools, buildings etc.
S_{order} = sales price
P_{profit} = overhead to cover profit
P_{sale} = overhead to cover sales cost, transportation, commissions etc.
n = number of units included in the order
C_{eng} = order specific engineering costs
C_{tool} = order specific costs for tools etc.

The type bound overhead is usually much higher for standard products than for tailor-made. The overhead on labor is often very high, several hundred percent, especially when expensive machine tools are used.

The costs were first calculated according to the described method as basis for the quotations. They were calculated again in the same way after delivery and the economical result of each specific order could be obtained from the difference between these two calculations. Post-calculations formed the reference for coming pre-calculations. This was of course difficult for the new, water-cooled generators and the accuracy was thus much lower. The real costs turned out to be much higher than the pre-calculated, see section 7.7.3.

Interviews with project managers Hans Klein and Bengt Alenfelt, as well as Lauri Piensoho, manager for the cost calculation office, clearly indicated that costs had much lower priority than technical considerations and delivery time. The main question was how to find a solution that could work satisfactory, and there was simply no time left for searching more economical alternatives. There were frequent discussions between the designers and the cost calculators, which often also involved the production specialists, but such discussions took place on a case-to-case basis; there were no systematic reviews.

7.6.4 Manufacturing and testing

The manufacturing of large turbogenerators is in many respects very demanding. It is a tricky combination of handling very large and heavy pieces and manufacturing small, sophisticated components e.g. for the cooling circuits. Rigorous cleanliness is required during the work with the high voltage coils and windings, and also in the case of details for the cooling system. Machining operations have to be performed with high accuracy in spite of the large size of many of the components. When Asea built these large generators, almost everything was made in-house. In principle, it was only a matter of purchasing material and some minor components or auxiliaries. The intention of this thesis is not to describe the manufacturing in any detail, just to give a brief overview and comment on a few details. The diagram in figure 7.42 illustrates the main steps in the production flow and none of these steps were outsourced.

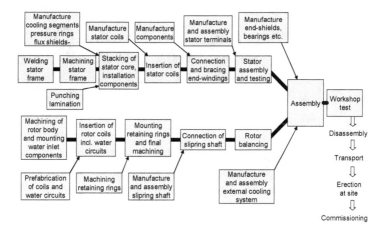

Figure 7.42 Simplified production flow diagram for water-cooled turbogenerator

The individual coil sides for the stator winding were manufactured in a separate workshop with clean room conditions. Such a coil side is a 6 – 8 m long composite structure with a long straight portion and accurately bent end parts. The challenges are to obtain the required geometric accuracy and proper quality of the high-voltage insulation. The process consists of forming the straight Roebel-bars, bending the end sections, fitting the connection pieces, machine taping the mica-glass tape, vacuum-pressure impregnation with epoxy resin, curing, application of corona protection and finally various inspections and tests. Figure 7.43 a shows the mechanized bending of the coil ends and figure 7.43 b is a photo from the machine taping. The whole process is sensitive and can easily get out of control. This happened and caused a lot of concern for the organization when the first GTD generators were to be manufactured. To identify and correct the problems required a lot of assistance from both the development department and the Central Laboratory.

Figure 7.43 Stator coil manufacture. a) Bending of coil ends and b) machine taping of coil end

In principle, the winding operation was a matter of manual work. A manipulator was used to facilitate the insertion of the heavy coil sides in the slots, but the rest was manual and required very skilled workers. The main operations included: wedging the slot portions, connecting the coil sides to each other and the terminals, applying all necessary supports to create a strong but flexible end section and connecting water hoses. A picture of an end-winding was already shown in figure 7.33.

The rotor body, i.e. the rotor core plus shaft ends, is made in one piece from a large forging. The weight of such a forging could be in the order of 50 – 90 tons for the actual sizes of turbogenerators. Each forging was subjected to a series of machine operations before it became a finished rotor body. It was a matter of turning, milling, drilling and grinding in several steps. The work required very experienced machine operators. A small mistake in the slot milling could destroy the rotor body and a new forging would cost at least a million SEK (in the 1970's) and take more than half a year to obtain. Figure 7.35 in section 7.5.3 shows slot milling of a large rotor.

Figure 7.44 Equipment for welding of cooling circuit components

Quite a bit of the production development was focused on the manufacturing of the stainless steel cooling circuits, especially welding operations for making joints between different components. Many small fixtures and TIG (Tungsten Inert Gas) welding automats had to be developed. The photos in figure 7.44 show a couple of examples where special welding tools are used. The welds were controlled by various methods such as ultrasonic tests, x-ray, or pressure tests followed by leakage detection.

Large 2-pole turbogenerator rotors are usually over-critical, i.e. the rotor has one or more lateral critical speeds below its rated speed. It is necessary to have a very well balanced rotor to be able to pass these critical speeds without severe vibrations. Such a rotor has different bending modes depending on its speed, and it is therefore necessary to balance the rotor with respect to all actual modes. It is not sufficient to balance the rotor in only two axial planes, as for stiff rotors. Balancing in five planes is more usual. Figure 7.45 contains a sketch and a diagram illustrating the bending modes and the position of the bearings.

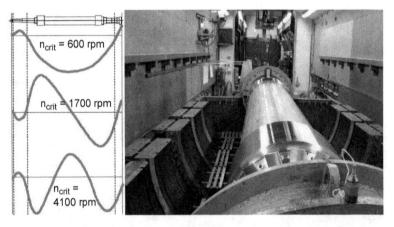

Figure 7.45 a) Example of 2-pole rotor bending modes and critical speeds b) GTD rotor in balancing machine. Note that lower halves of laminated flux shield yokes are in place

The rotor is mathematically represented by a matrix and the coefficients in this matrix are determined through vibration measurements during a number of test runs with different balancing weights. It is then possible to calculate a suitable final distribution of balancing weights. In order to check that there was no thermal unbalance, the rotors were also balanced with current. A laminated yoke was placed concentrically around the rotor to screen the surrounding structures from the rotating flux.

Of course, the manufacturing of a large generator required an extensive inspection program for successive verification of various components. Finally, the generator had to undergo acceptance tests, which were carried out in Asea's works before delivery to the power plant. It was impossible to perform a full load test of a 600 MW machine like Barsebäck in a test bay, but there were alternative methods to verify the performance through combinations of no-load and short-circuit tests. The test program included the following major points [313]:

- Resistance and impedance measurements

- Measurement of no-load and short-circuit characteristics

- Loss measurements

- No-load heat run test at rated voltage

- Short-circuit heat run test at rated current

- 2-phase short-circuit heat run test (to prove negative sequence current capability)

- Registration of voltage shape, shaft and bearing vibrations etc.

The Barsebäck generator represented a big development step and Sydkraft had requested, according to Gunnar Tedestål, extensive workshop tests. The test program contained therefore a large number of cycling tests, 70 for the rotor and 48 for the complete generator. The latter were both short-circuit and no-load cycles and each cycle had duration of four hours. Figure 7.46 shows a photo of one of the Ringhals 2 generators in Asea's test field.

Figure 7.46 Turbogenerator for Ringhals 2 at workshop test

7.6.5   Cooperation with partners

The development and manufacturing of these large machines created, in many cases, wide cooperation between the O-division and other Asea units as well as external organizations. It has been noted during many interviews that cooperation with the Central Laboratories, later Corporate Research, was the dominating one. It was not a matter of electrical or mechanical design of the generators, something which the O-division took care of itself. The Central Laboratories contributed instead to solve problems related to materials, insulation systems, sensors, special measurements etc.

In spite of some differences in opinions, there was good cooperation with Stal-Laval. Design and delivery of large turbine/generator units required a lot of coordination. Stal-Laval, as main supplier, had the responsibility for dynamic analyses of the complete rotating system including both lateral and torsional vibrations. All interfaces had to be specified and accepted; the foundation was also a common item, and so were installation questions. The latter were often quite complicated. It was a matter of handling parts with a weight of 200 – 300 tons and those could not just be put somewhere in a corner. The control and monitoring system was another area that required close cooperation between the two companies.

It has been mentioned in the previous section that Asea hardly outsourced any manufacturing to sub-suppliers that consequently did not participate in the development either. There were indeed many contacts with suppliers of important materials like forgings, steel tubes, insulations etc. but mainly with the purpose to obtain information on material properties and suitable methods for processing and inspection.

The end customers, on the other side, took a very active part in the development through their own specialists, but also with the help of some consultants. Vattenfall and Asea had already agreed, in 1968, in connection with the order for Ringhals 2, to set up a mutual group of specialists for regular reviews of the development and design. Vattenfall contracted, in order to strengthen the competence, the British Central Electricity Generating Board (CEGB) to provide experts as consultants for these review activities [278]. At that time, CEGB had a lot of experience from the operation of several tens of 500 MW turbogenerators supplied by different British manufacturers [13]. The review group was organized as "Project nr. 10" within the general frame work for mutual development work between Vattenfall and Asea. Representatives for Vattenfall were Peter Langer (1923-1995), Uno Jonsson and Sture Söderberg. Langer and Jonsson were specialists on generators and generator control while Söderberg was expert on mechanical engineering. From CEGB participated mostly Mr. John J. Arnold, head of their group of generator specialists. Questions, which received special attention, were of course those regarding the water-cooling circuits, but even the stator end-winding support and the retaining ring locking were much discussed. All drawings submitted to Vattenfall for review were passed on to CEGB for comments [314].

The review group was later extended to include a few persons from Sydkraft and Aroskraft, after they had ordered their corresponding generators. Sydkraft's representatives were Gunnar Tedestål, who had experience from the supply of large BBC turbogenerators to the fossil fuelled power plant in Karlshamn, and Curt Lindqvist. Tedestål mentioned in an interview that the review meetings gave a good insight in Asea's design work, but Sydkraft had no ambitions to influence directly on the design.

During this period, the end of the 1960's and the first couple of years in the 70's, there were hardly any direct contacts with competitors. Of course, BBC had a close cooperation with Stal-Laval concerning the turbines, but was very strict about excluding generator issues. According to a protocol from a meeting with the O-division board in May 1968, there had been discussions in Västerås with English Electric concerning possible cooperation and Eric Sjökvist was scheduled to continue these discussions in England [315]. However, no agreement was reached. The only forum for communication with colleagues from leading competitor companies was Cigrè's study committee for rotating machines. Swedish representative during these years was Claes Tengstrand (1922-2003), head of Asea's hydropower generator office during the 1960's. He was replaced in this Cigrè committee in 1975 by Sture Eriksson. Asea's turbogenerator development engineers had thus very limited direct competitor contacts during the critical years around 1970.

Asea did not engage any consultants in the development process, nor did they receive help from university researchers except from the MIT professor Herbert Woodson, mentioned in section 7.4.3. It can be concluded that the input of external knowledge was limited. This was probably not due to an underestimation of the need, at least not from the engineers concerned, but a lack of tradition. These engineers turned to specialists in the Central Laboratory for help with certain problems as they had usually done. Otherwise, they considered themselves to be the best domestic experts. Two important factors could be part of the explanation why there was so little input of external know-how. One was that the organization was overloaded with all the large orders and simply did not have time for any outlook. Another was that Asea did not have sufficient experience from building large turbogenerators to be able to approach the leading manufacturers. You must have interesting information to trade if you expect to obtain any. This was proved by the subsequent history when there were quite a few meetings with competitors. Nevertheless, Karl-Erik Sjöström mentioned as an early example that Asea had some contacts with BBC in the mid 1970's and received information on large sliprings and brushgears in exchange for information on water-inlet devices.

7.7 The initial results

7.7.1 Manufacturing and testing

The manufacturing of the first water-cooled turbogenerators, which had received the type designation GTD, was problematic. The machines were complicated. The workshop faced a lot of difficulties and the operations took much longer than expected. A message from the president of the O-division, in May of 1972, to Asea's executive management describes the situation as precarious [316]. Table 7.7 presents the delivery situation according to this message.

Table 7.7   Delivery situation for GTD generators

Generator	Delivery date	Reported delay weeks	Additional delay weeks
Aroskraft	feb-72	28	5
Ringhals 1	nov-72	12	6
Ringhals 2	dec-72	0	7
Barsebäck	jan-74	0	9

Things got much worse half a year later. The Aroskraft rotor winding had failed during a high-voltage test. The rotor had now an earth fault and needed a re-winding requiring many months. This was a disaster for Aroskraft but there was also a big risk that the same weakness had been built into the Ringhals rotors. Barsebäck was in the pipeline and the design office was occupied with design of the new generators for Ringhals 3 and Forsmark 1. This created a conflict of interests. Aroskraft wanted to use the rotor as it was for some time and carry out tests of other parts of the plant, primarily the turbine. The development department wanted to dismantle the winding and investigate the fault as soon as possible to be able to introduce necessary modifications on following rotors [317]. Vattenfall wanted their generators as soon as possible and was prepared to wait with high-voltage tests until a spare rotor could be ready. Vattenfall was also critical to Asea's development process and the technical director, Ingvar Wivstad wrote in a letter, concerning a new rotor, to Curt Nicolin in December 1972; *"The development work carried out has mainly focused on component design. No prototype rotor has been built, which often has been done in similar cases by other generator manufacturers. It has been proven that the designers have had too little experience to take into account the existing manufacturing problems."* (My translation) [318].

The reason for the fault could be attributed to the method for pressing the rotor coils after winding. The slot insulation cracked when the rotor was balanced at full speed. The remedy was a good example on how a radically new solution solved both the current problem and even another one. Kristian Dahl Madsen proposed that a flat copper tube should be inserted in each slot, just under the slot wedges. Epoxy resin should then be injected under high pressure so that the tubes would expand and compact the coils before it cured [319]. The coils became properly fixed and, at the same time, the cooper tubes formed a damper winding through bridging the axial gaps between the brass wedges in top of the slots. The photo in figure 7.35 b shows the expanded copper tube as well as all other details in such a rotor slot.

The risk for water leakages had been discussed as a possibility, therefore it was no surprise when the first leakage was detected at a leakage test of one of the Ringhals rotors in the autumn of 1972. This was the first in a long series of leakages, which led to a number of design modifications, but the balancing and workshop test of this Ringhals rotor passed without any further failures. Rotor balancing had been a major problem for the Oskarshamn generators and there was some concern over the risks for thermal unbalance. The experience from balancing of the first Ringhals rotors was positive in this respect and it was proved that the thermal balance status could be controlled by the cooling water distribution.

Also the stators caused difficult problems. The glued stator core for the Aroskraft generator had to be re-stacked due to detection of many hot spots during a so-called flux test. The glue was quite simply pressed away so that short-circuits occurred between adjacent sheets. Some serious problems were then encountered during the workshop test of the first Ringhals generator. There were further rotor leakages detected, but more severe, according to OK's annual report for 1973, vibrations and hot spots in the stator end regions. The end-windings had a number of vibration modes and resonances occurred with the excitation forces caused by the stator current. The leakage flux induced eddy currents in various components and created local hot spots where the cooling was insufficient. The problems required comprehensive design modifications and the corrective actions resulted in substantial delays and extra costs [320].

Except for problems like those mentioned above, the results from the performance tests of the first GTD generators showed good agreement with predicted values, and losses as well as temperature rises met the guarantees. For instance, the measured losses in the first Barsebäck generator was 6171 kW, which can be compared with a guaranteed value of 6400 kW.

7.7.2 Commissioning and operation

An important milestone was passed Saturday night on August 17, 1974 when the first large GTD generator, Ringhals 21, was connected to the grid and generated power for a few hours [321]. Ringhals 22 was commissioned later the same year, and Barsebäck 1 in the summer of 1975. The initial years of operation were characterized by frequent faults and disturbances. Some types of faults required immediate stop of the generators, others were found during inspections. Examples of the first type are leakages in water inlet devices, leakages in the rotor cooling circuits, short-circuits between sliprings and even leakages in the stator. The first rotor leakage in a running rotor occurred in Ringhals, in the summer of 1974, i.e. already during the start up phase. It was a fatigue crack in a water-inlet tube, just where it entered into the coil [322]. The reason for most of the subsequent water leakages was also fatigue cracks in the cooling tubes, but this will be dealt with in more detail in section 7.9.1. Faults found at inspections were, for instance: cracks in rotor slot wedges, broken flexible conductors in the stator terminals, loose stator end-winding supports as well as broken sheets and fastening screws in the flux shields. Common denominator for many of these faults was that they were caused by vibrations and thermal movements. Another irritating problem was excessive brush wear which resulted in deposits of carbon dust on sliprings and brush gear. The list of examples could be made much longer. Asea's specialists and design engineers became, during these first years of operation, much occupied with trouble shooting. However, the situation was gradually improved, a fact that can be illustrated by unavailability statistics. For instance, the records for Ringhals 21 shows an unavailability of 9.51 percent in 1975, 3.09 in 1976, 1.79 in 1977 and 1.58 in 1978 [323]. The situation was not much different for the other machines.

Figure 7.47 506 MVA, 3000 rpm turbogenerator at Ringhals 2, later up-graded to 540 MVA

7.7.3   Cost results

When Bertil I. Larsson took over as manager for development and design within the O-division in the summer 1970, almost immediately he initiated a review of the expected economical results from development and manufacturing of the first GTD generators. The result of the review, which was reported to the executive management, was an estimated deficit of 4 MSEK over the next four years [324]. A new review, performed one year later, showed instead an expected loss in the order of 9 MSEK for the four generators Aroskraft, Ringhals 21 + 22 and Barsebäck 1. Of this loss 7.5 MSEK was booked on the actual orders and the rest on the development account [325].

It was obviously difficult to make a reliable estimation of future costs. They became much higher than expected. The cost calculation office presented the final results for the first four orders mentioned above in a document dated September 1974. The result was the difference between the pre-calculated and the real costs, and was for these orders together close to 20 MSEK. The excess costs on the individual orders varied from 57 to 70 percent. Development and quality costs, which were carried by the expense budget, were not included in this amount [326].

All kind of costs became higher than foreseen: design, material, labor and overheads. Lauri Piensoho, who was manager for the cost calculation office, mentioned in an interview that the costs for new components often turned out to be three times higher than expected. The reason was that it took time for the workshop to find the best methods for various operations and this often resulted in a pressure from the workers on increased piece payment. In addition, the costs for expensive machine tools were added as a high overhead percentage on the direct labor cost. Therefore, the orders had to carry high machine tool costs even when most of the time they were used for trial runs. The cost calculation system was adequate for the ordinary products, but it probably resulted in too high costs for new products. This was, according to Piensoho, a problem not only for the GTD generators but also for other special products.

There were reasons to question the thoroughness in the quotation procedure. A message to the company's president in October 1972 contains information that the O-division had received a telephone enquiry on October 5th from Stal-Laval for a proposal on an 825 MVA generator to TVO. The specification arrived by courier the same evening. The quotation was sent by telex the next day! The message states that both generator size and cost had been determined in proportion to the Barsebäck generator. There was no indication that it should have been a preliminary proposal, but it took eight months until the contract was signed, and by then, the final order price was somewhat increased [327, 328].

7.7.4   Actions

It is not possible to identify any specific point in time at which the experiences from the first GTD generators could be summarized and transformed into specified design modifications or other actions. Various activities were instead overlapping each other during several years, from 1972 until 1977. It is nevertheless possible to find a number of measures aiming at an improvement of the situation. The descriptions in sections 7.7.1 – 7.7.3 above reveal clearly that such actions were urgently needed.

The design suffered from a number of problems. In many respects the generators were too difficult to manufacture and were not robust enough. The performance was generally good when the generators operated, but failures occurred far too often. This could partly be attributed to the concept chosen and partly to extrapolation difficulties. Some important design modifications were therefore introduced already during 1972 in connection with the development of the generators for Ringhals 3 and Forsmark 1. These are dealt with later in section 7.8.1.

The turbogenerators, especially the rotors, turned out to be one of the most critical items in the new large power plants. An onsite water leakage would require dismantling the rotor and transporting it back to Asea's factory in Västerås. Accurate fault investigation, repair and re-balancing could take anything from 2 weeks to half a year, then followed by transportation back to the power plant and re-installation of the rotor. The cost for power outages was very high; 1 – 2 weeks equalled the total cost for an extra rotor. This led, not surprisingly, to the manufacturing of a few spare rotors. Ringhals 2 received two extra rotors, Barsebäck one and Aroskraft one, which actually was air-cooled. The latter allowed operation with only 60 percent of rated power [329]. Thanks to all these rotors, the power production could, , be maintained at a fairly acceptable level.

During several interviews it has been learned that the ample resources for development and design were not approved during the late 1960's. The management did not realize the big challenge, which the large turbogenerators constituted. Their reference frames, hydropower generators and smaller turbogenerators, were not applicable. A drastic example was the assessment of the man-hours required for the design of the TVO generator. In an interview, Karl-Erik Sjöström remembered a meeting, in 1972, with the O-division management concerning the quotation of this generator. The preliminary estimation was 12 000 hours, which received strong criticism. The TVO generator was expected to be more or less a copy of the Barsebäck generator, only somewhat larger, so 3 500 hours ought to be sufficient for the design work. The final result, some years later, turned out to be 52 000 hours. Anyhow, the opinions changed quickly when the problems started to occur and Bertil I. Larsson has confirmed that all requests made for additional personnel were approved. The problems had become more complicated and required more advanced theoretical analyses, so consequently employment of graduated design engineers increased [330]. This can be illustrated by changes in the "Design and development department, OK1" staff during 1973. At the beginning of that year, 18 out of 87 were graduated engineers. At the end of the same year the corresponding figures were 27 out of 99, as shown in table 7.4. The situation was thus gradually improved for the GTD generators, but partly at the expense of smaller turbogenerators, at least temporarily.

Considering the amount of difficult problems, it would have been understandable if the company had tried to find external support and know-how. As mentioned in section 7.6.5, such contacts had been very limited and it remained so during the first half of the 1970's. The expertise on large turbogenerators was basically confined to the leading manufacturers and hardly any decision was taken to approach someone of them. Information on competitor's technology was mainly obtained through published reports and articles. The change that occurred in this respect during the latter part of the 70's will be referred to in later sections.

7.8   New generators

7.8.1   The second generation of GTD generators

The design of the first generation of GTD generators, which has been dealt with in section 7.5, represented a radical step in turbogenerator development and it is not surprising that there was room for improvements. There had been an almost fundamentalistic approach to the use of water-cooling for all components even when this was neither technically nor economically the best alternative. Some parts could preferably be air-cooled in a more traditional way. Besides this, a number of faults had occurred during manufacturing, testing and operation of these first generators. All this led to development and implementation of many new solutions in later generators and they were therefore considered a second generation. Table 7.8 contains an overview of all of Asea's 15 original turbogenerators type GTD. Later replacement machines, which often have been upgraded and modified to more recent designs, are not included.

The most obvious differences between the first and the second generation of these generators were the stator core cooling, the cooling circuits in the rotor and the excitation system.

Table 7.8   List of Asea's orginal GTD generators

Plant/unit	Deli-very	MVA	kV	Dr/L [mm]	Exciter	Stator core	Bearings
Aroskraft 4	1973	294.0	17.5	1150/2600	Static	Water-cooled	End-shield
Ringhals 21	1974	506.5	19.5	1150/4150	Static	Water-cooled	End-shield
Ringhals 22	1974	506.5	19.5	1150/4150	Static	Water-cooled	End-shield
Barsebäck 1	1975	710.0	17.5	1250/4900	Static	Water-cooled	Pedestal
Barsebäck 2	1977	710.0	17.5	1250/4900	Static	Water-cooled	Pedestal
Ringhals 31	1977	576.5	21.5	1150/4600	Brushless	Air-cooled	Pedestal
Ringhals 32	1977	576.5	21.5	1150/4600	Brushless	Air-cooled	Pedestal
Forsmark 11	1978	576.5	21.5	1150/4600	Brushless	Air-cooled	Pedestal
Forsmark 12	1978	576.5	21.5	1150/4600	Brushless	Air-cooled	Pedestal
Ringhals 41	1979	576.5	21.5	1150/4600	Brushless	Air-cooled	Pedestal
Ringhals 42	1979	576.5	21.5	1150/4600	Brushless	Air-cooled	Pedestal
Forsmark 21	1980	576.5	21.5	1150/4600	Brushless	Air-cooled	Pedestal
Forsmark 22	1980	576.5	21.5	1150/4600	Brushless	Air-cooled	Pedestal
Olkiluoto 1	1978	825.0	20.0	1250/5900	Brushless	Air-cooled	Pedestal
Olkiluoto 2	1980	825.0	20.0	1250/5900	Brushless	Air-cooled	Pedestal

All are 2-pole, 3000 rpm, power factor 0.85 and have water-cooled stator and rotor windings.

Unlike the water-cooled windings, which have a much higher current density than what is usual in conventionally cooled generators, the stator core has, due to the magnetic saturation, roughly the same flux density as in smaller machines. The specific losses could even be somewhat lower depending upon the high quality electrical steel used and the elaborate punching and stacking processes. From this point of view, it was therefore possible to cool the stator core with air in a large GTD generator also, provided a sufficient airflow could be circulated through the machine. The main advantages were reduced costs for the core and increased reliability due to absence of a lot of water-cooling circuits. Some water-cooling of the end regions remained due to higher losses in these parts. Such extra losses are caused by axial leakage flux. The transition to air-cooling was made during the design of the generators for the Ringhals/Forsmark series of eight units. The responsibility for the design had, at that time, been transferred from the special unit for GTD generators to the ordinary turbogenerator mechanical design office headed by Lennart Stridh (1936-1987). He was a strong advocate for using well proven concepts whenever possible. A 3-D sketch of the new design, actually the upgraded version, is presented in figure 7.49. Even the much larger TVO generators were provided with air-cooled stator cores, but due to the longer active length, the stators had to be divided in inlet and outlet zones as shown in the sketch in figure 7.50 [331].

Figure 7.48 825 MVA, 3000 rpm turbogenerator in Olkiluoto, Finland

Many of the early problems in the water-cooling components occurred in the rotors. The consequence of this was step-wise modification of these components. Some improvements had to be introduced immediately in the faulty rotors, but a more comprehensive redesign was made for the Ringhals/Forsmark rotors. An additional redesign was later made for the TVO generators. The problems related to the rotor cooling circuits and the measures taken are further dealt with in section 7.9.1.

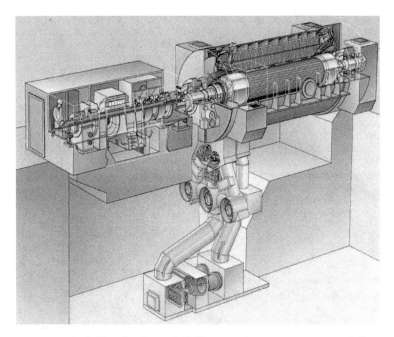

Figure 7.49 Sketch of Ringhals/Forsmark generator (upgraded)

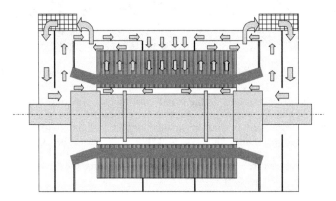

Figure 7.50 Sketch showing air-cooling of TVO stator core

The first generation of GTD generators had static excitation systems in which the DC current was provided from a thyristor rectifier controlled by an automatic voltage regulator. The excitation current, which as an example was around 7000 amperes at rated load in Barsebäck, was supplied to the rotor winding through carbon brushes and sliprings. There were 82 parallel brushes for each slipring. Resistors were connected in series with each brush in order to distribute the total current equally. Both sliprings and the resistor bank were water-cooled. The customers were not happy with neither the design nor the maintenance required, and the question of brushless exciters was raised. At that point in time, brushless exciters had, started to become common for smaller generators. KWU had even introduced them on very large turbogenerators [332]. Therefore, a development project started in 1972 aiming at a brushless excitation system for the next GTD generators. When later, in the winter of 1975-76, a couple of severe short-circuits occurred in the slipring unit of Ringhals 22, it was felt that it had been right to switch to brushless systems.

A brushless exciter consists basically of an outer pole, 3-phase synchronous generator and a rotating diode rectifier. The stationary field winding of the exciter is fed from a small static rectifier controlled by the voltage regulator. The excitation current for the main rotor is generated in the exciter's rotating armature winding and it can thus be supplied to the rotor winding without passing brushes and sliprings. Figure 7.51 shows a diagram explaining the principle.

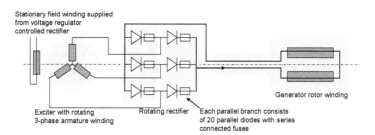

Figure 7.51 Principle of a brushless excitation system

To design a small brushless exciter for a couple of hundred amperes was not very difficult, but to develop one for 8000 amperes continuous current and 16000 as peak current was a challenge. It was necessary to connect 20 large diodes in parallel in each branch of the rotating six-pulse rectifier. The diodes had to withstand a centrifugal force corresponding to 3850 g at rated speed (5500 g at overspeed 3 600 rpm). A rapid fuse had to be connected in series with each one of the 120 diodes in order to prevent that short-circuit failures in single diodes would cause a total break-down. There was also a request that the system should be able to sense and keep track of blown fuses. Asea made an agreement with Vattenfall to change the specification for the Ringhals/Forsmarks generators from a static to a rotating brushless system. The brushless exciter project included tests to verify the mechanical strength of the diodes and fuses and development of an optical system for surveillance of the fuses. A photo of the rotating rectifier is shown in figure 7.52.

Figure 7.52 Rotating rectifier for 8000 A, 3000 rpm

A theoretical analysis of the shaft system showed that the transient torque, during an instant short-circuit, would become too high in case of a rigid coupling between the exciter unit and the generator rotor. The inertia of the exciter was simply too large. Therefore, it became necessary to design a torsional shaft, which was much weaker and allowed the shaft to twist without high stresses in the shaft system.

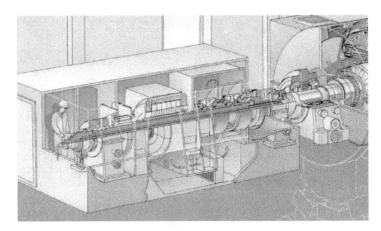

Figure 7.53 Outline sketch of brushless exciter

This complicated the mechanical design as the exciter shaft would consist of two concentric shafts. In addition to this, inside the inner shaft, the concentric inlet and outlet tubes for the rotor cooling would be placed. Furthermore, the outer shaft would also contain the terminal conductors for the excitation current.This torsionally flexible system had to be built and tuned so that it would be able to pass lateral critical speeds without problems, something which turned out to be the most difficult task to solve for the new excitation system. Figure 7.53 presents an outline sketch of the brushless exciter [333].

Another important design change for the second generation of GTD generators was an improved stator end-winding bracing, which allowed adjustments in order to avoid resonance vibrations.

7.8.2 1000 MW and more

The power generation capacity increased rapidly in the early 1970's and the growth projections were almost exponential. Studies of very large generating units were therefore undertaken in many countries. In Sweden generators with the output range 1000 – 1300 MW were foreseen for future nuclear power plants. Units above 2000 MW were considered in other countries, e.g by BBC and KWU according to Cigrè reports in 1972 [332, 334]. Asea and Stal-Laval decided the same year to look into turbine/generator sets of the 1000 MW size.

Asea started a pre-study with the intention to be able to make a customer presentation of a 1000 MW, two-pole generator in the spring of 1973. This study formed also, which was more important, a base for a development project called "Mille", which started formally in November 1973. The objective was to develop a design for turbogenerators from 1000 to 1350 MW and to verify this with a test rotor and a stator test rig. These would have full size diameter but a much shorter length than the real machines [335]. The approval of this expensive project represented a shift in policy compared with earlier development carried out on customer orders. An important reason for this was the technical problems that Asea had encountered during manufacturing and testing the first GTD generators, but it was also a matter of showing commitment to potential customers.

A special development team was set up for the "Mille project" and the design of two generators, one 1000 MW and one 1350 MW started. A forging for the test rotor was ordered from a steel mill. The time plan for the project stated that it should be finished during 1976, but it was stopped at the end of 1974 due to lack of engineering capacity. All resources had to be allocated to solving urgent problems with the generators already on order [336]. Asea's plan was undoubtedly very ambitious. A 1350 MW generator meant at least 1500 MVA and the largest two-pole generator in the world known of at that time was rated 1415 MVA, designed by the USSR manufacturer Electrosila. The most prominent design features of Asea's design concept were direct water-cooling of both stator and rotor windings while all other parts, including the stator core, would be hydrogen-cooled. However, the general development of very large units came to favour so-called "half-speed machines", i.e. four-pole generators, which in most respects are technically easier than the two-pole machines. Asea did never participate in that development.

7.8.3 Direct hydrogen-cooling

The new management for generator development and design, which took over during 1970, naturally questioned whether it had been a correct decision to choose complete water-cooling when practically all other manufacturers built directly hydrogen-cooled rotors. No doubt water-cooling had several advantages and furthermore, there were a number of such generators in the company's order book. BBC and KWU also developed water-cooled rotors, but they had directly hydrogen-cooled as their main alternative. Had Asea taken too large a risk and "put all their eggs in one basket" when it chose water-cooling for the rotors? In addition, such generators were not cost effective for outputs below 500 MW. However, Asea had no resources for developing its own system for directly hydrogen-cooled rotor windings, but become interested when an opportunity arose to acquire the technology from an external source.

Contacts had been established between Asea and the Hungarian company Ganz Electric in 1970. Its chief engineer, Mihaly Wallenstein, had invented and patented a "cross-flow gap-pickup" cooling principle, which Ganz had licensed to the French manufacturer Alsthom. A brief explanation of the system was given in section 7.1.4. Asea evaluated the system and found that it had good technical performance and seemed reasonably simple to manufacture. Ganz offered a license on favourable terms and Asea's executive management decided therefore to let the O-division proceed and sign a license agreement in April 1971 [337].

The first and only generator Asea designed and manufactured of this type, called GTH, had a rating of 284 MVA, 3000 rpm, 16.5 kV, i.e. it was very similar to the GTD generator for Aroskraft. The GTH generator was delivered in 1975 to a district heating power plant in Värtaverket, Stockholm. Manufacturing, testing and operation of this generator were all trouble free. It is interesting to make a comparison between the completely water-cooled generator for Aroskraft and the GTH generator in Värtaverket. The latter has a directly water-cooled stator winding while all other parts are hydrogen-cooled. Some key data for these two generators are given in table 7.9.

Table 7.9   Key data for GTD generator for Aroskraft and GTH generator for Värtaverket

Plant	Aroskraft	Värtan
Output [MVA]	294	284
Power factor	0.85	0.88
Speed [rpm]	3000	3000
Rotor diameter [mm]	1150	1100
Active length [mm]	2600	3400
Efficiency [%]	98.67	98.50
Utilization factor [kVA/rpm m^3]	20.7	17.0

Several years later, in 1978, Asea faced a very difficult situation with the delivery of the large generators for TVO. A study was then performed regarding the possibilities to re-design these generators from water-cooled GTD to hydrogen-cooled GTH. Asea received some support from Alsthom, which had built generators with this type of hydrogen-cooled rotors up till 945 MVA, 3600 rpm. The study showed that a switch could be possible, but the Finnish customer considered the idea as somewhat desperate and declined the proposal.

7.8.4   Foreign turbogenerators

The last nuclear units built in Sweden were Oskarshamn 3 and Forsmark 3. It was mentioned in section 7.2.3 that these were provided with 1150 MW, 1500 rpm units and the 1350 MVA generators were made by BBC. The orders for these large generators were placed in 1976 but they were not commissioned until 1985. They were designed in accordance with BBC's usual concept, i.e. with hydrogen-cooling of all components except for the directly water-cooled stator winding, and with a static excitation system.

The 1970's represented a peak in the development of large turbogenerators even on a global perspective. The unit sizes had increased rapidly but there were still optimistic projections for much larger machines. Oil crises, nuclear power accidents and revised energy prognoses brought the expansion to a halt a few years later. A presentation illustrating the state-of-art in turbogenerator technology during the most intensive period, was given at a VDE meeting in Berlin in November 1977. Friedhelm Heinrichs, technical director for KWU's generator division, presented at that meeting a comprehensive overview of the turbogenerator development, concepts for "border line" machines, limiting factors and growth projections. His presentation was partly based on documentation he had received from colleagues at other leading manufacturers [338]. The number of independent designs of these large generators was rather limited. Friedhelm Heinrichs mentioned GE and Westinghouse in the USA, GEC and Parsons in England, Alsthom, BBC and KWU in West Europe and Elektrosila and Elektrojashmasch in USSR. In addition to these, he also included Asea. There were other manufacturers of very large turbogenerators in Europe and in Japan, but with designs originally based on licenses from some of the companies mentioned above.

Basis for Heinrichs' overview was the cooling systems, in most cases water-cooled stator windings and hydrogen-cooling of other parts including the rotor windings. GE, Alsthom and Elektrosila had chosen gap pickup cooling for their largest 2-pole rotors while the other manufactures had various types of axial cooling of the rotor windings (see section 7.1.4). The largest 2-pole generator by then had a rating of 1333 MVA, 3000 rpm built by Elektrosila. Most of the manufacturers were also building 4-pole generators with axially hydrogen-cooled rotor windings, but KWU and to some extent BBC and Elektrojashmasch also used water-cooled rotors. The most powerful 4-pole generator in operation at that time was KWU's 1530 MVA, 1500 rpm unit in the German nuclear power plant Biblis. A cross section of this machine is shown in figure 7.54.

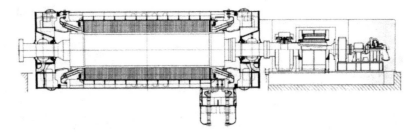

Figure 7.54 Cross section of Biblis 1530 MVA generator (KWU)

Elektrosila was the only manufacturer, except Asea, mentioned in connection with directly water-cooled 2-pole rotors. Elektrosila's concept was smart but somewhat drastic, typical for Russian designs in those days. Cooling water was sprayed into an open rotating chamber, from which it was pumped through the winding by the centrifugal force. The water, which had passed through the parallel circuits, was let out into a stationary water chamber and pumped back to the external system. The air inside the generator was replaced with nitrogen to avoid the risk for corrosion. Figure 7.55 illustrates this concept. There were also other manufacturers building prototypes with directly water-cooled rotors, e.g. Hitachi, but that was not included in the presentation [339].

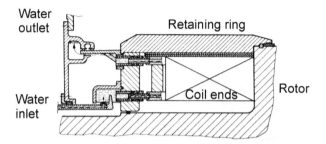

Figure 7.55 Sketch showing Elektrosila's water-cooling concept

The discussion on the possibilities to further enhance the generator size, including related problems, was based on a well known expression for the electrical machine power, namely equation 4-19. The diameter of the most critical component, the rotor, was limited by centrifugal stresses, and the length of how slim the rotor becomes. The stator dimensions were often limited by transport restrictions. The rated power could, except from geometric size, be raised through the increase of the linear current loading, which required improved cooling. It was then important to design the stator end zones with consideration to the large electro-dynamic forces and the concentrated losses created by the leakage flux. The 2-pole generators were, in almost all respects, more difficult than the 4-pole machines and therefore hit the "ceiling power" first. Mr. Heinrichs concluded that the maximum size for a 2-pole turbogenerator could probably be stretched to 2000 MVA while the corresponding value for a 4-pole machine would be 3000 MVA. He finally described a KWU project, which would lead to 2-pole generators rated 3000 MVA through the use of a superconducting rotor winding and a water-cooled airgap winding in the stator. Several manufacturers developed and even built experimental machines with superconducting rotor windings in the 1970's. A review made by Asea, in 1980, describes such projects in the USA, Japan, several West European countries and the USSR. Tests had already been performed with 20 MW, 2-pole machines and there were firm plans to build prototypes in the 200-300 MW size. Common for these demonstration projects were that the rotor windings were cooled by liquid helium and the stator windings were so-called airgap windings [340, 341].

It is now almost 30 years since Friedhelm Heinrichs, an internationally renowned generator specialist, presented this lecture. It was an excellent summary of the status and the possibilities in turbogenerator technology. It represents still state-of-art concerning generator sizes. The world's largest 2-pole generator, currently in operation, is claimed to be Leibstadt in Switzerland 2006, built in the 1970's but later upgraded to 1318 MVA. The largest generator on order, up till now, is the 4-pole generator, which TVO has ordered from Siemens, for the third unit in Olikiluoto. It will be put in operation in 2010 and it will be rated 1910 MVA, 1720 MW, 1500 rpm and 27 kV [342].

7.9 The major problems

The development, manufacturing and operation of the GTD generators initially created many problems, both technical and commercial. Many could probably have been avoided through a slower development pace and more comprehensive prototype tests, but several problems were shared with other, even larger manufacturers. The 1960's and 70's constituted a learning period for the generator industry and the knowledge increased partly through some generic failures, some of them very spectacular. One example is the British stator end winding problems mentioned in section 7.5.2. Asea contributed with some others.

7.9.1 Water leakages in cooling tubes

The first GTD rotors suffered repeatedly from water leakages, first in insulating sealings, but most of them from small cracks in the cooling tubes in the end section of the rotors. Analyses of different cracks showed that they were usually caused by mechanical fatigue. The tubes were subjected to both rotational speed and start-stop frequent deflections that initiated and propagated the cracks. The leakages started to occur in 1973 and continued frequently in 1974. It was mainly the rotors for the Ringhals 2 generators and the first Barsebäck generator, which suffered from leakages, a few during workshop tests and several during operation at site [343]. The design was improved, and the following generators for Ringhals 3 and 4 as well as Forsmark 1 and 2 experienced practically no rotor leakages.

A large but slim generator rotor will bend considerably due to the gravitation force, especially the weaker shaft ends. This means that the distance from a point at the shaft end to another point at a coil-end, carried by the retaining ring, will vary depending on if it is on the upside or downside, see figure 7.56. All the tubes going from a water manifold at a shaft end to various inlets somewhere at the coil-ends must withstand these deflections (see figure 7.58). If they fail, it is impossible to just make them strong enough; the only way is to make them more flexible. Asea modified the design of these cooling tubes in a number of steps trying to increase the flexibility and reduce the dynamic stresses as much as possible. There are actually two types of dynamic stresses. First the high frequent stresses, depending on the bending of the rotor as explained above. In addition, there is also a larger, start-stop frequent deformation, due to the radial expansion of the retaining rings and the coil-ends, caused by the centrifugal forces. Both had to be taken into account in the design. The photos in figure 7.59 show some alternative solutions that Asea developed for these cooling tubes. The last alternative with a skipping-rope shaped flexible tube eliminated all bending forces. Figure 7.57 also shows that the axial stresses in the rotor and the shaft ends will vary between compression and tension with the rotational frequency. An axial tube placed along the surface of the shaft end will hence be subject to a tensile fatigue stress, which magnitude depends on how the tube is allowed to move.

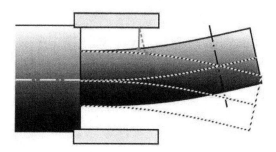

Figure 7.56 Rotational deflection between shaft end and retaining ring

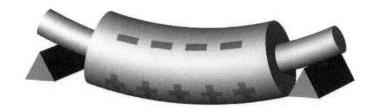

Figure 7.57 Dynamic stresses caused by bending due to gravitation

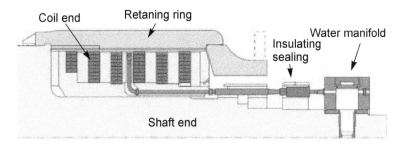

Figure 7.58 Tubes between water manifold and coil

Figure 7.59 Alternative designs for water connections to coil ends

A number of leakages occurred in the rotors for the TVO generators during a period from May 1977 until January 1979. The first leakages were found during balancing and workshop tests while the later ones were detected during operation in the power plant. Practically all of them depended on fatigue cracks in welded stub pipes at the circular water manifold as shown in figure 7.58. The reason for these faults was a combination of low quality welds and axial fatigue stresses as explained above in connection with figure 7.57. The remedy was a redesign of the stub pipes and introduction of curved tubes with more flexibility, which is shown in figure 7.60 [344]. Karl-Erik Sjöström reminded the author of a meeting in which an elderly Finnish professor in mechanics assisted TVO. Asked by TVO if he believed Asea could solve the actual problem, the professor replied: "Oh yes, provided Asea's young engineers realize that they cannot keep this large rotor straight by the help of those tiny tubes."

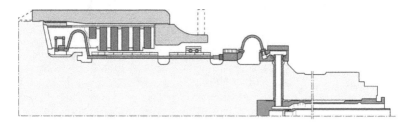

Figure 7.60 Water manifold, insulating sealings and cooling tubes with flexible loops

There were also a few rotor leakages in other parts of the system, and due to other reasons as well, but it will not contribute to this thesis to discuss these. The short review given above illustrates sufficiently a major problem with directly water-cooled rotors and that the development engineers gradually learnt from the mistakes by making the design flexible enough to avoid further fatigue cracks. The problems addressed in early test rigs focused on risks for erosion and galvanic corrosion, not at all on dynamic mechanical stresses. In addition, the test rotors were not big enough to cause representative dynamic stresses in the tubes.

7.9.2   Stress corrosion in retaining rings

Unit no.1 in Barsebäck was running normally with full load in the morning hours on Good Friday April 13, 1979. The operators on duty had, during 15 minutes, noticed a slight increase of the vibration level on the slipring side generator bearing, when the vibrations suddenly increased drastically. The unit tripped and a fire alarm was received from the turbine hall. After the fire had been extinguished, it was possible to see what had happened. The generator was severely destroyed and the generator end of the turbine hall was also damaged. The photos shown in figure 7.61 and 7.62 were taken the day after the failure had occurred.

Figure 7.61 Barsebäck turbine hall after retaining ring failure
(Figure 2.1 shows the same generator before the failure had occured.)

Figure 7.62 Barsebäck generator after retaining ring failure

An inspection would soon verify what had happened. It was a matter of a retaining ring explosion. The slipring-side retaining ring had broken into three pieces, which were thrown out through the stator end-winding and the generator end-cover. One of the heavy pieces hit the pedestal bearing so lubrication oil was sprayed around. The short-circuit of the winding caused arcing that ignited the oil and created the fire. The investigations would very soon focus on the reasons for the retaining ring failure.

The examination of the fractured surfaces revealed a primary crack caused by stress corrosion and secondary ductile fractures due to sudden overload as shown in figure 7.63. The retaining rings were made of a special, high strength, non-magnetic, austenitic steel and it was known that it could be sensitive to stress corrosion if it was exposed to water in combination with high stresses. The reason for choosing non-magnetic retaining rings was that they reduced the leakage fluxes in the end regions and hence stray load losses. Stress corrosion is inter-crystalline and exhibits a characteristic branching pattern. Figure 7.64 contains a photo from a metallurgical analysis of the actual material, which shows typical stress corrosion cracks.

Figure 7.63 Retaining ring fracture surface

Figure 7.64 Stress corrosion sample

The generator had been in continuous operation for almost one year when the failure occurred. No water leakages had been detected, but were nevertheless a matter for further investigations. The metallographic analyses indicated that the stress corrosion crack had grown over 6 – 9 months and the area where the crack started, on the inside of the ring, was not ventilated. A non-detectable micro leakage could have moistened the insulation material in contact with the ring. Extensive investigations showed, with a very high probability, that this had been the cause [345].

Many manufacturers of large turbogenerators used the same retaining ring material. It had almost become a standard and there were only a few steel works in the world that could forge such rings. Excerpts from the material specification are presented in table 7.6 (section 7.5.3). The high yield strength was obtained through a substantial cold expansion of the rings.

Asea was not the only turbogenerator manufacturer which experienced a retaining ring failure. The British manufacturer C. A. Parsons reported such a failure in 1974 in a 500 MW, 3600 rpm generator in Nanticoke Generating Station, Ontario [346]. It was a hydrogen-cooled generator and the retaining rings were made from a ferritic NiCrMoV steel with 3.2 percent Ni and 0.8 Cr, i.e. a "low-alloy" steel compared with the rings for Barsebäck. The failure in the Nanticoke generator depended on high local stresses in those parts of the retaining ring that belonged to the bayonet locking system, combined with a reduced crack initiation level, due to the hydrogen atmosphere. Failures like this enhanced transition from ferritic to the more expensive austenitic retaining rings.

BBC had supplied a completely water-cooled turbogenerator to Skaerbaek in Denmark. This machine had a catastrophic retaining ring failure in 1973, which was attributed to stress corrosion caused by water leakage. KWU experienced two failures, both in 1978. One of the rings burst at fairly low speed during balancing in the factory. Both BBC and KWU had used the same material as Asea. Even generators from other manufacturers such as English Electric, GE and Westinghouse had been destroyed by retaining ring explosions. A study presented at an EPRI workshop on "generator retaining rings" in October of 1982 reported 38 fractured rings [347, 348]. The matter of cracked retaining rings had, up till the Barsebäck accident, not received much public attention in the industry. The manufacturers tended to keep most information to themselves. However, Asea and KWU had, during the second part of 1979, two meetings were a few specialists from both companies exchanged information and discussed possible solutions [349]. Asea also turned to Krupp Metall- und Schmiedewerke in Essen, which had supplied the ring. In 1982, Krupp invited some generator manufacturers to a colloquium addressing the retaining ring failures. Krupp presented at this occasion its new, so-called P900 material, first developed for offshore applications. It was an austenitic material with 18 percent Ni and 18 percent Ma, which would be stress corrosion resistant in water and humid atmosphere. P900 became soon introduced as retaining ring material and has, since then, been widely used for such rings, especially by European manufacturers. For instance, Asea has successively replaced old rings with those made from the new material. Other steelworks have later introduced similar materials and it has now become a branch standard.

The severe failure in Barsebäck implied a repair almost corresponding to the manufacturing of a new generator, a process which would require at least 1½ year. The outage costs would be horrible. Anyhow, a provisional solution was found, which reduced the "out of service" period to five months, thanks to the political situation in Sweden at that time. Nuclear power was a huge problem for the government coalition and it had enforced a special law in 1977, the so-called nuclear power moratorium, which meant that the power companies were not allowed to put any new reactors in operation until they had demonstrated an acceptable method to dispose the nuclear waste. The new units delivered to Forsmark were consequently not in service. A quick study revealed that it could be possible to use a Forsmark generator in Barsebäck as a temporary replacement. It was indeed somewhat smaller, 577 MVA instead of 710, but a Forsmark generator, together with its transformer, could, by means of some mechanical adaptations, be used and produce around 550 MW. Sydkraft managed to negotiate a rental agreement with the Forsmark company and this saved approximately one year of operation.

7.9.3   Cracks in rotor bodies

Cracks in electrical machine rotors can be disastrous, especially in the case of large machines and high-speed machines, but cracks can also destroy smaller machines, e.g. standard industrial motors. A large turbogenerator is of course at extra risk in this respect, a fact that was clearly illustrated by the retaining ring failure described in the previous section. Cracks can also occur in rotor teeth, wedges, shaft ends, fan blades, rotor cores, rotor centers etc. A common denominator for these types of faults is that they usually result from mechanical fatigue. Examples of this are the fatigue cracks in various cooling tubes described in section 7.9.1. Asea also experienced some serious rotor body cracks in the generators supplied to TVO in Finland, a situation which required very special technical and commercial measures before it was solved. The solution involved the use of advanced and partly new theoretical tools as well as methods for monitoring and inspection. It is therefore motivated to address this problem in some detail. In order to understand the sequence of failures and corrective measures, it is helpful to know that three rotors had been manufactured for the two generators at Olkiluoto, but it is not necessary for the readers of this thesis to keep track of them.

The failures started with a water leakage, in a rotor terminal conductor, detected in the power plant in November 1979. During the repair of the broken conductor, ultrasonic and dye-penetrant tests revealed cracks in the wedges and in the corresponding slot sides carrying the axially inner section of the terminal conductors. Similar cracks, which were attributed to fretting fatigue, were found also in the other two rotors later on. The cracks seemed to be dormant, but the wedge profile and material were changed anyway and all the TVO rotors were successively modified in this respect [350]. Fretting fatigue in turbogenerator rotors had earlier been reported by other manufacturers. Late in 1976, Parsons' detected a large rotor crack in a 660 MW generator at Drax power station in U.K. The crack had been initiated by fretting fatigue starting at the ends of pole slot[*] wedges [351].

---

[*]. Pole slots are milled axially in the poles of 2-pole rotors in order to avoid double-frequent vibrations due to different stiffness in direct and quadrature directions. The pole slots are filled with magnetic material.

The first TVO rotor had been returned to Asea, in March 1980, for the modification described above. The inner coils were removed to enable the necessary work and this made it possible to perform further crack detection. These tests revealed new cracks, in this case located at the bottom of the winding slots right at the end of the rotor, as shown in figure 7.65. These cracks had been initiated from a sharp notch at the root of each tooth where a recess had been made for insertion of extra electric insulation at the slot opening. This was a design that Asea had used successfully for many years. The risk for fatigue had been considered non-existent because this was an area under compression from the retaining ring shrink fit. A method was quickly developed for ultrasonic crack inspection from the rotor surface. The inspections showed similar cracks in the rest of the teeth and also at a rotor still in use in the power plant. Two out of three rotors had these cracks, namely the two oldest that had the most operation hours and start-stop cycles. Both rotors with cracks had been in operation for a few thousands hours and had been subjected to more than 100 start-stop cycles. It was decided to repair the faulty rotors simply by removing the cracked zone through machining and modifying the design to get rid of all stress-rising notches etc. This shortened the active length by less than three percent and would not reduce the generator performance [352]. The repairs were quite time-consuming requiring manufacturing of new retaining rings, necessary machining, complete rewinding and balancing. In the mean time, one of the TVO units could be kept in operation with the third rotor. The same type of cracks could, however, be expected to occur in this rotor and it was therefore important to carry out ultrasonic inspections at regular intervals, and to carefully monitor the rotor vibrations.

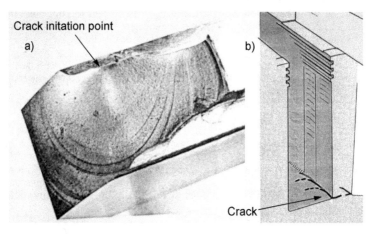

Figure 7.65 a) Sample of fatigue cracks removed from TVO rotor. b) Position of crack at rotor slot end

The turbogenerators in Olkiluoto were not the first that had this kind of problem. The British generator at Drax, already mentioned above, had a fatigue crack which covered almost half of the rotor cross-section, when it was stopped due to excessive vibrations after around 16000 hours of operation [353]. Another example was a 700 MW generator in the Aramon power plant in France which had been in operation for a year when it was stopped late in 1979 due to increased rotor vibrations. The shaft end on the exciter side was more than half broken by a fatigue crack that had started at a radial hole for a rotor terminal conductor. The generator had been manufactured by BBC's French subsidiary CEM. BBC had built many generators with the same design and the fault in Aramon was therefore attributed to a random material defect [354]. The most severe fault, however, was reported in 1980 from another French power plant, Porcheville, where a 666 MVA, 3000 rpm generator rotor had exploded and destroyed not only the generator but also major parts of the turbine and the building. Figure 7.66 shows the end section of the totally destroyed generator. The reason for the fault was, also in this case, fatigue cracks in the rotor body. The Porcheville generator was manufactured by Jeumont-Schneider [355].

Figure 7.66 Turbogenerator rated 666 MVA, 3000 rpm in Porcheville destroyed by cracked rotor

Asea had experienced the retaining ring explosion in Barsebäck in the spring of 1979 and reports were also received about other manufacturers' problems. The risks for catastrophic rotor failures were therefore a matter of great concern and a special study was initiated by the manager of the development and design department in October 1979, i.e. just a few months before the cracks in the TVO rotors were detected [356]. The study was never completed because the real events soon took over. The situation became extremely serious and comprehensive investigations were started in order to find a correct explanation for the initiation and propagation of the discovered cracks. Extensive know-how was built up, the rotor designs were improved and new methods for inspections and monitoring were developed. It can be argued that Asea had been too ignorant before, but it seems as if other manufacturers had acted in similar ways.

Why had the cracks in the TVO rotors occurred? It was obvious that they had started in a sharp notch, but no tensile stresses had been anticipated right there. FEM calculations were performed to determine both high frequent and low frequent stresses. It was found that the centrifugal forces actually changed the high compression stress (- 570 MPa), which the retaining ring shrink fit caused at stand still, into a tensile stress (120 MPa) at full speed. The latter was further enhanced when the retaining ring obtained a higher temperature during operation. The result of the stress analysis is shown in the diagram in figure 7.67. The question of whether these dynamic stresses could initiate a crack was subject to both theoretical and experimental investigations. Fracture mechanics had been developed during the 1970's as a new discipline within solid mechanics and Asea used it, for the first time for electrical machines, in connection with the TVO rotor problems. Janne Carlsson, professor in solid mechanics at KTH and specialist on fracture mechanics, was engaged as a consultant. Both theoretical analyses and tests on specimens showed that approximately 10 start-stop cycles were sufficient to initiate a crack in a sharp notch while it would require about 5000 cycles if there was a radius of 3 mm in the recess [357].

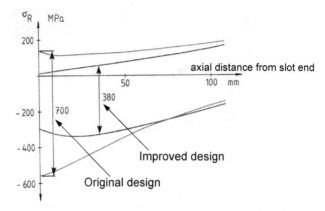

Figure 7.67 Low frequent radial fatigue stresses in slot end portion of TVO rotors

Fracture mechanics was used to calculate the threshold crack depth that was required for an initiated crack to start propagating due to the high cyclic stresses, which were roughly +/- 20 MPa. The threshold crack depth was found to be a matter of a few tenths of a millimeter. Attempts were also made to determine the crack propagation growth rate, and the critical, instable crack size by means of fracture mechanics. The stress pattern was complicated but the analysis indicated that it would take in the order of 10 months for a crack to propagate to critical size, and this size was slightly more than half of the cross section. The rotor stiffness would change significantly, long before a crack could become critical, so accurate vibration monitoring should prevent a dangerous situation from occurring [358, 359].

It was mentioned earlier that the third rotor kept one of the TVO generators in operation while the first two were modified. When this rotor was taken out of service in November 1980, it had developed the same cracks as the first two have had. A few months later, January 31st 1981, the repaired rotor no. 1 was stopped due to high shaft vibrations. An inspection revealed a large crack in the shaft end underneath the water manifold. The crack, caused by fatigue, had started from one of the radial holes for supporting bolts and had propagated half way through the shaft end, see figures 7.68 and 7.69. This was a fault, very similar to that which had occurred in the French Aramon generator referred to above. The investigations that followed had difficulties to prove why the crack had been initiated, but the propagation phase was pretty much a straightforward case [360]. In December of 1981, a French turbogenerator, similar to the Aramon, but installed in a power plant in Cordemais, was damaged in exactly the same way as the Aramon generator and very similarly to the last TVO failure. The Cordemais generator had been built by CEM, which at that time had been taken over by Alsthom. Alsthom's technical director for electrical machines, Gilbert Ruelle, contacted Sture Eriksson and invited Asea for a discussion about these failures. A first informal meeting was held in Paris in February of 1982 and a second was arranged in Stockholm at the end of June, with the purpose to exchange information between some key persons from Alsthom, BBC, EdF and Asea. It became evident that the cracks, in the French generators, had been initiated by fretting fatigue in some threads and then continued to propagate due to normal fatigue [361]. One conclusion was that Asea, at that point in time, seemed to have penetrated the actual problems more thoroughly than the others, especially the question concerning vibration monitoring. There were some strong reasons for this.

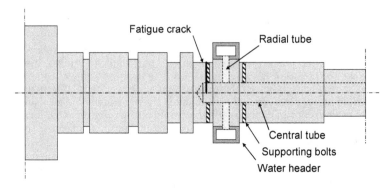

Figure 7.68 Position of crack near water mainfold

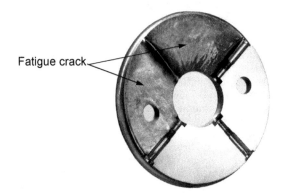

Figure 7.69 Cracked shaft end. The disc has been removed from the actual rotor

Asea had decided to build two new rotors, with all the necessary improvements, for TVO. They were under manufacturing when the last failure happened, but they were still several months from being finished. The rotor with the shaft end crack was beyond repair and the only available spare rotor was rotor no. 3, which still had the cracks described earlier. The question was: could this rotor be used for further operation thus avoiding a very costly down period? According to the fracture mechanical analysis, it should still have a few months of life left before the cracks became critical. Was enhanced vibration monitoring, combined with regular ultrasonic inspections, sufficient for avoiding a disastrous failure? Several opinions were expressed but the decision was finally made to use the cracked rotor. It was then in operation for several months until it had to be taken out of service due to increased vibrations. The cracks had then reached a depth of approximately 200 mm below the slot bottom [362, 363].

The vibration monitoring was a key factor for the operation. A system had been developed, by Stal-Laval, in which bearing vibrations in horizontal, vertical and axial directions, as well as touchless measurement of radial shaft end deflections, were recorded and processed online. Traditional vibration monitoring worked with one alarm and one trip level. This was not considered to be a safe method because a change of rotor stiffness, due to a large crack, does not necessarily increase the vibration amplitude but it will change the vibration vector, i.e. both amplitude and phase had to be considered. The established criteria for alarm and trip were, therefore, a change of any of the various vibration vectors recorded, outside the circles representing these two levels. The principle is illustrated in figure 7.70. An online communication link was set up between Olkilouto and Stal-Laval in Finspång, which had the main responsibility for the entire turbine/generator set [364].

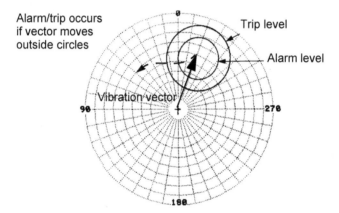

Figure 7.70 Vibration vector diagram with limits for alarm and trip

The vibration monitoring, introduced at TVO during this critical period, has later become a standard praxis for large power plant turbines and generators. It is of importance to note that the new rotors, which were installed later in 1981, have performed without any problems. The design was improved at a number of points resulting in much higher safety factors with respect to all types of fatigue stresses [365]. The cracks in the original rotors had practically nothing to do with the water-cooling of the rotors, but rather with traditional extrapolation difficulties.

7.9.4 Costs

The GTD generators, during the 1970's and the early 80's, created a huge cost problem for Asea. It has been mentioned in section 7.7.3 that the orders for the first four machines resulted in a deficit of 20 MSEK. Also the following orders gave negative results even if they were percentage wise significantly better. The pre-calculations had been improved based on experience from the first generators and the organization had also started to climb along the learning curve. It was instead the costs, caused by the major technical problems as those mentioned above, that became a serious matter. Examples of activities that caused such costs are trouble shooting, design and manufacturing of new rotors, as well as extra fast transportation and installation. The total costs reached such a magnitude that they required a special comment in Asea's annual report for 1980. The amounts were not specified, but the CEO's remark was *"The company's result has been considerably reduced by disturbances in the supply of the generators to the nuclear power plant TVO in Olikluoto, Finland."* [366]. There was never any accumulated amount presented of Asea's total costs for the GTD generators, but a rough estimation, made in the early 1980's, indicated around 200 MSEK.

7.9.5 Confidence

When Asea celebrated its 100 year anniversary, in 1983, it published a comprehensive company history, "Asea during 100 years", written by Jan Glete. He wrote in a chapter concerning the development and delivery of nuclear plants: *"…. This (the turbogenerator) was the only part of the company's nuclear deliveries, which the customers showed an explicit discontent for. "* (My translation) [367]. This is of course understandable taking into account all the serious technical problems, and considering that Glete did most of his research during the period when these problems were extra critical.

Since the beginning of the century, Vattenfall and Asea had developed a trustful customer - supplier relationship. Power plant generators were no exception in this respect. There had been problems from time to time, but these had been solved and Asea had, in reality, a position as most preferred supplier, not very unusual for a big domestic supplier. The GTD difficulties put the relationship under certain stress, but Vattenfall recognized Asea's determination and efforts to solve the problems. Uno Jonsson mentioned in a letter, as an example, Asea's willingness to supply a spare rotor [314]. If Vattenfalls decision, in 1975, to choose a 4-pole generator from BBC for Forsmark 3, shall be seen as lack of confidence in Asea is arguable, but the primary reason for this choice was that one large turbine/generator unit gave better economy than two parallel units. This excluded, in practice, a generator from Asea, which had no experience from large 4-pole generators. This is confirmed by a letter, dated Jan. 23, 1976, from Vattenfall's technical director Ingvar Wivstad to his Asea colleague Torsten Lindström [268]. Vattenfall was, of course, worried about the frequent failures in Ringhals 2, and decided together with Asea, in September of 1978, to set up a mutual working group for reviewing all the faults that had occurred and proposing and implementing improvements. The group, which was chaired by Bertil Nilsson, chief engineer within Vattenfall, presented its comprehensive final report in June 1979 [368].

A more significant proof of damaged confidence came in 1980 from the Finnish customer TVO. For several years, TVO's management had been worried by the Swedish generator problems and had even investigated possibilities to change the generators made by Asea. The final straw was when the rotor cracks were found late in 1979. In February 1980, TVO ordered a complete replacement generator from BBC and in August of the same year, they ordered a second one. Esko Haapala, former technical director for TVO, mentioned in an interview that TVO's board had been skeptical about the water-cooled rotors already from the beginning. Curt Nicolin had then assured TVO that he had confidence in the design, but a special clause was included in the contract which gave a possibility to switch to a hydrogen-cooled generator until the end of 1975. TVO later took up this issue, but a change would have resulted in too long delays, so it was instead decided to build a spare rotor. TVO's main concerns were the size of the generators, the largest ever built by Asea, and the water-cooling. Mauno Pavola, employed by TVO as mechanical specialist in 1975 and later the president for the company, explained that the generators were a permanent headache for several years. The consulting firm, Merz &McLellan, had even recommended a change to hydrogen-cooled machines. Due to the rotor leakages that had occurred, TVO engaged, from 1977 on, Professor Erwin Krämer from Darmstadt Technical University as a consultant and specialist on rotor dynamics and solid mechanics.

TVO took the retaining ring failure in Barsebäck very seriously and then, when fatigue cracks were found in TVO's own generators, the situation became almost desperate. For a long time, Asea and TVO had weekly review meetings in Västerås while temporary measures were taken to secure operation and new rotors were manufactured. Frequent meetings were also held with TVO's executive management in Helsinki in which both technical and contractual items were discussed. Asea delivered and installed the new rotors in 1981 and both Haapala and Paavola confirmed that TVO has since then been very satisfied, both with the generators and the cooperation with Asea. However, in order to obtain TVO's acceptance for the generator delivery, Asea had to sign an agreement in which it took some years of responsibility for payment and installation of the spare generators in the case Asea's generators should continue to have failures. Luckily enough, there were never any reasons for this.

If Asea's centennial history had been written a few years later, the judgment about the large turbogenerators would probably have been different, because the major problems had successively been solved and had become part of the company's history by the early 1980's. The generators have since then been very reliable and the customers have expressed their satisfaction.

7.9.6   Competitor contacts

The title of this section is perhaps misleading. Contacts with other turbogenerator manufacturers were not a problem, but they were, in many cases, established due to existing problems. It has been mentioned in section 7.6.5 that direct discussions with colleagues from other companies were rare during the development of the first GTD generators. The only forum for this was Cigrè's study committee for large rotating machines, and this remained an important contact arena, also during the later 1970's and through the 80's. Karl-Erik Sjöström, who replaced Sture Eriksson as the Swedish member on this committee, mentioned that the exchange of technical information with German and British specialists was especially valuable. A prerequisite for an open discussion about somewhat sensitive matters was the collegial atmosphere, built through participation in the committee meetings during several years. The relationships established in this way formed also a basis for bilateral contacts regarding certain subjects.

Some examples of bilateral contacts have already been mentioned. Stress-corrosion in retaining rings (section 7.9.2) was subject of discussions with KWU. Cracks in rotor bodies (section 7.9.3) were a matter for meetings with Alsthom and BBC. These meetings were based on mutual exchange of information without any commercial or legal implications. Asea also had written agreements with some of these manufacturers. In the end of 1975, Asea and BBC signed an agreement concerning exchange of technical information and a number of mutual working groups were established in 1977. This agreement included also an option for Asea to acquire a license on large hydrogen-cooled turbogenerators [369]. As mentioned in section 7.8.3, Asea had already, in 1971, signed a license agreement with Ganz concerning gap-pick-up cooled rotors. This technology was later further developed by Alsthom, and Ganz recommended therefore Asea to discuss an extended agreement with Alsthom. As a result, Asea sent in 1978 a "letter of intent", including a draft license agreement, to the French company whereupon further negotiations followed. Asea's intention was to receive an option, to have as a back-up, in case it should be necessary to convert the TVO generators to hydrogen-cooled machines [370]. Neither the agreement with BBC, nor that with Alsthom had any major impact on Asea's continued turbogenerator development, but the contacts with these manufacturers gave a better understanding for certain problems.

Asea also had a number of meetings with USSR manufacturers. In this case, the initiative came from the Russian side. A couple of their specialists were very active in Cigrè, but most of them had few possibilities for contacts with western companies. They therefore tried to arrange bilateral meetings in USSR with some foreign companies. Small delegations from Asea visited the

"All-Union Research Institute of Electric Machinery" and Electrosila's factory, both located in Leningrad, at two occasions, in 1975 and 1980 respectively. Specialists from Electrojashmasch in Charkov and Sibelectro in Novosibirsk participated also in these meetings. A small group from USSR visited Asea in Västerås in 1978. Subjects, which received extra attention during these discussions, were stator-winding vibrations, cooling methods and systems, brushless excitation systems, and retaining ring fractures [284, 371, 372]. Leader for the institute and the USSR delegation was Academician Igor A. Glebov (1914-2002), internationally recognized electrical machine specialist.

7.10 Changes during the 1980's and later

7.10.1 New company structures

Asea's O-division had kept its organization with only minor changes during the critical 1970's. Erik Lundblad (1925-2004), efficient troubleshooter but also scientist, had replaced Eric Sjökvist as president in 1975. Bertil I. Larsson became the sales manager in 1976, and Sture Eriksson took over as manager for the development and design department. New manager for the turbogenerator office became Karl-Erik Sjöström. Continuity was maintained as much as possible in the problematic turbogenerator field. Most of the engineers also remained during these years. The first changes came during the early 1980's when the difficult technical problems had been solved. The new Asea president, Percy Barnevik, had started to reorganize the company and the O-division was transformed to Asea Generation, first as a division and later as a daughter-company. The O-division's old functional organization was replaced with three parallel business units. Some of the managers moved to new positions but Sjöström remained in a key position, with respect to the turbogenerators, for more than two decades.

The situation changed radically in 1988 when Asea and BBC merged and formed ABB. Asea had had, for some years, a very positive development, while BBC had been less successful financially. The Swedish company therefore seemed to have a certain advantage when the merger was to be practically implemented. This was however not the case for large steam turbines and turbogenerators. The relationship between the two companies could instead be likened to that of a big brother versus his little brother. The former BBC unit could refer to a much more comprehensive reference list for large machines as well as larger engineering resources. Karl-Erik Sjöström, who participated directly in the coordination, has explained that in the case of hydropower generators, a working group tried to find the best solution based on the different local designs, but in the case of turbogenerators this was not considered necessary. The Swiss business unit management considered its own technology as superior and the product responsibility, for large turbogenerators, was allocated to Switzerland. The unit in Västerås was only entitled to deal with service and refurbishment of existing machines. Customer preferences have been decisive when ABB Sweden has received orders on complete GTD generators in connection with the upgrading of the power plants. The Västerås factory also obtained a limited market for its own air-cooled turbogenerators, < 200 MW. According to interviews, the technology developed in Switzerland was considerably more expensive.

A decade later it was time for the next big structural change. ABB and Alstom announced, in March of 1999, that they merged their power generation activities into a new company, ABB Alstom Power, owned equally by the parent companies. ABB's reasons for the merger were weak results for its large "Power Generation Segment" and a new strategy focusing more on expanding IT related business areas such as "Industrial Automation". Alstom, on the other hand, was particularly interested in accessing ABB's gas turbine technology because its long license cooperation with GE had been terminated by the latter [373]. Suddenly three different designs for large turbogenerators existed in the same company. The consequences were on this occasion larger for the Swiss and French units than for the Swedish one. It was necessary to reduce the number of generator designs. The former Alstom factory in Belfort was given responsibility for air-cooled generators and large 4-pole generators, while the factory in Birr, Switzerland kept all other types with the exception of the completely water-cooled GTD generators, which remained in Västerås.

Already the following year, in May of 2000, Alstom acquired ABB's part of the "Power Company" which thus became an internal part of Alstom. The relationship between ABB and Alstom had become tense, especially due to unexpected high costs for faults in delivered large gas turbines. ABB's CEO, Göran Lindahl, had, from the beginning, wanted to divest this business segment which did not fit into his new strategy. ABB therefore negotiated an agreement about a complete take over. Whether this strategy was right or wrong is up for discussion; several opinions have been expressed regarding this matter. Nevertheless, it is now only possible to conclude that a centenarian long tradition of building hydropower- and turbogenerators with Asea's, and later ABB's, trademark was broken. A discussion had started, already before ABB left the cooperation, that the Västerås factory ought to be closed. Alstom , wanting to further rationalize its production, fulfilled this intention and decided consequently to stop the manufacturing of generators in Västerås. Objections from the Swedish unit did not help. Alstom had serious financial problems and needed support from the French government, a situation which didn't favour keeping factories in Sweden. However, the factory in Västerås was taken over, in late 2000, by a new company, GenerPro AB, established especially for contract manufacture of large electrical machines and components. Alstom has, in spite of this, kept a unit in Västerås for service and rehabilitation of both turbo- and hydropower generators. This unit still has the responsibility for the GTD generators, not only for their service but even for the sale and design of new generators, as long as such are requested by the customers and can be treated as replacement for existing units. The market for GTD generators are thus in reality limited to those power companies, which already have such machines. These companies have used the possibility to increase the output of their nuclear power plants by upgrading the main components, i.e. reactor, steam turbine, generator and transformer. This has, in some cases, required acquirement of new generators and Alstom's Västerås unit has managed, in open competition, to secure these orders. The manufacturing of the new GTD generators is nowadays outsourced. The rotors are made by GenerPro in Västerås and the stators by Alstom's factory in Wroclaw, Poland. Both these workshops maintain, according to Karl-Erik Sjöström, a very high standard of quality. The outsourcing has not had any negative consequences in that respect.

GenerPro is a manufacturing company with no product development or design resources. The company has around 80 employees and is equipped for machining of rotors and other components, punching and stacking of stator cores, winding of stators and rotors, rotor balancing, assembly and testing. The production is focused on turbo- and hydropower machines though some other components are made as well. It is interesting to note that not only Alstom and ABB but also companies like Siemens, Toshiba and Brush belong to its clients. GenerPro has been successful in securing order from several leading OEMs in international competion. The president of the company, Gösta Hesslow, mentioned in an interview, that it is possible to operate at approximately the same cost level as e.g. competitors from Poland [374].

Another new electrical machine company in Västerås is VG Power, which was established by former ABB and Alstom employees in October of 2002. This company is, however, focused on service and retrofit of hydropower units and has nothing to do with large turbogenerators. VG Power has a staff of 50 persons out of which half of them in the engineering department. An interesting observation is that a German company, jointly owned by Siemens and Voith, has acquired a majority share of VG Power in the summer of 2006. In order to give a comprehensive picture, it should also be mentioned that ABB in Västerås is still developing and manufacturing salient 4- and 6-pole generators for geared gas- and steam turbine units in the power range 3 - 70 MVA. Among its important customers are companies like Siemens turbine factory in Finspång and Solar Turbines in California. ABB has a significant share of the global market for this kind of synchronous machines, indicating that it is still possible to successfully manufacture large electrical machines in Sweden [375].

7.10.2 Technical changes

The design of the GTD-generators was basically completed when the problems with the TVO rotors had been solved, i.e. around 1980 – 81. There has nevertheless been room for later improvements, especially after ABB was established and it became possible to adopt certain solutions developed by

BBC. A significant example was a new design of the stator core ends combined with axial bolts through the core for exerting the necessary axial pressure. The former, partly water-cooled, laminated end-sections were replaced by conical, air-cooled sections, which could be seen as an extension of the core itself as shown in figure 7.71. The new design has proved to be optimal for reduction of leakage flux and stray load losses. Even the bracing of the stator end-windings has been further improved.

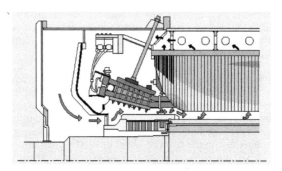

Figure 7.71 New stator end section design

Upgrading was briefly mentioned in the previous section. The TVO generators can be used as a good example of this. These were originally rated 825 MVA, 700 MW and were later up-rated to 905 MVA. An increase of power-factor from 0.85 to 0.9 simplified of course a substantial rise of active power. The higher rating was achieved through utilization of existing margins and certain improvements of the stator end cooling. New stators, with the conical end rings mentioned above, were installed in 1996. The rated power could then be raised to 950 MVA with a power factor 0.9. Both stator and rotor windings, as well as the magnetic circuit, remained unchanged from the original version. It has now been decided to further increase the output, this time to 1100 MVA, 990 MW and power factor 0.9. Stig Hjärne, who is responsible for electrical design, mentioned, in an interview, that this requires new generators, but the same physical dimensions as the present can be maintained. The stator and rotor windings must in this case be somewhat modified. Increased utilization is primarily achieved through higher linear current loading, which is facilitated through intensified cooling of the stator winding in combination with a somewhat larger copper area. Re-designed rotor cooling circuits allow shallower but slightly wider slots, which results in less saturation and hence lower no-load excitation current, making it possible for the rotor to accept the increased armature reaction caused by the higher stator current. In October of 2006, TVO placed a firm order with Alstom in Västerås for one such 1100 MVA, 3000 rpm generator and an option for another one. The main competitor was Alstom Power in Switzerland, which offered a generator with a directly hydrogen-cooled rotor [376].

The generators for Forsmark 1 and 2 as well as Ringhals 3 and 4 have, in a corresponding way, been upgraded from the original 577 MVA, powerfactor 0.85 first to 635 MVA, and in a second step to 677 MVA, powerfactor 0.9.

The design process has also been considerably improved during the last decades. Replacing drawing boards with interactive CAD systems has rationalized the design work. FEM analysis is no longer limited to mechanical problems. It is used for studies of a number of electro-magnetic problems, e.g. calculation of eddy current losses in different machine components. Risk analysis (FMEA) was introduced during the early 1990's as a quality assurance tool. Bengt Alenfelt mentioned, in an interview, that the design department also gained a lot from better communication with the suppliers, e.g. not specifying requirements which would only increase the cost of materials and components without affecting the final result. Transport and installation technology has also been improved. The time for replacement of a generator has been reduced from almost half a year to 2½ weeks.

7.10.3 A restructured industry

The market for very large turbogenerators has been weak for many years, at least in Europe and North-America. Major reasons for this are a slowdown in increased power consumption in the industrialized countries, the reluctance to build more nuclear reactors as well as very large fossil fuelled power plants. International statistics show a drastic reduction in nuclear power plant construction when comparing the periods 1970 - 1985 and 1990 - 2005 [377]. The overcapacity that followed the reduced market has resulted in a highly restructured turbogenerator industry.

The most pronounced changes have taken place in Europe. Table 7.2, in section 7.1.4, showed the situation as it was during the 1970's. The list contains more than 20 companies. In Germany, Siemens and AEG jointly formed KWU in 1969. England had the most fragmented situation with three independent manufacturers: AEI, English Electric and Parsons in the 1960's. The first two were acquired by GEC in 1967 and 1969 respectively, which then kept English Electric's generator factory in Stafford. BBC manufactured turbogenerators in four different factories: BBC in Birr and Mannheim, CEM in France and TIBB in Italy. The latter two had stopped manufacturing of large turbogenerators before BBC's merger with Asea.

Several companies ceased to manufacture turbogenerators during the 1980's. ACEC in Belgium, Elin in Austria, and Jeumont-Schneider in France are such companies. Parsons continued somewhat longer, but is now a service facility owned by Siemens. The Italian company, Ansaldo Energia, which has built turbines and generators in its factory in Genoa for very long, is still active. AEG ran into financial difficulties and withdrew from KWU, which thus became a part of Siemens. There are other European companies manufacturing smaller turbogenerators, i.e. Brush in UK.

In 1988 the fusion of Asea and BBC gave rise to ABB. This caused turbulence in the electric industry and soon after, in 1989, a merger between the French Alsthom-Atlantique and the British GEC was announced. The new company, GEC-ALSTHOM, continued to manufacture turbogenerators in two factories, Belfort and Stafford until 2004, when the latter was turned into a component and service site. (The company had much earlier changed name to Alstom.) For a short period followed, as mentioned in section 7.10.1, ABB's and Alstom's jointly owned company, which now is entirely owned by Alstom. All these changes mean that there are, at present, less than a handful of independent European manufacturers of large turbogenerators; the leading are Alstom and Siemens.

The situation for the East European manufacturers changed radically after the fall of the Berlin wall in 1989. Companies such as Ganz, in Hungary, and Skoda, in former Czechoslovakia, had not built really large generators but still were considered qualified manufacturers. These companies have been subjected to transformations and neither of them is manufacturing large generators any longer. Electrosila in St Petersburg, once established by Siemens, is now partly owned by this company, and has continued manufacturing of both hydropower and turbogenerators.

USA had only two companies, GE and Westinghouse, which built large turbogenerators during the 70's. GE is continuing to do so, while Westinghouse corresponding activities were taken over by Siemens in 1999. Japan had during the same period, i.e. the 70's, three major generator manufacturers; Hitachi, Mitsubishi and Toshiba and all three remain active in this field. The Japanese industry has been very stable compared to Europe and even North America. Large turbogenerators, at least up to 600 MW, are also made in China and India. The major Chinese manufacturers are located in Harbin, Dongfang and Shanghai and the largest Indian manufacturer is Bharat Heavy Electrical Ltd. These companies have large domestic markets, but are also exporting generators. Organized cooperation with western companies are common, e.g. between GE and Harbin Electrical Machinery Co, Alstom and Dongfang Electric Machinery Co. as well as Siemens and Shanghai Turbine Generator Co.

It is evident that the global turbine and generator industry is dominated by three multinational corporations: Alstom, GE and Siemens. Especially Alstom and Siemens have successively acquired other generator manufacturers, closed factories or turned them into service facilities, but also established new production, e.g. through joint ventures in China. All of them, even GE, have production in many countries around the world. Focusing on Sweden, it can be seen that Alstom, among other things, has engineering and service of generators in Västerås, and that Siemens has acquired the industrial gas and steam turbine factory in Finspång, which once belonged to Stal-Laval. GE has taken over factories for hydropower turbines in Norway and Finland.

7.11 The results

7.11.1 Operation records

It is evident, from what has been written earlier in this chapter, that the first GTD generators as well as those in Olkiluoto had a difficult time with serious teething problems. The second generation, which comprises the eight machines for Ringhals 3 and 4 and for Forsmark 1 and 2, has performed reliably from the very beginning. The others have done so after they have been modified or provided with new rotors. This means that the operation records from the early 1980's up until now have been very good. This is proved by statistics but perhaps more important by the fact that the power companies have chosen the same technology and supplier also in the case of new, upgraded machines. The diagram in figure 7.72 a shows how the unavailability for the water-cooled GTD generators has varied over the years. The peak at the end of the 1970's depends partly on the retaining ring failure in Barsebäck, but also on problems with rotor slot wedges in Ringhals. The result has become good, also in comparison with other large turbogenerators as shown in the availability diagram in figure 7.72 b. Both diagrams have been provided by Alstom.

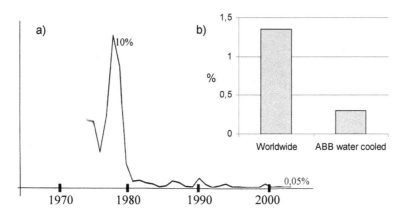

*Figure 7.72 a) Unavailability for GTD versus time b) Accumulated una-
vailability, 1981 - 95, in nuclear power plants caused by generators
(source IAEA)*

7.11.2 Competence development

The wide experience that the development, manufacturing and operation have given is a very valuable result, but it could also be fugitive. Even if the existing documentation is comprehensive, the know-how represented by a number of key persons is indispensable. A critical question is therefore: what competence and which resources remain in Sweden now and in the future? Alstom's engineering department in Västerås has 20 employees. Karl-Erik Sjöström estimated that five or six of these can be classified as real experts. Their competence and experience have to be canalized over to new persons if large turbogenerators shall survive as a Swedish activity. Deep theoretical competence, as such, is not the most critical; rather it is the combination of theoretical knowledge and practical experience that is of great importance. The possibility to obtain support from Alstom's larger engineering departments in Switzerland and France shall not be neglected, even if there is no direct cooperation in development projects right now. There are, however, regular meetings for exchange of information, and the key persons in Västerås view the contacts with their French colleagues to be especially positive.

7.11.3 Spin-off effects

Many years of advanced technical product development often give certain spin-off effects, i.e. results which can be used for other purposes or by other actors. In the case of the GTD generators, two examples are worth to be mentioned. The first is the use of stainless steel tubes in all cooling circuits, i.e. even inside the stator and rotor coils. Other manufacturers had instead used hollow copper conductors. Asea's technology to use stainless steel tubes was taken over by the other ABB/Alstom factories. The reason was that the steel tubes allows considerably higher flow rate than the copper tubes, and this helps to avoid clogging of water channels, which had previously been a problem.

The other example is the rotor vibration monitoring described in section 7.9.3. This sophisticated system, developed mainly by Stal-Laval, in connection with the crack problems in the TVO rotors, has received wide attention. The principle to base the monitoring on changes of the vibration vector has been adopted for large turbine/generator units all over the world.

7.11.4 Conclusions

The development of the GTD generators must be considered as a major technical achievement for a country like Sweden. The concept was at the beginning seen as daring and pioneering. It was different from what was usual in the industry and some viewed it as too risky. Many problems occurred and the GTD generators were, for a number of years, not only questioned but even regarded as a serious failure. Extensive development efforts solved the problems and the generators have for decades had a very good reputation. They have during the last 25 years generated around 30 percent of Sweden's electric energy. A corresponding figure for Finland is 20 percent. The concept has been maintained and the technology is in no respect obsolete.

Explicit conclusions concerning the Asea phase are:

- The Swedish development of large turbogenerators was basically market driven. Such machines were needed for the nuclear power plants.

- It was natural for Asea to develop the large turbogenerators in-house, even when it was necessary to obtain a license for the steam turbines. Licenses were always considered as an exception.

- Asea chose the fully water-cooled concept:
  -to avoid development of directly hydrogen-cooled rotors as an intermediate step,
  -because it was the most efficient cooling method,
  -and also due to synergies with large salient pole synchronous machines.

- Asea's immediate challenge became too large due to the drastic increase in machine size in combination with simultaneous orders for generators with different ratings.

- Asea underestimated the difficulties when they decided to make most of the development on customer orders. Manufacturing and testing of a full-size prototype as basis for the order design could have eliminated several problems and saved a lot of costs.

- The 1970's were characterized by severe technical problems due to new technology and extrapolation. Even other manufacturers suffered from similar problems. Asea's competence and resources were much improved from the work with trouble shooting, advanced analyses and improved designs.

- Test results show that the electrical and thermal design of the generators was correct. The major problems were mechanical fatigue and corrosion.

- Mutual exchange of technical know-how with other manufacturers increased substantially, as expected, when Asea had gained certain experience.

- Asea has chosen a technically elegant and efficient concept for large turbogenerators, but a design with directly hydrogen-cooled rotors would have been sufficient and less costly to develop.

- Asea tried to obtain a back-up solution through the license on a gap pick-up hydrogen cooled concept, but the timing was not right.

- The experiences from the final GTD design has proved that direct water-cooling of both stator and rotor is a viable and efficient concept for large turbogenerators.

- The market for large turbogenerators almost disappeared in the early 1980's of reasons outside Asea's control. The industry has since then been subjected to a radical structural rationalization.

No generators with this particular design have been built for other countries than Sweden and Finland. Very little of the technology has been taken over by others. This could indicate that the concept is not competitive enough, but on the other hand, it has been preferred in open competition for recently upgraded units. The cost-effectiveness is therefore probably not much different from other designs of large 2-pole turbogenerators and, besides, each manufacturer tends to hold on to their existing designs. The explanation is more likely to be found in the market and industry structure. The market for large steam turbines and generators had just started to shrink when Asea had solved the technical problems. There were simply no possibilities for Asea to sell GTD generators, together with Stal-Laval's steam turbines or in any other ways, when the international situation was characterized by large overcapacity. The conditions became different in ABB and later Alstom. These companies had access to the global market and could have decided to include GTD generators in deliveries of large steam power units, if advantageous. This has not been done, the concept with directly hydrogen-cooled rotors, emanating from BBC, has instead been chosen. Why, is it better? This question is not entirely relevant because it would have been possible to combine the best parts of each design into a common one. The uniqueness of the GTD generators is the water-cooled rotors and absence of hydrogen. The management of ABB's, and later Alstom's, turbogenerator operations located in Switzerland, has obviously come to the conclusion that the customers prefer more conventional, hydrogen-cooled generators and the possible incentive for a shift of concept will not outweigh the risks. It is, of course, difficult to motivate radical design changes when the market for really large generators has become very limited.

Some conclusions regarding the ABB/Alstom phase and the current situation are:

- The water-cooled GTD generators are at least as cost-effective as other turbogenerators of corresponding size, but this is highly dependent on the individual manufacturer's background.

- Direct water-cooling can be limited to the stator and rotor windings. Remaining parts can be air-cooled.

- Projects for upgrading existing units have shown that it has been possible to increase the utilization factor for the original GTD generators by 20 – 30 percent.

- The formation of ABB, and later transfer to Alstom, has effectively

limited the Swedish unit's possibilities to pursue development, manufacturing and marketing of large turbogenerators. It is, however, not
likely that an independent Asea would have been successful in the
export market, as the manufacturers of large steam turbines still build
their own generators.

• The global turbogenerator industry has become dominated by three
multi-national companies and a shift from production in Europe to
Asia is evident.

Do the GTD generators have any future and is there any future for the development and design of large turbogenerators in Sweden? These two questions
are strongly interlinked. The existens of GTD generators is definitely a prerequisite for Swedish engineering activities today. With the present owner
structure, this business and engineering activity is only accepted due to the
uniqueness of the design. Hydrogen-cooled machines would have been supplied from the large factories in Birr and Belfort. Existing Swedish and
Finnish power companies seem to prefer GTD generators in case of
upgrading, mainly because these machines are competitive and the reliability
has been high, but also due to the access to domestic expertise. The answer
to the questions at the beginning of this paragraph is that it depends to large
extent on the domestic customers. It is only if they see an advantage in the
GTD concept and explicitly press for this, that this activity will survive. The
fact that large actors such as Vattenfall have become multi-national can, in
this respect, be considered both as a possibility and a threat. There are few
reasons to be too optimistic about the long-term prospects.

Another, more universal, question is: has direct-water-cooling of turbogenerator rotors, as technology, any future? It does not look very bright at present
as practically all other manufacturers use hydrogen-cooled rotors, but there
are exceptions. The Russian company ELSIB in Novosibirsk builds turbogenerators, which are liquid-cooled in both stator and rotor. Other manufacturers, e.g. Siemens, claim that water-cooling is optimal for certain types of
large hydropower generators. The Swedish built GTD generators will be in
operation for several decades. There are, therefore, reasons to assume that
water-cooling of large rotors will continue to be used also in future machines,
but there are no indications that it will become a dominant technology. However, a break-through of super-conducting generators could turn everything
upside down.

8   Vehicle motors

New types of drivelines for road vehicles have become an increasingly important application for electrical machines. Many research projects at technical universities around the world deal with such machines and drive systems. The powerful automotive industry has, in several respects, set new, challenging targets for electrical machine development engineers and has also become an active partner in this process. It is no longer evident that the established electro-technical industry will maintain its traditionally leading position.

Sweden has, from an international perspective, a very strong automotive industry but it has not been in the forefront in the introduction of electrical drivelines. A few demonstration projects have nevertheless been carried out. Sweden is not a leader in the development of electrical machines for such drivelines either, and this chapter of the thesis applies therefore a more international view than the others. At any rate, some development efforts have been carried out and a couple of interesting examples will be described and discussed. ABB has made some attempts, in this new field, but it terminated them a few years ago.

It is of great interest to study the development in this particular area and the impact it may have on electrical machine development in general. A relevant question is; does the vehicle application lead to some sort of paradigm shift for the electrical machine industry?

8.1   "Electric vehicles"

8.1.1   The early years

Battery powered electric cars are not a new invention; several models were commercially available around the previous turn of the century. A rechargeable lead-acid battery had been built already in 1859 by the Frenchman Gaston Planté (1834-1889) and it was improved in 1881 by his fellow countryman Camille Alphonse Fauré (1840-1898). The same year, 1881, Siemens&Halske demonstrated electric motors driving trams in Berlin [378]. The necessary components for electric road vehicles were thus available and some documents claim that the first car was built as early as 1881 by another Frenchman, Gustave Trouvé (1838-1902) [379]. There had been earlier attempts, e.g. in 1870 by Sir David Salomon (1851-1925), but with poor results [380]. Other documents give priority to the English inventor John Kemp Starley (1854-1901) and to the American Fred M. Kimball, who both presented functioning electric vehicles in 1888 [381]. This took place more or less simultaneously with the appearance of the first cars driven by internal combustion engines (ICE). Famous examples of these were Karl Benz' (1844-1929) and Gottlieb Daimler's (1834-1900) cars, both presented in 1886 in Germany [382]. Another technology for vehicle propulsion, more commonly used during this period, was steam engine drivelines, which could benefit from the intensive development of such systems for railway traction, ship propulsion and industrial installations.

The electric cars became quite popular and were preferred over ICE-driven cars due to higher reliability, easier handling and better comfort. In the USA around 1900 almost twice as many electric cars as ICE-driven cars were manufactured. Most of the first taxicabs in New York were electrically driven, provided with interchangeable battery packs. The performance was also good which can be illustrated by the fact that an electric car, "La Jamais Contente" set a new world record in April 1899 when it reached a speed of 105 km/h at a track close to Paris [383].

Figure 8.1 Old electric cars, La Jamais Contente 1899 and Detroit Electric 1915

There was, for a period of two decades, a competition between the combustion engine and the electrical motor, which finally was lost by the latter due to well-known battery limitations. The problems, which have remained, were too short driving range, long re-charging time and high costs. Cadillac's introduction of electric start motors, for the combustion engines in 1910, did also favour the ICE driveline concept, so the electric cars disappeared more or less completely in the 1920's.

There had not been any production of electric cars in Sweden, but Asea actually built a small number of 2- ton distribution trucks, which were in operation during World War II. One of these trucks is shown in figure 8.2. The reason for using electrical trucks was, of course, the very difficult fuel situation during those years. These trucks were provided with series DC motors [384].

Figure 8.2 Asea's 2 tons electric distribution truck

8.1.2 The rebirth

For a long period during the previous century, there were practically no battery powered electric road vehicles other than a few ten thousand British milk distribution trucks. The electric drive concept was restricted to some off road applications such as forklifts and golf carts. None of these applications were very demanding. Lead-acid batteries and series excited DC-motors have been the standard concept in use.

Growing concern over air pollution stimulated a renewed interest in electric vehicles in the late 1960's. The first International Electric Vehicle Symposium (EVS-1) was held in Phoenix, USA in 1969. Some government funded development and demonstration programmes were carried out in Europe, Japan and USA during the 1970's. The oil crises in 1973-74 and 1979 temporarily increased the efforts, but the development slowed down again as soon as the crises faded. In addition, the introduction of catalytic cleaning of exhaust gases had improved the air pollution situation also contributing to a low interest in electric vehicles in the 1980's.

The turning point came in 1990. The Los Angeles metropolitan area had long suffered from severe air pollution caused mainly by the very extensive motor traffic in combination with a special location from climatological point of view. It became necessary to take some actions to improve the situation. In order to comply with Federal air quality standards, the California Air Resource Board (CARB) passed, in 1990, regulations with the intention to push electric vehicle technology forward. CARB's mandate required that all major car manufacturers producing cars for sale in California must offer so called zero emission vehicles (ZEV). The mandate stipulated that 2 percent of their sales, in 1998, should be ZEV, 5 percent in 2001 and 10 percent in 2003 [385]. This triggered of course the large American and Japanese manufacturers, which were the first in line to have to comply with the new rules, but it also served as an alarm bell for other countries and companies. By definition battery powered electric vehicles were the only vehicles that could be classified as ZEV. Most of the development activities therefore became focused on such pure electric vehicles. Several years later, in 1999, CARB changed the mandate opening up for other concepts.

8.1.3 Battery electric vehicles

The driveline in a battery electric vehicle (BEV), often referred to as an electric vehicle (EV), is principally very simple. It consists of a battery as energy source, a power electronic motor controller, an electric motor and some sort of mechanical transmission. In addition, it is usually necessary to have an onboard battery charger, a DC/DC converter for supply of all low voltage (12 V) loads and also electric drives of servo pumps and the air-condition compressor. The block diagram in figure 8.3 illustrates a BEV driveline including some auxiliaries.

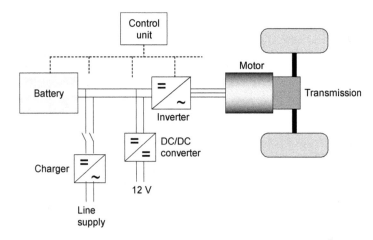

Figure 8.3 BEV block diagram

Estimations of the total number of registered BEVs in the world in the early 1990's showed some 4500 – 5000, an insignificant percentage of the total number of road vehicles, which was far above half a billion. These were basically two categories of electric vehicles. One was mini-cars from small, entrepreneurial manufacturers and the other was converted standard cars and vans from established automotive companies such as Fiat and Peugeot. Almost all available BEVs were provided with DC-motors and lead-acid batteries or, in some cases, nickel-cadmium batteries [386]. However, many different companies had started to develop new BEVs, either converted or so called purpose built cars. The latter can be illustrated by GM's prototype

Impact, which later was series manufactured as "GM EV 1". Another was BMW's E-1, see figure 8.4. Examples of converted cars were Ford Eco Star based on the Escort and Toyota's RAV4 Electric. Some new players saw a possibility to enter the automotive business as BEV manufacturers. A good example is the Norwegian company PIVCO, founded in 1990, which developed a small two-seat BEV for city driving. The 30 kW induction motor was supplied by Siemens. However, the company run into financial difficulties and was bought by Ford in 1999 and changed name to "Think Nordic". The production was increased and most of the cars were exported to California. Three years later, Ford changed its strategy concerning BEVs and the Norwegian company was sold. It is now, after some detours, owned by a Norwegian investment group and the aim is to restart building and marketing small electric cars [387].

Figure 8.4 A Norweigan Think and Dr. A. Goubeau, BMW with the E-1

The DC-motor and often also the lead-acid battery were eventually abandoned for most of these new prototype and demonstration vehicles. The development of new solutions with better performance was necessary. Size and weight are very critical in the case of vehicle applications, which require motors with high specific torque (Nm/kg) and power (kW/kg), and in the case of batteries high energy and power density (kWh/kg and kW/kg). The efficiency is also very important, as road vehicles have to carry their energy storages with them. Thus, higher efficiency will result in a longer driving range. AC induction motors and even PM motors therefore replaced the traditional DC-motors in the new BEV projects. The battery development focused in the early 1990's on alkaline batteries such as NiCd and NiMH, high temperature batteries like NaS and NaNiCl but also on metal-air batteries, e.g. Fe/air [388].

ABB was actively engaged in the development of equipment for BEVs in the early 1990s as were several other large electro-technical companies. Long before the merger, which led to the formation of ABB, BBC had started to develop a high temperature battery with sodium and sulfur electrodes and a ceramic electrolyte, the NaS battery. It had excellent performance data compared with other battery types. ABB therefore built a pilot production line in

Heidelberg, Germany and supplied batteries for a number of BEV projects. The company also delivered DC drive systems for BEVs, first from factories in Germany but the manufacture of the DC motors was later transferred to one of ABB's motor factories outside of Paris. The main customers, for the drive systems, were Volkswagen and Renault. The vehicles were Volkswagen's Golf based "City Stromer" and Renault's electric Clio, shown in figure 8.5. ABB supplied hundreds of DC drive systems during the first half of the 1990's. The main data for these systems are shown in table 8.1 [389]. During that period, ABB also got involved in the development of AC drive systems for automotive applications. The company stopped its battery development and manufacture in 1994, partly due to safety reasons (risk of fire in the batteries) but mainly due to the financial prospects.

Table 8.1   Data for ABB's DC drive systems

Continous power [kW]	18	19	28
Maximum power [kW]	27	32	38
Speed range [rpm]	0 - 2000 - 6700	0 - 900 - 3000	0 - 2000 - 6700
Voltage [V]	72 - 144	108 - 144	144 - 240
Maximum current [A]	300	330	250
Motor weight [kg]	76	165	84
Converter weight [kg]	13	17	17

Figure 8.5 Renault Clio Electric and VW City Stromer

In spite of all the efforts, there has not been a break through for the BEVs. Much of the development has ceased and the BEVs seem to remain a small niche product. The focus has shifted to hybrid electric vehicles (HEV) and fuel-cell vehicles (FCV). The reason for this is the battery, the same situation as one hundred years ago. The energy density is too low, the charging time is too long, the lifetime, in number of discharge cycles, is too short and the costs are too high. It is true that some types of batteries would allow a driving range of 150 – 200 km, which is quite sufficient for most daily driving, but such batteries are very expensive. Affordable batteries limit the range to < 100 km which does not give the driver sufficient flexibility. The batteries are, nevertheless, important also in terms of HEVs and FCVs, but then the power density becomes more important than energy density. Table 8.2 below shows characteristic data for some battery types [390]. Presently, the most promising battery technology is the Li-ion type.

Table 8.2   Comparison of battery data

Battery type	Pb/PbO	NiCd	NiMH	NaS	NaNiCl	Li-ion	Li-polymer
Energy density [Wh/kg]	35	55	65	120	100	75 - 150	75 - 160
Power density [W/kg]	150 - 400	250 - 400	400 - 900	200	150	800 - 1200	500 - 1200
Cell voltage [V]	2.0	1.4	1.2	2.0	2.6	3.5 - 4.0	3.5 - 4.0
Cycle life (80 % DOD)	300	1000	1200	1500	1000	2000	> 1500
Operating temp. [°C]	- 30 / + 60	- 20 / + 55	- 10 / + 45	+ 300 / + 350	+ 300 / + 350	- 10 / + 45	- 10 / + 45
Relative cost [%]	100	300	800	Not manufactured	500	500 proj.	500 proj.

8.1.4   Hybrid electric vehicles

A hybrid electric vehicle (HEV) has, by definition, two different power sources for propulsion, normally an ICE and one or more electrical drive units. The corresponding energy storages are the fuel stored in a tank and usually some type of battery. The basic idea is to reduce the fuel consumption and even the emissions by avoiding running the ICE in an inefficient way. It can also be to facilitate ZEV operation, i.e. pure electric driving, in environmentally sensitive areas. HEVs offer such possibilities, while they, in other respects, maintain the performance and flexibility of conventional vehicles.

This concept is not new. Dr. Ferdinand Porsche (1875-1951), the legendary automotive engineer and industrialist, built the world's first hybrid car in Vienna in 1898. It was a so-called series hybrid with electric motors contained within the hub of the front wheels. A couple of years later, he developed even a hybrid car with four wheel motors. Such a car, known as a Lohner-Porsche, is shown in figure 8.6. Ferdinand Porsche, who had started to work for an electrical machine factory, held in fact a patent for such wheel motors [382]. An early patent application for a parallel hybrid car was filed in the USA, in November 1905, by a Belgian engineer, Henri Pieper, who was granted US Patent No. 913,846, titled "Mixed Drive for Autovehicles" in 1909. The concept is presented in figure 8.7 [391]. It is not, however, until recently that HEVs have received wider attention.

Figure 8.6 Lohner-Porsche hybrid car from 1902

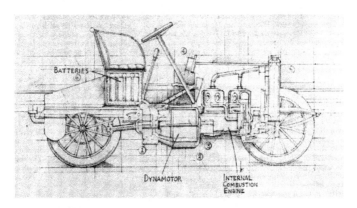

Figure 8.7 Sketch of H. Pieper's hybrid car from 1905

Some different types of HEVs can be identified. Each of these has its advantages and disadvantages and hence more or less suitable application areas. Two basic types have been mentioned in the previous paragraph, the parallel hybrid and the series hybrid. The definitions refer to the power flow through the drive motors. The ICE and the electric motor can, in the first alternative, propel the vehicle in parallel, while in the second case the ICE drives a generator that provides the electric motor with power via a power electronic controller. The engine and the motor are thus, in the latter case, in series with respect to the power flow. Figure 8.8 shows the principal outline of these two hybrid systems.

Parallel hybrids have, from a system point of view, most in common with the

conventional ICE driven vehicles. They have full-sized, or slightly decreased, engines and often multi-gear transmissions. The operation point of
the engine depends, to a large extent, on the actual driving conditions because
there is a mechanical connection between the engine and the wheels. This
complicates the possibility to run the ICE under optimal conditions and to
avoid transients. One important advantage with a parallel hybrid is that the
electrical system can be fairly small and contain few components. Only one
electrical machine is needed, which is used both as a motor and a generator.
The other main advantage is a better overall efficiency than the series hybrid,
especially at highway driving, due to the direct power transfer from the ICE.
A disadvantage is, on the other hand, that the mechanical transmission becomes more complicated. Parallel hybrids can be made in different ways and
with different degrees of hybridization, i.e. the ratio between electrical and
mechanical power. One is the so-called mild hybrid that has a very limited
electrical power where the electrical machine is placed in the driveline between the engine and the gearbox as shown in figure 8.9 or is belt-driven.
This type of electrical machine replaces the start motor and the generator and
its rating is limited to 5 – 15 kW for a passenger car. It can be used for limited
regenerative braking and can also act as a booster. Most important is that it
facilitates a stop-and-go function, i.e. automatic ICE stop when the car stands
still and automatic start when the driver takes off, thus avoiding fuel consuming idling. Small versions of this system are also referred to as micro hybrids,
ISG (Integrated Starter Generator), or ISAD (Integrated Starter Alternator
Damper) [392].

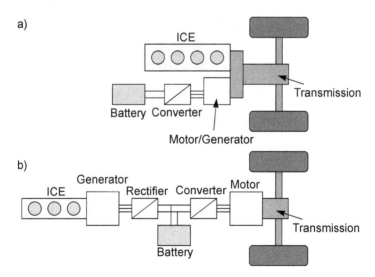

Figure 8.8 Diagrams showing a) parallel hybrid and b) series hybrid

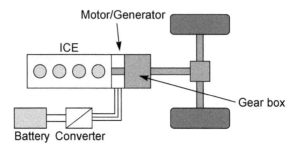

Figure 8.9 Typical "mild hybrid" with motor/generator directly coupled to the ICE

A hybrid car can also be built by having the front wheels driven by the ICE while an electric drive system is connected to the rear axle. There are examples of such cars where the road is the only mechanical connection between the two drive systems, but there are also examples on hybrid cars having more advanced four-wheel drive systems [393, 394]. Parallel hybrids, with somewhat larger electrical drive systems than the mild hybrids, can usually be used for limited ZEV operation, but to what extent depends primarily on the battery.

The parallel hybrid is considered as cost effective and has therefore become introduced in commercial cars. Examples are Honda Civic Hybrid and Citroën C3 (Stop & Start) [395, 396]. Several others are under way.

The series hybrid is in principle very similar to the BEV, but it contains an ICE-driven generator that supplies power to the electric drive system. Such a system has the advantage of making the ICE more or less independent of the actual driving conditions. The ICE, combined with the generator, runs at an optimal working point when the traction load is large enough or the battery needs to be charged. The ICE is switched off when the load is low and the power is then supplied from the battery. The ICE can also be somewhat smaller than in a conventional vehicle as extra power for accelerations etc. can be provided from the battery. A series hybrid is suitable for ZEV operation as the electrical drive system is dimensioned for full power, how long is determined by the capacity of the battery. The diagram in figure 8.10 illustrates the operation of a series hybrid.

The series hybrid also has some distinctive disadvantages. The electrical system becomes fairly large, as it must be dimensioned for maximum power, and it is requiring two electrical machines and two converters while the parallel hybrid only needs one of each. The efficiency becomes, in most cases, lower because the power has to be converted in many series connected components on its way from the ICE to the wheels.

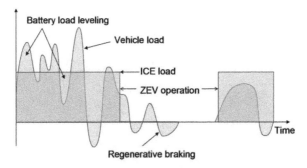

Figure 8.10 Diagram illustrating operation of a series hybrid.

The series hybrid is therefore best suited for such urban drive cycles, where a conventional vehicle would have a great deal of idling. Another suitable application can be plug-in hybrids, i.e. those where a substantial part of the necessary energy is provided from the grid through battery charging. Hybrid vehicles with individual wheel motors would preferably also be built as series hybrids. An interesting example is a six-wheel military terrain vehicle developed by the Swedish company "Hägglunds" (BAE Systems Hägglunds). This illustrates that an important advantage of series hybrids is that they offer possibilities for a much more flexible mechanical outline and packaging of the entire vehicle. Figure 8.11 a shows a picture of prototypes of Hägglunds' armored terrain vehicles, which exist in one track- and one wheel-driven version. A system diagram for the six-wheel version, provided with two diesel-driven generators, is presented in figure 8.11 b. [397]

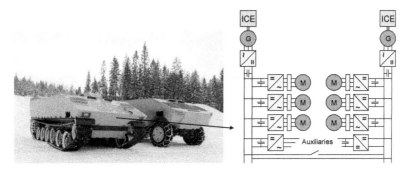

Figure 8.11 a) Terrain vehicles with electrical propulsion systems.
b) Driveline system for six-wheel vehicle

Hägglunds' vehicle is an example where the power units (ICE+generator) are dimensioned for full power. It can be considered a vehicle with diesel-electric transmission more than a hybrid. The other end of the scale is an electric vehicle with a range extender, i.e. a small ICE powered generator set for battery charging on board. A series hybrid can, in principle, be anything between these two extremes.

Figure 8.12 Volvo ECC, 1992 with gas turbine-driven high-speed PM generator and water-cooled induction motor below the turbine

Figure 8.13 a) Series hybrid 12 ton distribution truck in operation from 1998-2000
b) Parallell hybrid vehicles presented in 2006 (Volvo)

A number of concept and demonstration projects were carried out during the 1990's in Sweden by Volvo, most of them in cooperation with ABB. All of these were series hybrids but with different types of power units, from ICE- to gas turbine-driven PM generators. The drive motors were, in most cases, water-cooled induction motors. Figures 8.12 and 8.13a show a couple of these vehicles [398, 399]. Volvo has, in the beginning of 2006, presented new HEV prototypes, now with parallel hybrid systems, see fiure 8.13b.

The world's most well-known hybrid car is the Toyota Prius which was introduced in 1997 and has now, ten years later, been sold in more than 600 000 units globally. It is a so-called power-split hybrid that can be considered a combination of a parallel and a series hybrid. It has a mechanical connection from the ICE to the wheels through a planetary gear, but also an electrical connection via a generator, converters and an electric motor. The electrical components are not dimensioned for transfer of full power, but the system nevertheless allows the ICE to operate more independently than a corresponding parallel hybrid. Figure 8.14 contains a photo and some main data for the Toyota Prius and figure 8.15 presents a schematic sketch of the hybrid system. The Toyota Prius is not intended for any ZEV operation, even if it has a limited capacity for that [400, 401].

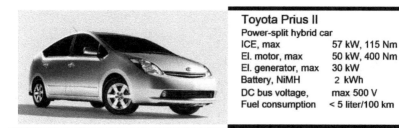

Toyota Prius II
Power-split hybrid car

ICE, max	57 kW, 115 Nm
El. motor, max	50 kW, 400 Nm
El. generator, max	30 kW
Battery, NiMH	2 kWh
DC bus voltage,	max 500 V
Fuel consumption	< 5 liter/100 km

Figure 8.14 Toyota Prius

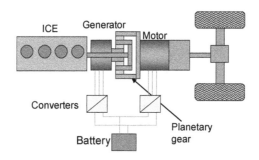

Figure 8.15 Principle sketch of Toyota Prius drive system

Another power-split concept has been developed at KTH during recent years. It has, in many respects, similar properties as the Toyota Prius system but it is based on an electrical transmission instead of a planetary gear. The system is a good example of advanced applied research in electrical machines currently carried out in a Swedish university. Besides the ICE, the system consists basically of two electrical machines, integrated in the same unit, two converters and a battery. One of the electrical machines consists of two concentric rotors and the other one of a rotor and stator. The inner rotor is driven by an ICE and has a 3-phase winding connected to a converter via sliprings and brushes. The outer rotor is combined between the two machines and is provided with permanent magnets both on its inside and outside. The stator too has a 3-phase winding connected to a converter. The outer rotor is coupled to the wheels via a reduction gear. Figure 8.16 shows a sketch of the system.

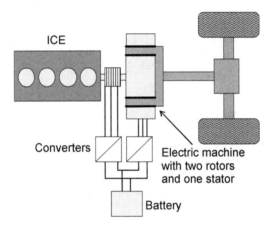

Figure 8.16 Sketch of 4QT system

The main objective of this system is to reduce the fuel consumption by letting the ICE operate along an optimum operation line independently of the vehicles operation point. The double rotor machine transmits the torque from the ICE, but it can increase or reduce the speed and thus the power. The stator can increase or reduce the torque transferred to the reduction gear making it possible to move from an arbitrary operation point to an optimal one by circulating a certain electrical power in the system including using the battery. The "vehicle" can operate in all four quadrants around such an optimal point and the system is therefore called "four quadrant transducer" or simply "4QT". The diagram in figure 8.17 explains the principle. Other than improved fuel economy and lower emissions, the 4QT system also gives a possibility for a certain ZEV operation. How much depends on the battery size. The research project at KTH has not only included the presented radial/radial flux topology but also axial/axial flux machines as well as combinations of these [402, 403, 404].

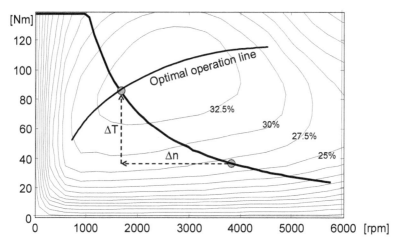

Figure 8.17 ICE efficiency diagram showing the 4QT operation principle

8.1.5 Fuel-cell vehicles

Many consider the fuel-cell as the ultimate power source for road vehicles. It is efficient, clean and quite. It has an energy-efficiency higher than 60 percent and the exhausts contain nothing else than water if it is fueled with hydrogen. A fuel-cell, which simply can be described as a fuel-powered battery, consists of electrolytic cells that generate electric power when the electrodes are supplied with the reagents, hydrogen and oxygen.

This invention is not new. Sir William Grove (1811-1896), a British jurist and physicist, discovered the fuel cell principle in 1839, when he series connected some cells containing hydrogen and oxygen and succeeded to produce electric power. The practical development was, however, slow until the 1960's, when fuel-cells began to be used as power units on spacecrafts. A number of different fuel-cells have been introduced during the later decades. Table 8.3 presents a list of these and some of their main data. It can be seen that most types operate at high temperatures, which makes them less suitable for vehicle applications. The Proton Exchange Membrane (PEM) fuel-cell dominates completely the efforts to develop commercially viable fuel-cells for road vehicles [405].

Table 8.3   List of fuel-cells with main data

Fuel Cell	PEMFC	AFC	PAFC	MCFC	SOFC
Electrolyte	Solid organic polymer	Potassium hydroxide	Phosphoric acid	Molten carbonate	Solid zirconium oxide
Operating Temperature	60 - 150 °C	90 - 100 °C	175 - 200 °C	600 - 1000 °C	600 - 1000 °C
Application	Electric utility Transportation Portable power	Military Space Transportation	Electric utility	Electric utility	Electric utility
Advantages	Quick start-up, simple construction Low temperature	High efficiency	Generate heat and power	Cheap catalyst fuel flexible	Cheap catalyst fuel flexible
Disadvantages	Expensive catalyst sensitive to fuel impurities (< 100 °C)	Removal of CO_2 from air and fuel required	Pt catalyst Large size Weight	Corrosion Material stability	Material stability

These days the automotive industry invests huge amounts on development of PEM fuel-cells and fuel-cell powered vehicles (FCV). The technology has been successfully demonstrated in numerous projects but the costs are still much too high. Prognoses indicate that the commercial breakthrough will come between 2010 and 2020. Another major problem is the access to hydrogen. To develop the necessary infrastructure for production and distribution of hydrogen is a gigantic undertaking. The storage of hydrogen onboard vehicles is also problematic. It can be stored either as compressed gas at a pressure in the order of 300 – 700 bars, or in liquid form, which requires a temperature below 20 K (– 253 °C). There are also possibilities to store hydrogen chemically in metal-hydrides, but this method has not yet been used for FCVs.

An alternative is to provide the vehicle with a reformer that converts a suitable hydrocarbon fuel to hydrogen, but such a process also creates some carbon oxides. This technology simplifies the infrastructure but complicates the vehicle and increases its costs. Most FCVs, presented in various demonstration projects, are provided with tanks for compressed hydrogen. A battery or super capacitors are required in parallel with the fuel-cell to give the system sufficiently good dynamic properties. Figure 8.18 shows the system layout of an FCV and figure 8.19 a picture of Honda FCX, one of the FCVs most close to market introduction.

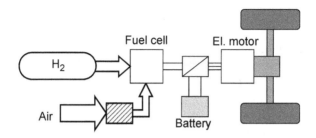

Figure 8.18 Sketch showing system layout of FCV

Figure 8.19 Honda FCX fuel cell car

The electrical drive system in an FCV is, in principle, not different from that in a BEV or a series HEV. The electric motor has to be dimensioned for the full traction power. It also opens up the possibilities of using individual wheel motors. The application as such does not therefore require development of new and different electric motors. However, a major shift from ICE-driven vehicles to FCVs will create a quantum leap in the global electric motor production. Imagine an annual production of 30 – 50 million motors in sizes from 30 kW and larger compared to the current production of electrical motors in this range, which ABB estimates to roughly a million; the industrial impact would of course be very significant. An important prerequisite is, however, that abundant quantities of hydrogen can be produced by means of renewable energy, e.g. solar power, at reasonable costs.

8.2 The challenges

8.2.1 A new application

It can be argued whether road vehicles can be seen as a new application for electrical machines, but the radically increased importance of HEVs and FCVs, justifies such a view. From a position, shadowed by numerous industrial applications, the automotive electrical drive systems are nowadays in focus for much attention. The reason is of course the need to replace the traditional ICE drivelines with cleaner and more energy efficient alternatives.

Road transportation by cars, buses and trucks is fundamental for the modern society and has contributed immensely to the standard of living, at least in the industrial world, but also in developing countries. These forms of transportation are dependent, almost 100 percent, on fossil oil as fuel, which has lead to well-known, negative consequences. Fossil fuel combustion results in different emissions, mainly carbon oxides (CO and CO_2), nitrogen oxides (NO_x), hydrocarbons and certain particles. All of them, except CO_2, have created health risks in densely populated regions, but these emissions have been almost eliminated by new technical solutions such as improved combustion, catalytic exhaust cleaners, filters etc. CO_2, which is a so-called "greenhouse gas", remains a huge problem due to its major contribution to global warming [406]. The other fundamental problem is that fossil oil is a limited resource and the current consumption exceeds substantially the discoveries of new reserves. The cost for crude oil is therefore predicted to increase in the future, but the deviations between various forecasts are big [407]. The situation will be even more critical as the population in countries like China and India, but also many other countries, increase their use of ICE-driven road vehicles. This is the background to why the automotive industry, although initially somewhat reluctantly, began looking into new alternatives. The urgency is emphasized by the fact that more than half of the current global crude oil production is used for the transportation sector [408].

What are the alternatives? One obvious option is a radical reduction of the fuel consumption, which can be achieved by the use of much smaller cars, more efficient engines (e.g. diesel instead of otto engines) but also through introduction of hybrid electric drivelines. The other alternative is to substitute the fossil oil with various renewable and CO_2 neutral fuels such as ethanol, methanol and biogas. Hydrogen and electricity are energy carriers which can be seen as alternative fuels, provided that they are produced by renewable energy resources, preferably sun-, wind- and hydropower, or by nuclear power. In case of hydrogen and electricity, FCVs and BEVs are likely solutions, but instead of the latter, "plug-in" HEVs, equipped with batteries that can be charged from the grid are also an option. HEVs can, of course, be used to improve the fuel economy also in the case of ICE-driven vehicles fueled by the renewable fuels mentioned above. Many of these options require electrical

drive systems, much larger than those used until now in traditional road vehicles. The extent of electrical drive systems can become enormous and it is
therefore relevant to look upon this as a new application for somewhat larger
electrical machines.

8.2.2 New customers

Electrical machines, in sizes varying from a few kW and up, have traditionally been manufactured by large and medium-sized electro-technical companies. Thousands and thousands of industries and other customers all over the
world use these products. The individual users have, however, had a very
limited influence on the machine development, which mainly has been gov-
erned by the manufacturers themselves. The situation becomes very different
in the case of large-scale use for automotive applications.

The automotive industry is the largest in the world dominated by a dozen
large, multinational car manufacturers such as General Motors, Toyota and
Volkswagen. It also includes manufacturers of heavy vehicles, e.g.Volvo and
Scania, as well as suppliers and sub-suppliers of different sizes. Examples of
very large suppliers are Bosch, Delphi and Denso.

The leading vehicle manufacturers are very demanding and specify in detail
the systems and components they want to have; it is seldom a matter of buy-
ing something from the shelf. This demand initially caused confusion and
disagreement between automotive and electrical companies. The former did
not have sufficient competence regarding electrical drive systems but knew
what was necessary to achieve; the latter were unwilling to go far beyond
their established technical solutions. One example can serve as illustration.
BMW, with its headquarter and development centre in Munich, developed,
during the early 1990's, BEV prototypes with drive systems from the small
American company Unique Mobility. Why did BMW not choose Siemens, a
very competent supplier also located in the Munich area? The project manager Dr. Andreas Gobeau explained in an interview, "*Siemens just told us
what we should have instead of listening to what we wanted!*" He added that
he first later realized the fundamental difference between the large electro-
technical industries and the automotive suppliers. The first both specified and
developed their products, which they offered to a wide and diversified market, while the latter developed products specified by a very few, big car manufacturers. The powerful automotive companies had more ambitious targets
than using existing technology. They have consequently become much more
competent in electrical drive systems and the electro-technical companies
have realized that it is impossible to compete with industrial motor technology for this special application.

It is not just a matter of tougher technical specifications; it is, in some aspects,
a new way of doing business. Electrical machines for drivelines are vital
components with a big influence on vehicle performance. The development

of these therefore becomes an integrated activity in the whole vehicle development and it is necessary for the machine supplier to adhere strictly to the vehicle manufacturer's project plans, quality assurance procedures etc. The automotive companies have their own general purchasing conditions [409], which contain several important deviations from the electro-technical suppliers' normal sales conditions [410]. An example of this is the requirement for "open books", i.e. complete presentation of all costs for material, labour and overheads as well as profit. Another example is product liability, which includes responsibility for consequential damages to an extent that could ruin many suppliers. Nevertheless, many manufacturers are attracted by the potential size of the automotive business and are therefore willing to take the risks. The large, established electrical machine manufacturers seem to be most reluctant, perhaps because they consider that there can be more to lose than to gain. Table 8.4, which contains a list of leading electrical motor manufacturers indicating those that actively promote automotive applications, supports this opinion.

Table 8.4   Electrical motor manufacturers - automotive drivelines?

Manufacturer	Electrical machines for automotive drivelines?	Other automotive products?
ABB	No	No
Alstom	No	No
Baldor Electric Co.	No	No
Brook Crompton	No	No
GE	No	Yes
Hitachi	Yes	Yes
Leroy Somer	No	No
Siemens	Yes	Yes
Toshiba	Yes	Yes
WEG	No	No

8.2.3   New requirements

A company that plans to start to develop and manufacture electrical machines for automotive drivelines faces a lot of new requirements. Far-reaching commercial conditions have been mentioned above. Low cost is not a unique target, it is common for almost all product development, but the automotive companies are known for being extra fierce in this respect. Other important requirements concern performance, environment and reliability.

The required performance for a vehicle with an electrical driveline is often expressed "that it shall be at least as good as for a corresponding conventional vehicle" which means that it shall have torque and power characteristics that could match what an ICE driveline would have had. The electrical motor can, in most cases, meet this requirement even combined with a single step reduction gear instead of a multi-shift gearbox. The motor must, however, have a large field weakening range as indicated in figure 8.20. Weight and dimensions are critical for mobile equipment especially in cars. High specific torque and power are therefore requested. High speed and forced cooling are means to reduce the physical size of the motor. Typical data for vehicle and industrial motors are presented in table 8.5. It is obvious that the vehicle motors must be designed with a much higher utilization factor than industrial motors of similar size. Another important performance criterion is the efficiency and also in this respect, the demand is very high on automotive drive systems. Higher efficiency results in a longer driving range as a road vehicle carries a limited energy source determined by the size of its fuel tank or battery.

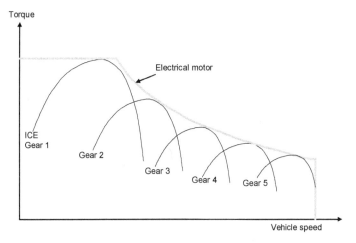

Figure 8.20 Torque and power characteristics for conventional and electrical drive system

There are two kinds of environmental requirements, one constituted by the ambient conditions under which the machines have to operate and another is the impact these products have on the environment. A vehicle motor must be able to function over a wide ambient temperature range, withstand mechanical vibrations and shocks and be protected against dust, water and other hazardous atmospheres. It shall also be designed with consideration to stringent EMC requirements, both for received and radiated electromagnetic fields. There are of course industrial motors, which also have to operate under difficult conditions, but standard motors are subject to less severe stresses than automotive motors. A comparison of some typical data is also included in table 8.5 [133, 411]. The request that the machines shall be free from hazardous materials and be recyclable is not different from other industrial products.

Table 8.5   *Specific performance data for standard industrial and automotive drive systems*

Application	Industrial	Automotive driveline
Specific output [kW/kg] Continous	~ 0.2	~ 1
Specific torque [Nm/kg] Maximum	3 - 4	4 - 25
Max speed [rpm]	< 4500	> 10000
Efficiency [%] Rated load	91 - 95	92 - 96
Coolant temperature [°C]	< 40 air	80 - 100 liquid

Reliability has an impact on safety, which is a core issue for the automotive manufacturers. The motors must function in such a way that people are not exposed to risks to life or health. Safety is more emphasized in case of vehicles, which are mostly driven by private persons in public streets and roads as opposed to industrial motors which are usually handled by skilled personnel and are placed in restricted locations. Another important reliability requirement is the minimum "mean time between failures" (MTBF) that determines the risk for standstill caused by these specific products. According to information from one of the Swedish car manufacturers, they specify instead a maximum number of "repairs per thousand vehicles" during the guarantee period, RPTV = 20. This was a total number for complete cars, while the requirement on critical sub-systems was no failure at all during 160 000 km, i.e. half of a car's entire lifetime [412].

There is also a request on delivery reliability. The vehicle manufacturers operate their factories with "just in time" deliveries that do not allow delayed supply of components. Neither can any faulty components be accepted because they would also stop the production line. Supply contracts usually specify large penalties both for delays and faulty components.

8.3   The electrical machines

DC motors have long been used for electric vehicle drive systems and they are still used for some BEVs. The development, during the last 10 – 20 years, has however been focused on AC machines including so-called brushless DC motors. The traditional DC motors cannot be considered as an important base for the more recent development and they are not used in any of the modern HEV and FCV projects. They will therefore not be subject for explicit study in this part of the thesis. Neither will electrically excited synchronous machines be included, even if they nowadays are used as alternators for cars and trucks. The reason is that they have not, so far, been in practical use as traction motors for road vehicles in spite of the fact that studies have been performed and prototypes have been built [413, 414]. A recent example is a research project carried out at the university in Lund [82].

8.3.1 Induction machines

Induction motors have a number of properties that make them suitable for electrical drivelines. A fundamental prerequisite is of course the availability of good inverters enabling variable speed operation (see section 6.7.7). The squirrel-cage motor is cost effective and very robust. It can be built for high speeds and can operate at comparatively high temperatures. Both properties contribute to reduced physical dimensions and weight. The induction motor is a reliable machine type and its efficiency is better than that of DC motors. It has a suitable torque characteristic, which means that it can be field weakened even if the constant power region is somewhat limited, as shown in figure 8.21, and it does not need any speed or rotor-position sensor for its control. [415]

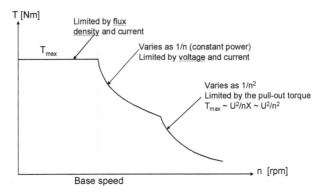

Figure 8.21 Torque and power characteristics for field weakened induction motor

The induction machines have no significant disadvantages for this particular application. They are good but simply not the best when comparing performance data such as specific torque, efficiency and power-factor. Nevertheless, they have been the preferred machine type in many vehicle projects, especially in the USA. The connection to production of industrial motors has for obvious reasons been particularly strong in the case of induction machines. It has been economical to utilize parts from the large-scale production of industrial standard motors.

Induction motors started to replace DC motors in railway traction applications during the 1970's. The first serious attempts to use them in road vehicles came nearly one decade later though singular test vehicles had already been demonstrated in the late 70's [416]. Professor Hans-Christoph Skudelny, at the well-known Rheinisch-Westfählischen Technischen Hochschule (RWTH) in Aachen, published for instance an investigation on this type of drive systems in 1982 [417]. The same year GE and Ford commenced the development of an induction motor drive system for an electric car, ETX-I, funded by the U.S. Department of Energy (DOE), which was completed in 1985 [418].

An early example of an induction motor driveline is a project for a couple of electric cars built and demonstrated in Finland around the mid 1980's. These BEVs were based on chassis and bodies of a Talbot Horizon passenger car manufactured in Finland by Saab-Valmet. The electrical drive system used was developed by Strömberg. The induction motor was a standard enclosed squirrel-cage motor normally used for industrial applications and it was flange mounted to the cars standard clutch and gearbox. It had a rating of 20 kW, peak power of 27 kW and a maximum speed of 4800 rpm. The motor weighed 125 kg, which means a peak power density of 0.22 kW/kg, i.e. considerably less than two of the DC-motors listed in table 8.1. It was obvious that standard induction motors were insufficient for automotive drivelines [419].

The next step was to take advantage of the expensive tools used for the manufacture of standard motors, but to perform modifications necessary to achieve a lighter and more compact motor for a certain output. Such a motor was built by ABB Motors in Västerås in 1991. This motor is actually an example of an activity, which started as a "skunk work", i.e. not an officially authorized job, which later became a showpiece. The motor was intended for a series HEV provided with a gas turbine-driven generator. The base for this motor was a standard 4-pole induction motor, frame size 132, with a nominal rating of 7.5 kW at 1500 rpm that was redesigned so that it could develop 40 kW continuously and 70 kW in peak. The maximum torque was 180 Nm, the base speed 6 000 rpm and the maximum speed 12 000 rpm. The increased output was achieved through higher speed, water-cooling of the stator and closed loop air-cooling of the rotor, much thinner lamination sheets in the core and an improved stator winding fill-factor. The motor also had to be provided with special bearings for higher speed. The weight of the motor was 50 kg. The peak power density was, through the modifications, raised to 1.4 kW/kg, i.e. far more than 0.5 valid for the corresponding standard motor. The maximum torque density was 3.6 Nm/kg [154, 398]. The hybrid car as such was first presented at the International Automobile Exhibition in Paris in 1992 as Volvo's Environmental Concept Car (ECC) and it received huge public attention for a long period.

High rotational speed and liquid cooling were the keys to small induction motors. There are many examples of this from the years around 1990. Two interesting cases are the motors for GM's BEV "Impact" and a motor developed primarily for Ford's "Ecostar".

GM presented its futuristic concept car "Impact", in January 1990, at the Los Angeles Auto Show. The car was originally provided with two 42.5 kW induction motors, one for each front wheel. The concept with two motors was later abandoned after one car crashed as a consequence of a sudden failure in one of the motors. The new motor, which was developed by GM's daughter company Delco, had a rating of 103 kW and 141 Nm in peak at 7000 rpm. Its maximum speed was 14000 rpm. This 4-pole induction motor was liquid-cooled with a copper squirrel-cage winding in the rotor. It had a weight of 68 kg including an integrated reduction gear and differential. Assuming 2/3 of the weight can be attributed to the motor itself, the specific peak output and torque can be estimated at 2.3 kW/kg and 3.1 Nm/kg respectively. These values are quite high for an induction motor [420]. As a curiosity, the "Impact" set a speed record for electric cars in 1994 when it reached 295 km/h. This required a special gear ratio and crushed ice in the cooling system to enable the power overloading.

The "Impact" was a forerunner to GM's "EV1" that was intended for series production, primarily in order to meet the Californian mandate in 1998. "EV1" had practically the same motor as the "Impact". Around 1150 units were manufactured before the production was stopped in 1999. Figure 8.22 shows a photo of an "EV1".

The man behind the Impact drive system, Alan Cocconi, established in 1992 a small California based development company, AC-Propulsion, which later has developed a forced air-cooled induction motor with high performance data. It is rated 50 kW continuously and 150 kW in peak. It has a maximum torque of 220 Nm, base speed of 5 000 rpm, maximum speed of 10 000 rpm and its weight is only 50 kg. This means that the specific peak output and torque are as high as 3.0 kW/kg and 4.4 Nm/kg respectively. A general comment concerning many of the motors used as examples, both in this section and the following, is that the given peak torque, peak power and base speed values are not consistent. The explanation is that there is simply not just a constant torque and constant power region.

In 1990, after the earlier research efforts mentioned above, Ford and GE started a development program aimed at creating a few, commercially viable, drive units for electrical cars. One of them, a 75 hp (55 kW) unit, was intended for the "Ecostar" and specialists from GE documented the development in a conference paper in 1992 [421]. Two years later, a drive system that was, in principle, the same as this one, was presented in a common paper from Ford and Siemens [422]. Siemens had replaced GE as partner to Ford. The system had to be cost effective, have high performance and should also have good manufacturability for both low and high production volumes. The concept chosen was a transaxle induction motor integrated with a planetary reduction gear and differential as shown in figure 8.23.

Figure 8.22 GM EV1

The frame was made from cast magnesium in order to reduce the weight as much as possible. The motor was liquid cooled by means of the lube oil, which thus was used for double purposes. Ford's motor was a 4-pole squirrel-cage motor with a rated output of 44 kW, maximum power of 55 kW and maximum torque of 190 Nm. The base speed was 3 000 rpm and the maximum speed 13 000 rpm, which meant a very large field-weakening range. An assessment of the physical size, from published material, indicates a similar stator diameter as for the ABB Motor described above, but 40 percent larger active length. The difference in utilization can depend on the fact that the Ford motor was oil-cooled where as ABB's was water-cooled. Ford studied different manufacturing strategies and decided to outsource the motors to a supplier with more flexible production facilities compared with a big car manufacturer. The projected production volumes were too small and thus Ford entered into a partnership with Siemens. It is not evident how they shared the development work, but it is probable that Siemens focused on the active parts of the motor while Ford took care of the transmission and the case. The latter looks very typical for automotive design.

Figure 8.23 Ford/Siemens transaxle motor with reduction gear

Toyota is a company that has played a major role in the development of electrical drivelines for automotive applications and it is therefore of interest to follow its activities. At the International Electric Vehicle Symposium (EVS-10) in Hong Kong in 1990 Toyota contributed a paper on an induction motor drive system which it had developed for passenger cars and light trucks [423]. The paper describes an air-cooled 20/30* kW motor with an integrated reduction gear. The data presented does not indicate that it had extremely high performance. It was physically larger than the more powerful ABB motor mentioned above, but it has to be taken into account that it was air-cooled. Nonetheless, Toyota could hardly be considered a leading company in the development of electrical motors at that point in time.

Several large electro-technical companies developed and built prototype induction motors for electric vehicle drive systems in the early 1990's. ABB, GE and Siemens have already been mentioned. Additional examples are AEG, Westinghouse and Hitachi. AEG's motor had a maximum output of 30 kW and 160 Nm at 1900 rpm. Its maximum speed was 9000 rpm and the weight 80 kg. The specific output was thus very low, 0.38 kW/kg in spite of the fact that it was liquid-cooled. This depends of course to a large extent on the low base speed, but even the specific torque was fairly low, 2 Nm/kg, which indicates that it was in reality only a slightly modified standard motor [424]. Westinghouse and Hitachi had gone somewhat further in their development projects, which can be seen in their presented papers [425, 426]. Westinghouse had for instance, like GE and Ford, chosen to make the housing of magnesium. It is of interest to note that out of the six companies mentioned above, only Siemens and Hitachi are manufacturing electrical motors for automotive propulsion today. This will be further discussed in section 8.4.

Several drive systems, with induction motors, were developed and demonstrated during the first half of the 1990's. Some examples were presented at EVS 12 in Anaheim, California 1994. Two of GM's divisions had developed a drive system for shuttle buses, which had two liquid cooled wheel motors and a total peak power of 2 x 70 kW. Each motor had two parallel 3-phase windings, fed from different inverters, and a squirrel-cage rotor with brazed copper bars [427]. GE also presented a hybrid bus with induction wheel motors rated at 75 kW, with oil-cooling and planetary reduction gears. The design was based on the motors originally developed by GE in cooperation with Ford as mentioned above [428]. Two Austrian papers were presented as well, one which was from AVL describing a hybrid drive system, functionally similar to Toyota's Prius system, but with induction machines [429]. The other was from Steyr-Daimler-Puch describing an integrated unit, demonstrated in a Volkswagen Golf, consisting of an induction motor, an inverter and a reduction gear [430]. Both projects were focused on system design, while the induction motors seemed to be fairly normal. This is natural as both companies were suppliers of more traditional automotive components.

---

*. Continous power/peak power

Siemens has maintained a focus on induction machines, especially for HEVs and FCVs in which the driving range is less dependent on the drive system efficiency compared to BEVs. In a paper published in EVS 18 in 2001, Siemens claimed a cost advantage of 30 percent for induction motors compared with corresponding PM motors. Data was presented for a 75/120 kW, 150/240 Nm motor with a base speed of 5000 rpm and a maximum speed of 12500 rpm. It had a weight of 82 kg, which gives a specific peak output and torque of 0.91 kW/kg and 2.93 Nm/kg respectively. The specific power is difficult to compare due to differences in base speed but the specific torque is a relevant figure for comparison. It is worth noting that the torque density is not any higher than corresponding data previously mentioned for 10 years older induction motors [431]. Even if it isn't evident what the prioritized development objectives were, the performance data quoted show that the development pace for induction motors has been slow during recent years. Another paper, published by GM at the same occasion, confirms this conclusion. It describes and presents data for an induction motor drive system of similar size as the one referred to above. The development focused more on the integration between motor, reduction gear and differential, which all were built into one unit [432].

A review of the proceedings from the EVS conferences in 2003, 2005 and 2006 as well as from the IEEE Vehicle Power and Propulsion conferences in 2004, 2005 and 2006 reveals that most papers dealing with induction motors are focused on control issues, especially sensorless direct torque control (DTC). The use of induction machines as integrated starters and generators (ISG) is covered by a few reports. Only a couple of papers give any information on motor design and technical data, but these do not indicate any extensive ongoing development. Most interesting in this respect is a contribution on improved efficiency in an ISG with a die cast copper rotor [433].

8.3.2 PM machines

The access to new, powerful permanent magnet materials, especially NdFeB, during the 1980's led to an intensive development of PM machines for different applications. One of these was electrical drive systems for vehicles. The PM machines have a number of advantages, which are important for this application. Most significant are their superior torque and power density and the high efficiency. They are also suitable for several interesting machine topologies, partly because the magnets high coercivity force allows fairly large airgaps. They can be built for both high and low speeds and be accurately speed controlled, even though the field weakening requires special measures. The most common method is to control the phase angle of the stator current so that it will have one component that gives an mmf, which counteracts the mmf from the magnets. This principle is illustrated in the diagram of figure 8.24. More detailed explanations can be found, e.g. in the PhD theses included in the reference list [434, 435].

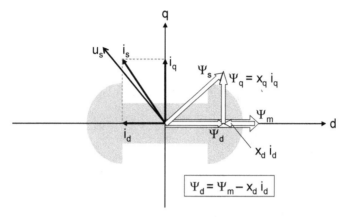

Figure 8.24 Phasor diagram illustrating field weakening

There are two major disadvantages with PM machines containing NdFeB magnets. One is the high cost for the magnets and the other is that they are relatively temperature sensitive. The magnets will experience a partial, irreversible loss of their flux density above certain temperatures, usually in the region of $120 - 180$ °C depending on material grade. Another disadvantage is that the PM motors usually need expensive resolvers for sensing the rotor position during operation at very low speeds. It must also be taken into account that it is more difficult to handle magnetized components during production.

Development of PM machines for electric drivelines began in the second half of the 1980's in America, Europe and Japan. A number of different concepts were introduced, some of which will be briefly presented below. These concepts illustrate how the border lines for electrical machine development were expanded due to the access of new material.

The Colorado based R&D company Unique Mobility, now known as UQM, had already been building simple BEVs for some years when it started to experiment with electric drive systems 25 years ago. Customer financed development of high performance PM motors with controllers soon became the core business for the company. Prototypes were sold to different vehicle manufacturers. A contract from BMW, in 1990, for a transaxle PM motor was extra important. UQM happened to pay a visit to BMW, which had started its BEV development. BMW's project manager, Dr. Andreas Goubeau found UQM's concept very promising and the American company was also very sensitive to BMW's requests. The motor was to be rated 32 kW, maximum torque 140 Nm, base speed 2200 rpm and maximum 8000 rpm. UQM chose to develop a motor with a high pole number, 18 poles, in order to reduce the weight as much as possible. The flux paths become short and the flux per pole small, which resulted in a very thin stator core and rotor ring. The iron losses were therefore small in spite of the high frequency, especially since the rotor had surface mounted magnets, which acted as a large airgap reducing armature reaction flux harmonics.

Figure 8.25 Unique Mobility's PM transaxle motor for BMW's E1

The end windings in the stator also became very short thanks to the high pole number which helped to reduce both weight and copper losses. The magnets were made from NdFeB and they were secured by a thin titanium ring. Water-cooling was provided through the stator casing. A picture of the motor is presented in figure 8.25. The weight of the motor was 38 kg resulting in a specific power of 0.84 kW/kg and maximum torque density of 3.7 Nm/kg [436, 437].

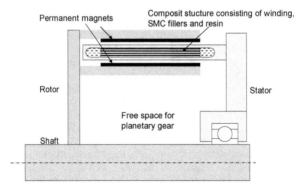

Figure 8.26 Sketch of Unique Mobility's double rotor topology

UQM tried, for a period, to use a topology in which the rotor consisted of two concentric rings with the stator placed in between. It had both inner and outer magnets and the stator did not need a "core back", i.e. it consisted only of slots and teeth. This concept reduced the iron losses to a minimum, but it required a special technology for manufacturing the stator. The concept was abandoned due to insufficient mechanical stiffness as it required an overhang design of both rotor and stator, which is shown in the sketch in figure 8.26.

PM motors with surface mounted magnets have inherently low inductances (both L_d and L_q and consequently X_d and X_q) and are therefore difficult to field-weaken by means of a component of the stator current (I_d) as seen in the phasor diagram in figure 8.22. UQM nevertheless managed to obtain a large constant power range, from 2200 to 8000 rpm, by substituting the field-weakening principle with a so-called "phase advance control". This method presumes six-step wave form and fairly low motor impedance; so that a certain winding can be "pre-charged" with current until the induced voltage becomes larger than the battery voltage. The current continues to flow for some time until commutation to the next winding occurs. More and more of the current is used for "pre-charging" instead of torque production as the speed increases and the result is roughly constant power [438].

UQM was a typical development company and BMW therefore wanted also to involve a larger company with adequate manufacturing resources. This desire led, in 1992, to a three party agreement between ABB, BMW and UQM, according to which ABB, as a first step, should evaluate UQM's technology. The result of the evaluation was that ABB did not find sufficient synergies with existing production and lacked also the capacity to start something radically new. The cooperation was therefore terminated [439]. This case illustrates the difficulties well established manufacturers often have to adopt new concepts. UQM has continued its activities with development of vehicle drive systems using the concept described above, mainly for American customers.

The German company Magnet Motors belonged to the pioneers for development of PM machines for vehicle propulsion. Its activities have mainly been focused on development and prototype manufacture of equipment for military and other heavy vehicles. Around 1990 the company presented a PM machine design with very high specific torque. The machine can be described as a "step motor" and as such has a different number of poles in the rotor and in the stator. It is an outer rotor design with NdFeB magnets in the rotor. The pole number is high and the rotor ring becomes very thin resulting in a large airgap diameter compared with more conventional topologies and hence also a larger torque. The inner stator consists of a number of salient poles provided with coil windings. Liquid cooling of the stator contributes to a high utilization factor. Figure 8.27 shows a cross section of a Magnet Motors machine with x poles in the rotor and y in the stator. In a paper from 1993, the authors claim that such a machine has a maximum specific torque as high as 16 Nm/kg and a maximum specific power of 2.4 kW/kg. It is obvious that these motors could be classified as high-torque motors operating at fairly low speeds; the maximum speed was in this case 3200 rpm [440].

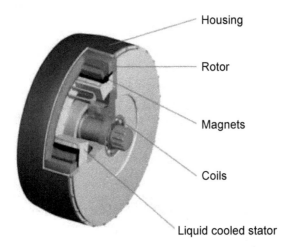

Figure 8.27 Sketch of Magnet Motors outer rotor PM machine

Magnet Motors machines require very special converters. Each stator pole is fed from a separate four-quadrant converter, which means that the total number of power switches is very high. The machines are usually not built for high speed. The concept is suitable for wheel motors, but also for more traditional installations. One example is that Volvo used such motors in two concept vehicles built in 1994; the environmental concept truck ECT and the corresponding bus ECB. This motor had the following key data: peak power 150 kW, peak torque 2850 Nm, maximum speed 2950 rpm and weight 100 kg [441]. Magnet Motors has licensed its technology for series production to ZF Sachs, an established supplier of transmissions, steering systems etc. to the vehicle manufacturers.

In this thesis, it is of particular interest to review a couple of Swedish efforts to develop PM machines for automotive applications. The first example started as a research project at KTH as early as 1987. The intention was to develop a small gas turbine-driven high-speed generator (HSG). A pre-study had shown that a 2-pole PM machine was the best alternative. The project included development, manufacturing and testing of a 20 kW, 100 000 rpm synchronous generator. The rotor consisted of a cylindrical NdFeB magnet, contained inside a non-magnetic steel sleeve, and shaft ends. The stator had a 3-phase, toroidal airgap winding made from litz wire[*]. The stator core was only a steel ring stacked from thin lamination. The bearings were ceramic ball bearings. Figure 8.28 shows a simple drawing of this high speed generator. It had a sinusoidal flux, very low inductance and comparatively low losses. It was tested in a special test cell in the electrical machine laboratory driven by a primitive gas-turbine. The generator was used as a motor for starting the gas turbine. Figure 8.29 contains pictures from the laboratory tests [442].

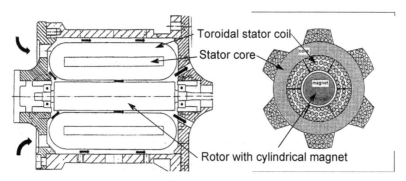

Toroidal stator coil

Stator core

Rotor with cylindrical magnet

Figure 8.28 Drawing of first HSG generator

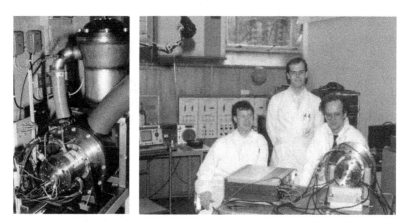

Figure 8.29 High speed generator with gas turbine and first run of HSG

The results were positive and a decision was made to continue and make a full scale demonstration in a HEV application. The development of a 40 kW, 90 000 rpm unit started therefore in 1990 in cooperation with ABB, Vattenfall and Volvo Aero. The initial idea was to demonstrate the power unit in a converted Volvo station wagon, but after an agreement with Volvo

---

*. Conductor consisting of several hundreds or thousands twisted parallel very thin copper strands

Cars, it was decided to implement the system in a new concept vehicle, the Volvo ECC mentioned in section 8.1.4. This series HEV was presented publicly for the first time in October 1992 at the Paris International Automobile Exhibition. It worked well for being a concept car and it rendered wide public attention. It was therefore followed, a couple of years later, by a truck, Volvo ECT, and a bus, Volvo, ECB. The high speed generator developed for these vehicles was rated at 100 kW at 70 000 rpm. Both the 40 and the 100 kW generators had the same conceptual design as the first 20 kW generator mentioned previously [398, 443]. The smaller turbine/generator unit was also supplied to Renault for installation in a French HEV project.

The HSG was an advanced concept and the inventors, Peter Chudi and Anders Malmquist, were awarded the prize "Designer of the Year" in 1993 from the Swedish National Board for Industrial and Technical Development. The development and manufacturing was in many respects supported by expertise and resources from ABB. In spite of successful demonstration projects, displaying in total six Volvo HEVs with both sizes of power units mentioned above, the concept was disbanded and not used for further automotive development. The main reasons were timing and costs. Automotive manufacturers were not seriously interested in HEVs until the end of the 1990's and the gas turbine had no chance to be cost competitive with conventional ICEs as prime mover for this size of generators. Another drawback was that this type of power unit is only suitable in series HEVs. Instead a redesigned version of the larger HSG came later to be used for stationary combined heat and power generating units.

In 1997 ABB received an order from BMW for development and supply of a drive system for BEVs, which were required so that the company could meet the Californian mandate in 2003. It has already been mentioned that BMW, some years earlier, worked in cooperation with UQM and, to some extent, ABB. BMW had also, for a period, worked with the French company Auxilec, but was searching for a larger manufacturer as its supplier of series produced drive systems. Contacts had been maintained between ABB and BMW, but the final development contract was a result of a formal evaluation procedure with several competing quotations.

BMW's specification required a PM motor rated 32 kW continuously and 50 kW in peak, 192 Nm maximum torque, 2500 rpm base speed and 10000 rpm maximum speed. It should be a water-cooled transaxle unit with an integrated inverter. The requirements on dimensions and weight were tough. ABB decided to develop an 8-pole motor with buried NdFeB magnets in a laminated rotor. A cross section of the motor is shown in figure 8.30. The inverter unit was placed directly on top of the motor as the photo in figure 8.31 shows. A major difficulty for the design was to obtain high performance over the large field-weakening range, combined with a large voltage range, 200 - 340 V, and the latter depending on the state of charge in the battery. Figure 8.32 contains an efficiency diagram, which illustrates one aspect of the motor performance [444].

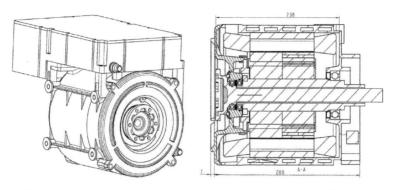

Figure 8.30 3-D and cross section drawing of ABB's BMW motor

Figure 8.31 Photo of ABB's BMW motor with integrated inverter

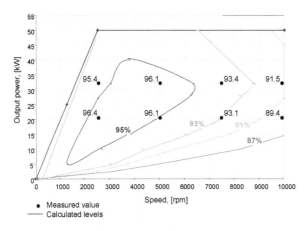

Figure 8.32 Efficiency map for ABB's BMW motor

ABB supplied a number of these systems to BMW for bench testing and for installation in cars. The project never resulted in series deliveries because BMW decided, in 1999, to terminate the development when CARB changed the Californian requirements. It had been a fairly large project for ABB with participation from a number of units including ABB Hybrid Systems, ABB Motors and ABB Corporate Research. The project gave useful technical knowledge both concerning PM motors and the control of such motors. It also showed ABB how difficult it can be to enter into a new market, which the company has no control over.

There are other Swedish examples of development of PM machines for automotive applications, e.g. the 4QT machine at KTH mentioned in section 8.1.4. PM machines have also been developed in research projects at Chalmers and, during recent years, primarily LTH. However, none of these have, as of yet, been installed in real vehicles and they are therefore left without further comments.

The recent international development of PM machines for vehicle drivelines shows interesting examples of different machine topologies. Most important, or at least most produced, is Toyota's "Prius motor" and its successors. It is conventional in the sense that it is a radial flux, inner rotor, 3-phase, 8-pole PM machine. It has buried NdFeB magnets in the laminated rotor, which has resulted in a certain reluctance torque contribution. The stator has a distributed winding and is liquid-cooled by both oil and water. Lube oil is sprayed over the end-windings and flows to the lower part of the casing which is water-cooled. Figure 8.33 contains a photo of an exhibited Prius driveline. The motor in the present Prius model is rated 50 kW and 400 Nm in peak. The base speed is 1200 and the maximum speed is 5600 rpm. The active weight of the motor is estimated to be about 35 kg. It is worth noting that the concept of this motor and that of ABB's "BMW motor" are fairly similar. Both motors have buried magnets, the same pole number and the same number of slots. This is clearly shown in figure 8.34 containing photos of the stator and rotor laminations for both motors. An attempt has been made to assess various data for the Prius motor and compare them with corresponding data for ABB's motor. The result is shown in table 8.6. It is evident that Toyota's motor has a much higher linear current loading and hence utilization factor. This fact can mainly be attributed to the efficient oil-cooling of the inner parts, but also to the fact that the Prius motor has more favorable operating conditions than the "BMW motor". The latter should have full performance in a wide voltage range, 180 – 320 V DC, depending on the battery's state-of-charge and the load. Toyota uses a DC/DC converter between the battery and the DC link resulting in reduced requirements on field weakening and a possibility to control the motor voltage so that the motor losses can be minimized. This is illustrated by the diagrams in figure 8.35. A comprehensive report on PM motors for HEV applications has been issued by Oak Ridge National Laboratory. It contains, among other things, a detailed analysis of the Prius motor [445].

Figure 8.33 Prius drive unit

Figure 8.34 Core lamination of ABB's BMW motor and Toyota's Prius motor. Both motors have the same airgap diameter

*Table 8.6   Comparison between Toyota's Prius-motor and
ABB's BMW motor*

Alternative	Toyota	ABB
Maximum power [kW]	50	50
Maximum torque [Nm]	400	192
Maximum current [A]	220	290
Base speed [rpm]	1200	2500
Maximum speed [rpm]	6000	10000
Pole number	8	8
Stator core diameter [mm]	269	232
Rotor diameter [mm]	160.2	158.8
Outer diameter [mm]	~ 285	250
Active length [mm]	83.5	125
Airgap [mm]	0.9	0.6
Slot number	48	48
Core lamination thickness [mm]	0.35	0.35
Utilization factor [kW/rpm m3]	19.44	6.34
Cu fill factor [5%]	50.6	41.1
Cu diameter [mm]	0.9	0.75
No of parallel strands	13	16
Cu area	8.27	7.07
Max current density [A/mm2]	26.6	10.3
Weight, active parts[*] [kg]	34	36
Magnet thickness [mm]	6.5	4

[*].   Estimation made by the author

Toyota used the "Prius motor" as base for the development of a more than twice as powerful motor for the SUV hybrid Lexus RX400h. The Toyota engineers claim that they were aiming at building the world's top-level motor with respect to compactness and weight. This new motor is rated 123 kW, 333 Nm in peak and the base and maximum speeds are increased to 3530 and 12400 rpm respectively. The torque is a little less compared to the Prius motor so the high power density is obtained through the increased speed. The magnets are placed in V-formation in each pole, which enhances the reluctance torque. The new motor has thinner lamination to compensate for the higher frequency. The weight is not published but can be assumed to be slightly less than for Prius. This gives a specific peak power around 3.0 kW/ kg and torque 8.3 Nm/kg [446]. Toyota developes the motors in-house and the company is today one of the worlds leading manufacturer of electrical machines with an annual volume of approximately 500 000 units or 30 000 000 kW. The corresponding figures for ABB's LV motor factory in Västerås are in the order of 100 - 150 000 units and 1- 2 000 000 kW.

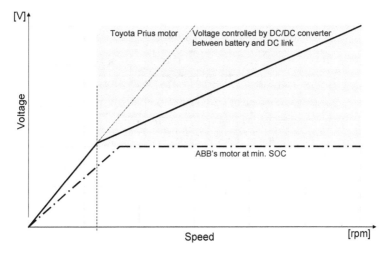

Figure 8.35 Diagrams explaining control of ABB's BMW motor and Toyota's Prius motor

The use of electrical propulsion opens good possibilities to drive and control each wheel individually through wheel motors. The advantages with this are improved control and more freedom in vehicle layout and packaging. Wheel motors need to be compact and PM machines are therefore the main alternative. One suitable concept is to use some type of outer rotor, radial flux machine like Magnet Motors' described above. Figure 8.36 a shows an example of a similar wheel motor with reduction gear developed by TM4, a Canadian development company owned by the power utility Hydro Québec [447]. The length of these motors is usually critical and axial flux motors have therefore been developed in some projects. A couple of papers presented at the EVS 21 conference in Monaco 2005 deal with axial flux wheel motors. One paper, from the American bearing manufacturer Timken, contains a comparison between axial and radial flux motors designed to meet the same specification [448]. The peak power was 100 kW, which is rather high for wheel motors, and the radial flux motor turned out to be 30 percent heavier than the axial flux variant, but the volume was almost 70 percent larger. Axial flux motors can be built very compact as shown in figure 8.36 b. The peak torque density was 13.6 Nm/kg. Another axial flux wheel motor has been developed by GM in cooperation with the University of Rome. It has a toroidal stator winding placed between two rotating discs containing the magnets. The winding is placed in radial slots in the spiral wound stator core. The stator is liquid cooled. This motor is intended for direct drive, i.e. there is no necessity for a reduction gear [449]. GM's motor is conceptually very similar to an experimental "Axial flux Torus machine" developed as part of the 4QT project at KTH in 2000 – 2002 [450]. A comparison of these two motors is included in table 8.7. GM's motors have been demonstrated in a Chevy van.

Table 8.7   Data for toroidal winding, axial flux PM machines

Developer	GM	KTH
Power cont/max [kW]	16/25	29/not specified
Torque cont/max [Nm]	200/500	70/not specified
Speed base/max [rpm]	750/1200	4000/6000
Outer diameter [mm]	390	320
Total length [mm]	95	115
Weight [kg]	30	22
Number of poles	24	6
Specific power [kW/kg]	0.53/0.83	1.31/-
Specific torque [Nm/kg]	6.7/16.7	3.2/-

Figure 8.36 a) Radial flux and b) axial flux wheel motors

GM also presented, at the EVS 21 conference, a report on a study of a trans-versal flux, 3-phase PM motor. This project was primarily aimed at a high torque density drive for hybrid buses. The paper reports that SMC is used as material in the magnetic circuits and that the major problem is low power factor, which is typical for transversal flux machines [451]. Interesting, from a Swedish perspective, is the use of SMC as Höganäs AB is a world-leading supplier of this kind of material. Furthermore, a couple of research projects at KTH are based on transversal flux technology for HEV applications.

Radial flux PM machines can be built with surface mounted, inset and inte-rior (buried) magnets. The latter two provide a certain saliency and can there-fore produce a reluctance torque in addition to the torque generated by the magnet flux. They also have a wider field weakening range than motors with surface mounted magnets. A recent American study compares machines with different combinations of surface mounted and interior magnets as well as distributed and concentrated windings. The machine is intended as a direct-driven starter/generator with a very high constant-power speed range; 10:1. The report does not nominate any "global best" solution, but shows that the interior magnet machine has some important merits, even cost wise. The ma-chine with surface mounted magnets is lighter but needs a concentrated wind-ing to meet the tough speed range requirement [452].

The major drawback for the PM motors is their high costs, mainly caused by the price of the magnets. It is of course also a matter of production volumes. The diagram in figure 8.37 shows anticipated costs versus volume for a 50 kW motor. The source is an American report and the motor is probably an UQM design [453].

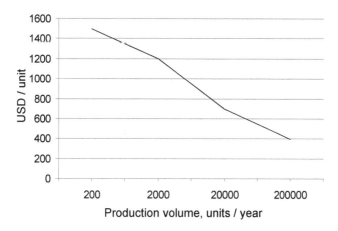

Figure 8.37 Production costs vs. volume for 50 kW PM motor

8.3.3 Reluctance machines

There are two different types of reluctance machines, the synchronous and the switched reluctance types (see section 4.8.2 and 4.8.4 respectively). Both can be used for vehicle applications and there are a number of examples available. Most of them are switch reluctance (SR) machines and the following comments are therefore focused on these.

The SR machines have a simple and cost effective design. The rotor consists only of a shaft and a laminated steel core with saliencies. These machines are robust and can be designed for operation at very high temperatures. They are similar to the induction machines in the sense that the magnetic flux is created by the stator current, but an important difference is that a SR machine has no rotor winding and hence no large rotor losses. It can easily be designed for different speed ranges and the speed control principle is simple, though some experts claim the implementation to be difficult. Some sort of accurate rotor position control is usually required. They need special converters instead of normal 3-phase PWM inverters. The SR machines have also been known for having high torque ripple and noise level but seem to have improved in these respects. In order to obtain good performance, a SR machine must be designed with small airgaps, which can be a complicating factor depending on possible bearing arrangements. Figure 8.38 shows a photo of the active components in a SR machine.

Figure 8.38 Active parts of SR machine (courtesy SR Drives Ltd)

SR machines have received more attention, during recent years, in relation to automotive applications. Three papers on SR motors were presented at EVS 21, all of them European. The papers describe motors with water-cooled stators [454, 455, 456]. The chosen number of stator and rotor poles are high; 24/18, 18/12 and 24/16 respectively. They have rather high field weakening ranges, in the order of $4 - 5$. Two of the papers focus on the different switching modes used to cover such large speed ranges. A typical torque and power characteristic is presented in figure 8.39. The specific performance is not consistent. One of the machines has a peak torque density of 4.2 Nm/kg and the next one has 15.2, both of them based on active weight. None of the commercially available BEVs or HEVs is equipped with SR motors, in spite of the wide attention these motors have received. The American automotive supplier Dana Corporation does, however, offer SR type starter/generators.

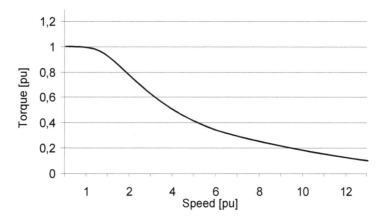

Figure 8.39 Torque and power vs. speed diagram for SR motor

There is, at present, no manufacture of SR machines in Sweden, but the company Emotron, located in Hälsingborg, is supplying drive systems for certain stationary applications with this type of motor. Emotron started this activity in 1984 and has developed both the SR motors and the converters with motor control. The motor production is nowadays outsourced to foreign manufacturers. Emotron has also participated in international vehicle projects with its SR technology.

Synchronous reluctance machines have also been used in a few demonstration projects, e.g. in Italy, but they have never received wide attention and are therefore left without further comment in this study.

8.3.4 Comparisons

The electrical machines used in conventional modern road vehicles are more standardized than most of the other components. Generators, start-motors (starters) and small motors for all kind of functions are standardized by manufacturers like Bosch, Valeo and others. When a vehicle manufacturer needs to increase the generator capacity, he often prefers to take two or three standard units in parallel instead of specifying a larger one. The situation is completely different for larger machines, which are to be part of drivelines. No standard products exist as of yet, and the possibilities of finding and reusing suitable existing machines are limited. Therefore it is rather a matter of choosing the best machine type and topology for a certain driveline and then trying to find a supplier that can design and build it. The previous sections show that there are many concepts to choose between.

A comparison of different machine types must be based on a number of factors that are relevant for each type of vehicle. Such factors are listed in table 8.8, in which the importance for various applications is also indicated. This list can be criticized for being subjective, but it has been reviewed during interviews with a number of specialists from automotive companies. It is obvious that factors as cost, performance, size and reliability are rated with high importance.

Table 8.8   Rating of important factors for electrical machines in different vehicle applications

Type of vehicles	Passenger cars	Heavy, commercial vehicles
Technical performance	1	1
Weigth and dimensions	1	2
Efficiency	2	1
Controllability	1	1
Robustness	2	1
Cost	1	2
Reliability	1	1
System safety	1	1
Reliability of delivery	1	2
Supplier service	2	1

Specific power and torque expressed in kW/kg and Nm/kg are essential. It can be argued that volumetric numbers such as kW/liter and Nm/liter are equally, or even more important for evaluation of the physical machine size, but they are unfortunately more difficult to find due to lack of published information. The comparison, based on weight, gives however a good indication of the relative performance of different machine types. The specific peak torque, given in different sources referred to in previous sections, and the corresponding figures for specific power are plotted in the diagram in figure 8.40. It is obvious that the specific power is fairly similar for the different machine types while PM motors are superior in the case of specific torque. The explanation is that they can more easily be adapted to low-speed, high-torque topologies compared to induction motors.

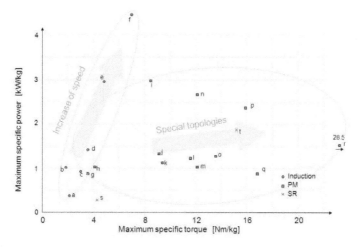

Figure 8.40 Specific power and torque for different machine types

Motor and converter efficiencies are important for fuel consumption and, in case of BEVs, also for the driving range. Power electronics have high efficiency that does not vary much with motor type, provided that the power-factor is reasonably high. The motors can show larger differences in maximum efficiency, but this is a difficult criterion to use for relevant comparisons. A vehicle motor normally operates with a highly varying load, which requires that the total efficiency map has to be considered. The driving cycles are very important for the energy efficiency as they take into account the time the motor operates at different load points. A German study, comparing three different motor types, reported the best fuel economy in five driving cycles out of six for an induction motor in spite of the fact that it had the lowest maximum efficiency [457].

A comparison of costs, for the different types of drive systems, must include both motor and converter as power electronics still are more expensive than the corresponding machines. As with other products, the motor cost depends on production volumes. In the case of motors produced in larger quantities, tens of thousands or more, the costs for material usually represent at least 70 percent of the total production cost. Permanent magnets are expensive and PM motors therefore often become 30 – 50 percent more expensive than induction and SR motors of similar size. It is outside the scope of this thesis to include accurate cost information so table 8.9 below, which compares a number of properties for different machine types, includes only a qualitative, relative rating. It is not possible to use simple arithmetic on the table for appointing the best motor, because each application would require individual weighting factors to be applied on the different properties. The table is nevertheless informative.

Table 8.9   Comparison of different electrical machines for vehicle applications

Motor type	DC	Induction	PMSM	PM spec.	SR
Size	5	3	2	1	2
Weight	5	3	2	1	2
Efficiency	5	3	1	1	2
Controllability	1	2	3	3	4
Robustness	4	1	2	2	1
Temp. sensitivity	1	1	3	3	1
Cost	5	1	3	3	2
Converter requirem.	1	2	3	4	5

It is possible to notice regional differences in the choice of electrical machines for vehicle applications. American companies have had a certain preference for induction motors while PM motors have been preferred by Japanese manufacturers. DC motors have remained longer in Europe and the interest in SR motors seems more pronounced in Europe. What are the reasons behind this? Is it a matter of traditions, company strategies, vehicle categories, timing or something else? Timing has probably been an important factor. The American development, which DOE supported during the first part of the 1980's (see section 8.3.1), was carried out when induction motor variable speed drives had begun to spread in industrial applications. It was natural for the big companies to choose a new, promising technology. It was too early for PM machines. The NdFeB magnets were not available until the mid 80's. The situation was different for the Japanese manufacturers, which started more intensive development around 1990. By then it had already been demonstrated that PM motors had superior performance even though they became expensive. Japanese companies are known for putting long-term strategic goals ahead of short payback and Toyota, Honda and others have probably decided that they needed the best possible technology to become market leaders. This will be further commented on in section 8.4.2.

The situation in Europe was more fragmented. Old cities with narrow streets have favored the use of small cars with rather moderate requirements on performance. A number of niche companies started to build small and simple BEVs. They looked for the cheapest available electrical drive system, which was DC motors with chopper controllers. Some established car manufacturers participated with converted standard cars and they also preferred DC systems as it was a quick and acceptable solution at a fairly limited cost. After some years the situation changed and both induction and PM motor systems are now used in projects all over the world. None of the Swedish HEV prototypes have had anything other than these two AC technologies.

8.4   Development policy and process

8.4.1   The automotive development process

The development of a new car model represents a huge cost, sometimes of such a magnitude that failure could put the existence of a manufacturer at risk. The customers have many attractive cars to choose between and it is necessary to be competitive. Great efforts are therefore put on product planning and specification. Process effectiveness and timing are crucial for the development project, of which quality assurance and testing are important parts. The large automotive companies have very formalized processes for development of new products and they usually expect their suppliers to follow similar principles. This has been emphasized by the fact that the vehicle manufacturers expect more and more of the development to be performed by major system and component suppliers, while they concentrate their own efforts on total vehicle design and system integration. In this respect it does not matter whether the supplier is internal or external. Most parts are tailor-designed for certain car models (or platforms) and were developed earlier by the vehicle manufacturers themselves. However, certain components like batteries, generators, start motors, etc. have normally been standardized and developed by specialized suppliers. The automotive companies have therefore had limited competence for design of such electrical components. This situation has changed during recent years, however, as many vehicle manufacturers today have very large electrical and electronic engineering departments.

The diagram in figure 2.4 in the second chapter indicated how the number of electric motors in cars has increased during the last decades. Most of them are of course small and are used for driving fans, pumps, windshield wipers, windows, etc. and to adjust mirrors, chairs and headlights to mention some functions. It is possible to see a corresponding increase in number of processors; modern premium cars have in the order of 30 micro processors installed. Another important observation is the increase of engineers with electric and electronic background in the automotive companies. Volvo Cars, as an example, had less than 50 engineers of this kind in 1980. The number has now increased to over 1000 [458].

The automotive industry has, within a couple of decades, gone from common mechanical mass-production to development and manufacture of frontline high-tech products. Jan Glete, currently professor at the University of Stockholm, wrote in an article in 1983: *"High-technical usually refers to technology, which requires qualified engineering with substantial elements of theoretical work. ... ... The most pronounced high-technical industries are probably the electrical industry (... ...), the aircraft-, space- and weapon industries and the advanced chemical industry. ... ...Not even such a, for the consumer expensive, product as the car can be considered as high-technical – the car industry requires qualified management rather than high-technology.* " (My translation) [459]. The situation is different today. A recent article in IEEE Spectrum listed the worlds 100 largest R&D spenders during 2004. Number one on this list was Ford closely followed by Daimler-Chrysler, Toyota, Pfizer and GM. Four out of five are automotive companies [460]! However, development costs in this context also includes, in many cases, tools and other type of investments so the comparison is to be used with care. Vehicle manufacturers are advanced system integrators and much of the advanced engineering is made by qualified suppliers. Gunnar Larsson, former R&D director for the VW-group, who also has worked for Audi, Volvo and Saab, explained in an interview that environmental requirements and safety issues have triggered much of the automotive industry's high-tech development.

The large automotive companies have corporate research organizations that deal with long-term R&D within various fields of technology. Combustion technology is an obvious example, but even such subjects as vehicle dynamics, telecommunication, IT and fuel cells can be mentioned. These research units also often have responsibility for development of advanced concept vehicles. Development of new car models for the market is handled directly by the individual product companies. However, the policy to use common platforms within a large corporation allows substantial parts to be developed by a sister company. The automotive industry often makes a clear distinction between pre-development and product development. An example of pre-development, which can be quite extensive, can be development and demonstration of a prototype HEV driveline. Pre-development often includes study and evaluation of different concepts. Product development means "sharp" projects, e.g. development of a new car model or a new series of engines. Such a project applied to the development of an electrical driveline could consist of a number of activities and gates as the diagram in figure 8.41 illustrates.

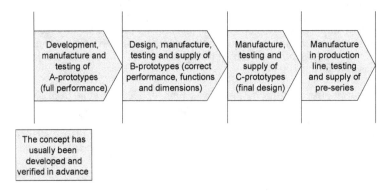

Figure 8.41 Main steps in development of a vehicle driveline

The experience has shown that it is not quite easy for a traditional electro technical company to adapt to the principles of the automotive manufacturers. Indeed, the development has become more formalized also in the heavy electro industry, but it is much more fragmented. A vehicle manufacturer focuses on a few large projects of relevance for the entire company, while a company such as ABB is working on many independent projects. The shift of product generations can, as evident from chapter 6, take place successively and the border line between planned development and order design is many times floating. The cultural differences between the automotive and the electro industry are significant.

8.4.2 Actors and competence centers

The development of electrical machines in the multi-kW size and larger has been led by the major electro-technical companies for more than a century. Medium-sized and smaller manufacturers have, with few exceptions, been followers. Universities have contributed with theoretical studies and graduated engineers. This traditional allocation of roles has changed during the last decades, much due to the development of automotive drive systems. Much was as it used to be during the initial period, i.e. the 1980's and early 90's. Companies like GE and Westinghouse in U.S., ABB, AEG, Alstom and Siemens in Europe and Hitachi in Japan participated, as mentioned in section 8.3.1, in projects for development of electrical drivelines, but most of them have now withdrawn from these activities. The "second league" motor manufacturers were more or less invisible as far as automotive drivelines are concerned.

The need for electrical machines for vehicle propulsion has instead initiated the establishment of a number of new, fairly small, actors which are developing new motor concepts together with suitable power electronic controllers. Early examples are UQM and Magnet Motors, companies established by entrepreneurs without electrical machine background. Both companies have developed innovative motors based on the new rare-earth magnets (see section 8.3.2). Examples of other, more recently, established electrical machine and drive system manufacturers focusing on automotive applications are AC Propulsion, Enova, TM4 and WaveCrest; all based in North America. All except the first one have expressed ambitions to become leaders in electrical drivelines supplying major OEMs. They are concentrating on R&D, prototyping and small scale production. Two of them have chosen induction motors and two have preferred PM motors. Cooperation with leading universities has been explicitly mentioned, but otherwise the know-how has been brought into these companies by key persons who had worked in other organizations [461].

Traditional automotive suppliers of electrical systems and components such as Bosch, Delphi and Denso have neither played any major role, so far, nor having contributed much to the development of electrical drivelines. Bosch was not at all interested in developing a motor for BMW's BEV in the early 1990's. Furthermore, they have hardly participated in recent EVS conferences. It is instead the vehicle manufacturers themselves who have taken a leading position in development of electrical motors for this application. Toyota is the most pronounced example but Honda and to some extent GM and Daimler-Chrysler are also developing motors in-house. In spite of what has been written above, automotive industry experts believe that major system suppliers like Bosch and others will enter the electrical driveline business when the volumes get large enough.

Toyota has become, as mentioned in section 8.3.2, a big manufacturer of electrical machines and is therefore of particular interest. The company was established as car manufacturer in 1937 by the Toyoda family and is still ruled by this family, even though today it only holds a minority share. Toyota, which has been profitable for more than 50 years, acts in accordance with a far-reaching long-term strategy that explains their pioneering role in the field of HEV. Toyota's chairman, Eiji Toyoda expressed, already in 1990, his concerns for the automotive development. He pleaded that a new type of car was required for the 21st century, a car that could meet the requirements on environmental sustainability. Eventually, in 1993, Toyota started a pre-project aiming at a concept for a radically new car, but it was not until November of 1994 that the decision to build it as an HEV was made. Starting from scratch, Toyota had built some competence in electrical machines during the early 1990's, but it was an open question whether the motor for Prius should be purchased or manufactured in-house. The alternatives were either to purchase induction motors or to develop and build PM motors. After extensive prototype tests, it was decided to choose the PM motor due to its smaller size and higher efficiency. The cost of the motors was less of an issue at this point of time. Toyota's management looked more towards better development potential as opposed to higher costs when it decided to use PM machines. A visible milestone was when Toyota's new chairman, Soichiro Toyoda, in a speach at the EVS 16 conference in Osaka 1996, stated that future automotive drivelines would be electric. This could be seen as a potential threat to the company because of its dependence on ICE technology at that time. The ICE had always represented core technology for all major car manufacturers. Toyota's management did not want to lose control over the important driveline development to electrical companies and thereby reduce its own future role. That is probably why Toyota decided to build the necessary competence to develop the electrical drive systems in-house and also to manufacture the machines in its own factories. This has been done with determination and Toyota is today also a leading electrical machine company [462, 463]. Toyota's release of Prius, in the end of 1997, shook the rest of the automotive industry which, in comparison, had been inactive in the case of HEV development [464].

Another group of new actors in this field are some leading suppliers of automotive transmissions, e.g. ZF in Germany, Allison Transmission in the USA and Aishin in Japan, though they often outsource the manufacture of the electrical machines. One obvious motive for their interest is that the electrical drivelines could be a threat, which could replace substantial part of the mechanical transmissions. The electrical part of an HEV driveline is often referred to as an "electrical transmission".

Research at technical universities, especially the applied research, is usually focused on areas that are in strong development, i.e. on frontline technologies. A review of the current research programs, at a number of leading universities, reveals that a remarkable share of the electrical machine oriented projects are related to automotive drivelines which can be verified by reading lists of projects and recent publications at a number of university websites. A few examples of leading universities in this specific area are:

- Laboratories d'Electrotechnique in Grenoble and Toulouse, France
- RWTH in Aachen and the technical university in Munich, Germany
- University of Rome, Italy
- Tokyo University, Japan
- Technical University of Eindhoven, Netherlands
- Imperial College in London and the Universities of Manchester and Sheffield, UK
- Illinois Institute of Technology, Texas A&M University and University of Wisconsin in Milwaukee, U.S.

The leading Swedish universities in this particular field are KTH and LTH. It is no big surprise that automotive applications are the focus of much university research, but it nevertheless sends an important message about the shift in the center of gravity in the electrical machine discipline. The fact that a number of actors with limited background in electrical machine technology have emerged has also increased the need for assistance from the universities.

8.4.3 OEM policies

The ICE and the transmission have long represented core competence for vehicle manufacturers and have, to a large extent, been manufactured in-house. A BEV or HEV driveline became more complicated. How should the electrical parts be treated, as batteries and generators or as the engines? So far, there is no uniform policy in this respect. Toyota and Honda are obvious examples of the latter alternative. European companies like P.S.A. and Volkswagen are instead outsourcing development and manufacture of electrical machines. Manufacturers of large vehicles such as buses and trucks are all purchasing their electrical drivelines from various suppliers. It is, however, important for all kinds of vehicle manufacturers to build up extensive system knowledge and to carry out system engineering in-house. They definitely do not treat the electrical machines as "black boxes".

According to several representatives for vehicle manufacturers, it is obvious that outsourcing is made on strictly commercial grounds; there is no room for any national preferences. The author has asked a few key persons representing the Swedish vehicle manufacturers for their comments on the supply of electrical machines and drives for hybrid vehicles. Keeping in mind that is a heterogeneous group of companies, the comments can be summarized as follows:

• Electrical machines are not considered as core technology today, but will probably become core (or key) technology in the future. It is important for the vehicle manufacturers to increase their competence in this field; it is not sufficient at present.

• Electrical machines and drive systems will be purchased from external suppliers provided that they are available. Need for large volumes in a distant future may change the situation.

• Taylor made systems would prevail due to the complicated mechanical integration, but standard components would be used whenever possible.

• Strategic partners are preferred as suppliers, and the requirements are that such suppliers should have: product development, approved quality, reliable deliveries, global presence, and financial resources to survive and support its products.

• Swedish companies have possibilities to become suppliers of electrical machines and drives, at least for sub-systems and auxiliary systems; i.e. tier 2 suppliers. ABB has been mentioned as a company that has the necessary qualifications, if they would be interested.

8.4.4   ABB's experience

ABB, like some other leading electro-technical companies, has had an hesitant attitude as a supplier of electric motors and converters for road vehicle drivelines. Several attempts have been made but have not been very successful, at least not commercially. A natural question is, why has this world-leading manufacturer of industrial drive systems not also become an important supplier of automotive drivelines? Is it a matter of bad experience or doubts on the potential market? ABB Sweden has been deeply engaged in most of the Group efforts and it is thus of interest to include a brief summary and analysis in this study.

ABB started to deliver DC drive systems around 1990, first for Volkswagen and later for Renault (see section 8.1.3). An induction motor was developed at the same time by ABB Motors for the Volvo ECC project (see section 8.3.1). This was during the period when ABB in Germany tried to become a supplier of NaS batteries for BEVs. ABB therefore had extensive contacts with a number of automotive companies and decided to also aim at drive systems. A special unit, ABB Automotive Drive Systems, was established in the beginning of 1992 in Mannheim, Germany with responsibility for ABB's global activities in this field. An important part of the strategy was to promote AC drive systems with induction motors from Västerås and inverters from Helsinki. The first prototype system of this type was supplied to Renault. In parallel, even more effective systems came into discussion as both UQM and Magnet Motors, independently of each other, approached ABB and offered licenses on their PM technologies. It has been mentioned in section 8.3.2 that ABB, BMW and UQM signed an agreement and started to evaluate UQM's PM motor technology [465]. ABB refrained, after a year, from this cooperation due to lack of synergies with its AC drive systems. The product responsible units, especially the Finnish one, did not want to split forces by working with a foreign technology. There were also doubts on the system safety for field weakened PM motor drives due to the risk of high over-voltages in certain fault situations.

The unit in Mannheim was closed in 1994 as a result of reorganization within ABB's management and business area structure. The new managers, at the headquarter in Zürich, had a different view on ABB's possibilities as supplier of automotive drive systems and decided to close all such activities [466]. However, certain cooperation had been established with Volvo Car Corporation (VCC), which planned a large HEV demonstration program. The relationship between Volvo and ABB Sweden had, at that time, become very close because the president of ABB Sweden, Bert-Olof Svanholm, had also been appointed chairman of the Volvo Group. Furthermore the CEO of VCC, Per-Erik Mohlin, had earlier been a division manager within Asea. ABB's Swedish management therefore decided to proceed with the cooperation with Volvo and it even established a special development company, ABB Hybrid Systems, as a local activity [467].

A small number of induction motors, inverters, PM generators and DC/DC converters were delivered to Volvo's hybrid car project until the project suddenly was closed by VCC's new CEO for economic reasons. ABB Hybrid Systems continued instead with development and supply of electric drivelines and auxiliary systems for large HEVs to Volvo Truck and Volvo Bus Corporations; specifically two 12 ton distribution trucks and two city buses intended for demonstration in regular, commercial operation. Each vehicle was provided with two water-cooled induction motors, based on a standard 30 kW 4-pole motor, but in this case rated 150 kW in peak and with a speed range $0 - 4\,800 - 8\,000$ rpm.

Figure 8.42 ABB Hybrid Systems personnel and series hybrid car with gas turbine-driven high-speed generator

The earlier contacts with BMW had not been completely broken and ABB Hybrid Systems got an opportunity to offer the development of a drive system with PM motors for a planned BEV including delivery of a number of systems. ABB received the order for this project and the development started in August 1997. Dr. Andreas Goubeau mentioned in an interview that ABB Motors' production resources and ABB's deep understanding of the motor technology were important factors. The motor has been described in section 8.3.2. The project was difficult technically, commercially and administratively, partly due to a difference in corporate culture. BMW "placed the stick high" and ABB learned a lot from this. BMW criticized ABB for not responding efficiently enough to avoid delays. ABB's intellectual level was excellent but the implementation did not live up to this, again according to Andreas Goubeau. BMW terminated the project in 1999, when it became clear that the amendment of the Californian ZEV mandate released the manufacturers from the requirement to sell ZEVs starting in 2003. Another important factor was that the experience from California showed that the market did not accept the BEVs.

The Swedish management had realized that ABB Hybrid Systems needed to be part of a global business area and initiated such discussions with the management in Zürich. A few options were investigated, e.g. cooperation, or even a joint venture, with GM on drive systems for heavy vehicles. The first contact was taken by GM and it was not insignificant that ABB's chairman Percy Barnevik also was a member of GM's board of directors. The final result was, however, that ABB Hybrid Systems and ABB's automotive activities were closed in 2000. At that point ABB had, over a period of 10 years, made several efforts to get into this market and had spent substantial money on development, somewhere between 80 – 100 MSEK. The reasons for the termination were primarily:

- The time before the market would need production volumes that could give an acceptable payback was too long.
- The company felt a lack of control and became too dependent on a few car manufacturers decisions.
- The company was too inexperienced in operating commercially as an automotive supplier.
- The limited development resources needed to be allocated to the company's core business.

ABB has, during the last 5 - 6 years, limited its vehicle drive system activities to the support of a couple of university projects.

(Part of the facts in section 8.4.4 is based on the author's personal notes and correspondence.)

8.4.5   Swedish efforts

Sweden has, in relation to its size, a remarkable large automotive industry with five vehicle manufacturers: the car companies Saab and Volvo Cars, the truck and bus manufacturers Scania and Volvo and also Hägglunds, which builds military vehicles. The export of road bound vehicles represented around 13.5 percent of the country's total export in 2006. The importance of this industry for the country is obvious and it is a key issue to maintain an healthy automotive industry also in the future. The international competition is very tough. It is a commonly accepted opinion that the best way to preserve a front-line position is through high competence in specific areas and HEV technology has been identified as one of these [468].

All five vehicle manufacturers mentioned above have built and demonstrated a number of HEVs. All of them also have programs and projects for further HEV development. Part of their efforts are through participation in common national R&D programs for HEV and FCV. One program is called "Energy systems in road bound vehicles" and is financed by the Swedish Energy Agency" and another is part of a very large program co-financed by the government and the vehicle manufacturers, usually called "The Green Car". These programs, started in the beginning of this decade, contain in total more than 30 post-graduate university projects. Seven of these projects are focused on electrical machines and they are carried out in close cooperation with some of the vehicle manufacturers [469, 470]. The Swedish Energy Agency has, in 2006, initiated preparations for a long-term national competence centre "Swedish Hybrid Centre", allocated to the major technical universities and with support from the automotive industry.

Sweden has a large number of companies that are established as automotive suppliers, but none of them build electrical drivelines. ABB is the only motor manufacturer, which has been engaged, but is currently inactive in this field. The vehicle companies have expressed a desire to have domestic suppliers of electrical motors and drive systems and have therefore initiated activities in that direction. It is, in spite of modern communication, considered as an advantage to have partners nearby for development of highly integrated systems. The question mark is whether volumes can be sufficiently large to enable a competitive production. As mentioned in section 8.4.3, contracts for series deliveries are placed on strictly commercial terms. The possibility for a new supplier to reach production volumes enabling competitive prices can easily put him in a "Catch-22 dilemma".

8.5   Achievements and prospects

8.5.1   Comparison with traction motors

Electrical motors were used for propulsion of both rail and road vehicles back in the late 19th century. The road application was soon placed in backwater while the development of electrical trains and trams continued. The traction motors for these kinds of vehicles have therefore undergone a steady development compared to the intensive process, which the road vehicle motors have been subject to in recent years. There are several similarities between these two applications in spite of the difference in size. A comparison could hence be of interest.

Asea built its first electrical locomotive, a very small one designed by Jonas Wenström, in 1891. It was powered by a DC motor like all other electrical rail vehicles in those days. International development was however faster, especially in the USA and Germany, and the first serious attempts made by the Swedish State Railways (SJ) in the beginning of the 20th century were based on locomotives from Siemens, AEG and even Westinghouse. A breakthrough for Asea was the electrification of the railway for transportation of iron ore from Kiruna in northern Sweden to Narvik in Norway in 1910-1915. Asea formed a consortium, together with Siemens, for the supply of all electrical equipment including the locomotives. A new type of traction motor, the single-phase commutator motor, originally developed by Westinghouse, was now installed. The advantage of an AC motor was the possibility of using a high voltage for transmission of power on the long overhead line (16 kV in Sweden) and then use transformers in the locomotives for supplying the motors with a suitable voltage; a few hundred volts. The single-phase commutator motor can be described as a series DC-motor. When the current changes direction the flux is also reversed, which means that the torque maintains its direction. Time constants and commutation properties limited the frequency for these motors which explains why the Swedish railway lines still has a frequency of $16^2/_3$ Hz. Several Swedish cities built tramlines in the beginning of last century. Asea soon became a major supplier of the tramcars, but these were all powered by series wound DC motors [471].

Asea eventually obtained a very strong position as supplier to SJ. Asea and SJ became, over a long period, a "development couple" as Asea and Vattenfall were in the field of power generation and transmission. Commutator motors remained the main option for the Swedish locomotives during half a century while the series DC motors were used for lighter rail vehicles. The big change came around 1960 with the introduction of power electronics. The locomotives could be provided with diode rectifiers, which implied a reentrance of the DC motor. Asea presented, in 1965, the first locomotive in the world with a thyristor rectifier, see figure 8.43, but now with separately excited motors. It was a success and it opened the export market for the company. The new technology made it much easier to adapt the equipment to various voltages and frequencies while the old machinery was very much confined to Swedish conditions [472].

Figure 8.43 Asea's thyristor locomotive

The next step came around 1980 when Asea built a test locomotive with inverter fed induction motors. The company was fairly late with the testing of this technology. BBC had demonstrated such a drive system in the early 1970's and Strömberg had chosen it for the Helsinki metro around 1975. The French manufacturer Alstom started to supply high-speed trains with synchronous motors in 1978. Tore Nordin, chief engineer for Asea's Traction Division, mentioned that he proposed development of induction motor drives several times but Asea's management was reluctant, the inverter technology was not considered sufficient. Furthermore, the existing thyristor based systems with DC motors were very successful. It was not until the late 1980's that the induction motor was accepted for Swedish traction drives. The reason was that the X2000 high-speed trains needed lighter motors than the old technology could offer. The development has then been focused on traction systems with induction motors both for main line and commuter applications. ABB's traction business has been sold in two steps, first to Daimler-Chrysler and then to the Canadian company Bombardier. The division in Västerås is still developing traction motors and drives, but the motor manufacture is outsourced to ABB's electrical machine factory. Most of the products are exported and the engineering force in Västerås is the largest within Bombardier for traction drive systems.

The requirements on traction motors and motors for road vehicles are, in many respects, similar in spite of the difference in size. Both have to perform heavy starts and be capable of operating for longer periods at high speeds, which results in the same type of torque and power characteristics even though road vehicle motors have a much larger field weakening range as shown in figure 8.44. The environmental conditions are also similar as they must withstand rain and snow, mechanical shocks and vibrations, low and high ambient temperatures etc. Even the traction motors must be very compact in order to fit into the bogies. Figure 8.45 shows an example on this. The size is effectively limited by the railway gauge [473].

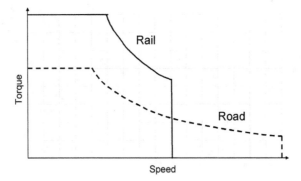

Figure 8.44 Example of torque characteristics for rail and road vehicle motors

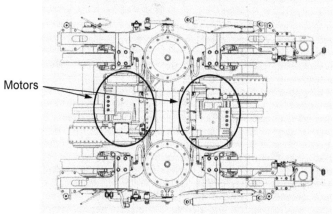

Motors

Figure 8.45 Bogie with traction motors

The development of traction motor performance during the 20th century can be illustrated by the graph in figure 8.46, which contains specific power and torque densities for both locomotive and light rail vehicle motors [474]. Corresponding values for road vehicle motors are included for comparison. The cooling is essential for the motor performance and it has to be noted that even the modern traction motors are air-cooled while the road vehicle motors usually have liquid-cooled stators. Smaller traction motors are often closed while somewhat larger motors are open-ventilated or even forced ventilated.

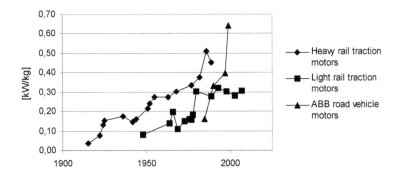

Figure 8.46 Specific performance for traction and vehicle motors

It is difficult to compare electric car motors with traction motors for locomotives, but the automotive and the track bound applications come much closer in case of city buses and trams. Trolley buses are especially interesting in this respect. Sweden had not had any trolley buses since 1964, when Stockholm and Gothenburg closed their systems with such buses, actually made by Asea, until the city of Landskrona opened a trolley bus line in 2003. These modern buses have drive systems supplied by the Hungarian company Ganz Transelektro. Each bus has one 4-pole induction motor, which is open ventilated and rated at 167/270 kW, with a maximum torque 2060 Nm, maximum speed 3500 rpm and it weighs 780 kg [475, 476]. The specific peak power and torque become 0.35 kW/kg and 2.64 Nm/kg. It is a traction motor that also can be used for trams and is therefore relatively heavy. Motors designed directly for bus applications have usually higher specific performance. Corresponding values for the water-cooled induction motors in Volvo's hybrid buses mentioned in section 8.4.4 were 0.9 kW/kg and 2.7 Nm/kg and the PM motor for the concept bus Volvo ECB had as high values as 1.5 kW/kg and 28.5 Nm/kg as mentioned in 8.3.2. Traction motors for light rail vehicles and even trolley buses are based on the technologies developed for main line trains and locomotives. The requirements are thus severe, a fact which results in very robust but also expensive motors. Such motors are far from optimal for automotive drivelines and this is probably the most important explanation why railway traction motors and automotive motors are developed by different actors.

8.5.2 Comparison with industrial motors

Industrial motors do not constitute an homogenous group of products. It could be relevant to choose standard induction motors for stationary industrial applications for the comparison, i.e. the type of motor, which has been studied in chapter 6. Interesting to compare are specific performance, efficiency and costs. Table 8.5 in section 8.2.3 contains a few comparison except for the costs, which still have prototype level for most vehicle motors. This will change in favour of the vehicle motors when the production volumes increase. A comparison with high-torque, industrial PM motors would considerably decrease the difference in specific torque.

It is obvious from the table that the specially designed vehicle motors have a considerably higher performance. Liquid cooling, special concepts and topologies etc. are means to achieve this. It has earlier been concluded that industrial motors are not suitable for vehicle drivelines. An interesting question is whether the automotive motors will be used for industrial applications? This has been considered as a risk, especially in a situation where automotive motors are produced in large numbers at low costs. Traditional manufacturers of electrical machines have sometimes applied a defensive strategy; they participate in the automotive field in order to avoid a backlash in their own core business. The fear for such a scenario seems to have been reduced in recent years. One reason is the difference in production volumes and number of variants. The quantities needed for automotive drivelines grow rapidly to levels, which require extremely specialized production lines while the manufacturers of industrial motors must secure the flexibility to make hundreds of variants. Another reason is that it would require a quite different marketing organization.

8.5.3 Future for automotive motors

Electrical machines for automotive drivelines can be expected to have a bright future, but are still just out of the starting blocks. The situation can be expected to remain turbulent until the market has grown and a number of leading actors have been established. Production volumes may become very large, but the number of manufacturers will be much less than for industrial motors. The reason being that the automotive market will depend on a limited number of large OEMs while the industrial market is very fragmented.

The pace of market growth for vehicles with alternative drivelines will depend on regulations and cost factors. The first is a political issue based on environmental concern. The second is mainly a matter of crude oil prices and costs for substitute fuels. Both are important but the latter will probably have the most impact on market development. Doubled fuel costs will probably affect the type of vehicles to be used. A number of institutes and business consultants make regular forecasts on the global market growth of HEVs. Table 8.10 contains a couple of recent forecasts. The discrepancies used to be large, but the recent prognoses are in much better accordance. All of them show a significant increase of vehicles, so the question is not "if" but "when". There are no reliable FCV prognoses, but annual volumes above two millions have been estimated from 2020.

Table 8.10 HEV forecasts

Year	Institute	Volume	Market
2009	M. Anderman	~ 780 000	Global
2010	ABI Research	1 000 000	Global
2012	Frost & Sullivan	150 000	EU
2012	J D Power	780 000	USA
2012	Nomura Research Institute	2 200 000	Global
2015	Freedonia Group	3 900 000	Global
2020	Freedonia Group	7 500 000	Global

Electrical drivelines are already used for different vehicles, both small and large, for private and for commercial use. Several degrees of hybridization exist. It is common that new technologies spread and find more applications and this will probably happen also for these drivelines. The variation in machine types and sizes can be expected to increase, because there is no global best solution, which suits all cases. ISGs and perhaps even wheel-motors will most likely be standardized while for larger machines a similar situation as with ICEs can be expected.

A look back reveals that the basic electrical machine concepts are old, even from the 19th century, but the implementation has been subject to continuous development. This process has been particularly intensive for vehicle motors in recent years. There is no reason to foresee any drastic trend change in the near future, but it is plausible that the focus will shift towards production development when the volumes increase. Machine performance can be expected to improve, but not radically as it is already very high and a further decrease of weight and dimensions could have an adverse effect on the efficiency. Electrical machines have been influenced by technical development in several other fields and this will undoubtedly continue. Progress in material development has a major impact, both directly and indirectly, on electrical machines. One example is better and cheaper permanent magnets. Soft magnetic composite (SMC) with considerably higher permeability would be another. SiC power electronic devices will simplify integrated solutions but at the same time probably increase the request for operation at higher temperatures also for the machines. Inexpensive, room temperature superconductors would entirely revolutionize the electrical machine technology but this is only a utopia. The diagram in figure 8.47 illustrates briefly the major development steps for electric machines for vehicle propulsion and points at some possible future directions.

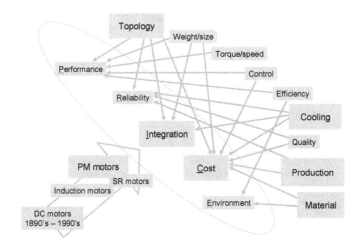

Figure 8.47 Electrical machine development for vehicle propulsion

Considering the successful production of ICEs in Sweden, it is relevant to discuss whether Swedish manufacturers could even build electrical vehicle motors. Looking back, there is no doubt that the country has sufficient competence to develop and manufacture electrical machines of many different types. Asea/ABB has, for long periods, belonged to the internationally leading companies in this field. An important question is whether a competitive domestic production can be achieved without the participation of this company? This depends mainly on the vehicle manufacturers. It is difficult to identify any other actors with sufficient industrial power. The globalization with competition from low-cost countries makes such decisions difficult. The author's opinion is that the best chances for Swedish industry is to focus on larger, high-performance machines produced in limited numbers and leave mass-produced ISGs etc. to the low-cost producers.

8.5.4 Conclusions

Road vehicles with electrical drivelines may have a long history, but not very interesting from electrical machine point of view. The traditional DC motors dominated until 15 – 20 years ago. What has happened then is mainly a question of developing induction and PM motors for this application. Initially, the market for electrical drivelines had a slow growth, but it has become very interesting in recent years. The conclusions summarized below are focused on the period from the mid 1980's until today:

- Environmental concerns and shrinking oil reserves have triggered the global efforts to develop alternative drivelines for road vehicles, most of them with electrical drive systems. Essential milestones were the Californian mandate, in 1990, and the launching of Toyota Prius in 1997.

• Automotive drivelines need special electrical motors with considerably higher performance than industrial standard motors. These vehicle motors have been developed by various actors such as large electro-technical manufacturers, small development companies, but also by some automotive companies.

• In many cases PM motors are preferred due to better performance in spite of their higher costs. Especially the torque density (Nm/kg) can be much higher for PM motors compared to induction motors while the power density (kW/kg) is fairly equal for these types of motors. The efficiency is also better for the PM motors.

• The use of PM motors has opened the possibilities for development of several unconventional topologies, such as axial flux machines and outer rotor machines, many of these used as wheel motors.

• Important targets for the development have been small dimensions and low weight, high efficiency, large field weakening range, ability to withstand both low and high temperatures as well as other difficult environmental conditions. Low costs and high quality have also been essential targets.

• In spite of several technical similarities, no signs have been found on coordination or cooperation between the development of electrical motors for automotive drivelines and the development of traction motors for railway applications. This depends probably on commercial factors.

• The application and the customers were new to the established electro-technical industry and the market situation is not yet settled. Some automotive companies are developing and manufacturing electrical machines themselves, others are purchasing. A number of small new companies have been attracted by the new market opportunities.

• Large differences in business culture and policy are often a more severe obstacle than the technical requirements, for the possibilities to create a customer/supplier relationship between automotive and electro-technical companies.

• Sweden has, in relation to its size, a very strong automotive industry. All five vehicle manufacturers have carried out HEV prototype projects, in several cases with participation from the electro-technical industry and the universities. However, no commercial development or production has been started in spite of the available competence.

• Various types of PM machines have been developed and tested at KTH, LTH and Chalmers. ABB has built both induction and PM machines as well as complete drive systems for a number of demonstration projects. The ABB's motors have, from an international perspective, been somewhat conservatively designed with moderate specific performance, especially with respect to torque density. This depends on a strong influence from the design of industrial motors and the use of the traditional radial flux, inner rotor topology.

- ABB withdraw from its attempts to become a supplier of electrical drive systems for automotive applications, primarily because the company wanted to focus its resources on its traditional core business, but also due to the commercial difficulties in the automotive market.

- The Swedish government, vehicle manufacturers and universities have, for several years, cooperated in R&D programs aiming at development of HEV technology, including electrical machines, and enhancing the national competence in this particular field.

The future of electrical machines for automotive drivelines has been discussed in the previous section. An obvious conclusion is that this will remain an expanding application for a long time, but "how and where" is partly an open question. A few points that summarize the author's view on the future development are given below:

- Practically all major vehicle manufacturers plan to introduce HEVs, and in some cases also FCVs, within the next five years. The market will expand and the automotive market for electrical machines will eventually be much larger than the industrial.

- Several types of cars and commercial vehicles, and different concepts and hybridization degrees will require electrical machines of many different sizes and topologies. Standardization will, therefore, be limited to certain applications and basic solutions.

- Induction and PM motors will remain dominant, but challenged by SR and electrically excited synchronous motors.

- The electrical machine development is influenced from the development of other technologies. Material development, e.g. magnetic materials, will be of great importance also for future electrical machines. Introduction of SiC semiconductor components is another example which will have a big impact also on the machines.

- The performance of the electrical machines will continue to be improved, but at a slower rate than before. The focus will shift more towards cost reductions. System control will be more important for the total efficiency than somewhat reduced electrical motor losses.

- The electrical machines for automotive drivelines will be manufactured by larger companies having sufficient resources and financial strength for product development, series production, and sustainable customer support. These manufacturers can be part of the automotive industry, vehicle manufacturers as well as major suppliers, or belong to the electro-technical industry. Small development companies will have to license their technologies to larger manufacturers or be acquired by such larger companies.

• Current Swedish R&D activities have explicitly shown an ambition from the participating vehicle manufacturers to enroll potential suppliers in these activities. This wish is also expressed by the vehicle manufacturers in the 10 year plan for a new Swedish HEV competence centre. Electrical machines have been identified as one of the most interesting components, and it can be assumed that the interest for national development of these machines will remain for some years.

• The Swedish car manufacturers Saab Automobile and Volvo Cars are owned by GM and Ford respectively, and they will primarily use components that are common for certain vehicle platforms within these large corporations. The possibilities for supply of electrical machines from Sweden are therefore very small. It is only in case Saab and/or Volvo will be allowed to develop certain niche vehicles that such deliveries seem realistic.

• The Swedish manufacturers of heavy vehicles are in the position to decide by themselves which components and suppliers they want to use. The production volumes are limited compared to the car industry, the vehicle manufacturers need external support, and the application is more related to industrial systems which are more familiar to potential Swedish suppliers. There are possibilities for future domestic development and supply of electrical machines for heavy vehicles provided certain conditions can be met.

• There will be no national preferences; the Swedish manufacturers have to offer competitive products. The best chances represent probably large machines and certain auxiliary machines. The machine supplier must fulfill some requirements, a fact that leaves only two alternatives. The first is that ABB reconsider its position and decides to supply machines, at least for heavy vehicles. The second is that an established automotive company acquires a suitable small electrical manufacturer, or a part of it, and helps it grow.

It is not within the scope of this thesis to propose specific actions, but a conclusion is that the government should preferably act as a catalyst if Swedish development and manufacture of electrical machines for this expanding application shall materialize.

9   Powerformer® - a machine for very high voltages

Large electrical machines have long been viewed as fairly mature products. A lot of attention was therefore aroused, both publicly and professionally, when ABB presented its radically new high voltage generator in the late 1990's. The so-called Powerformer® had originally been developed within ABB Corporate Research in Västerås, Sweden, and the company claimed that this machine would revolutionize the generator market. The concept could also be applied to large motors, which consequently were named Motorformer™.

The Powerformer project is an interesting example of a comprehensive, recent electrical machine development. The technical concept was unconventional, the development as such was in some respects extraordinary and the market projections were great. It was a truly Swedish development project and it is therefore appropriate to include a study of Powerformer development in this thesis as it quickly became a part of the Swedish electrical machine history.

9.1 Introduction of an invention

ABB had invited to an important press conference on February 25th, 1998. R&D achievements were to be presented and information on a radically new product was to be released. The headlines in Swedish papers the next morning read: "*World news from Västerås – the new ABB generator obtains only praise, can create an enormously profitable business*" and "*ABB has found its Losec**" [477, 478]. It was very unusual for ABB to get such big media coverage for a new product, coverage which was generally reserved for important business news such as annual results, major company acquisitions, criticized divestments etc. This particular day, the message was that ABB had a new invention, an electrical machine, which would take over the future global market for power plant generators.

The inventor of this particular high voltage generator was Mats Leijon, manager for the High Voltage Electromagnetic Systems Department within ABB Corporate Research, later appointed professor in electro technology at the University of Uppsala. Leijon joined ABB in 1987 after having finished his PhD studies at Chalmers University of Technology. Mats Leijon has, in sev-

---

*. Losec®, the world's leading ulcer drug, has been the major cash-cow for the large Swedish/British pharmaceutical company Astra Zeneca.

eral interviews, mentioned that his first ideas for a generator for very high voltages date back to the beginning of the 1990's. The major advantage with such a generator would be the possibility to connect it directly to a high voltage transmission line without any bulky and expensive transformer. He showed his idea in May 1994 to Harry Frank, at that time president of ABB Corporate Research in Västerås, who immediately decided to support it. Four months later, Frank and Leijon presented the invention to the executive president for ABB in Sweden, Bert-Olof Svanholm (1935-1997), who realized its huge potential and gave the proposal his full support. A special development team was set up with Mats Leijon as manager and the development work started and proceeded under rigorous secrecy. Four and a half years later, time had come for a public release [479].

The premier presentation of the new generator, Powerformer, had been carefully planned in order to make the best possible impact. The meeting, with invited journalists, was held in Dättwil, close to Zurich, the day before ABB's annual big press conference for the presentation of results from the previous year (1997). Present were not only Harry Frank and Mats Leijon but even the CEO Göran Lindahl and other high-ranking ABB representatives. A big model of the prototype Powerformer had been transported from Sweden to be exhibited together with some other of the company's other recent R&D results. The message was clear; ABB had a new product which would radically change the market for large power plant equipment. Powerformer would reduce investment costs and improve operation economy [480]. The response was a success. ABB's new generator received a lot of positive attention in both professional and general news media. A suspicion has sometimes been expressed, that this press conference was staged so that favorable technical news should somewhat balance the negative fact that ABB's result had fallen from 6.3 percent, in 1996, to 3.6 percent in 1997, but this rumor is most probably untrue. Nevertheless, ABB's shares went up and the company experienced a period of positive development for what ever reason [481].

The inventor, Mats Leijon, received public attention far beyond what has been normal in this kind of manufacturing industry. He was awarded KTH's Grand Prize for Powerformer development in 1998, i.e. the same year the invention was presented. He received the Finnish academy of science "Walter Alström prize" in 1999, Chalmers' "Gustaf Dalén Medal" in 2000 and the Swedish Association of Graduate Engineers' "Polhems Prize" in 2001 to mention some of the awards.

Figure 9.1 Swedish delegation at Powerformer launching in Dättwil, 1998. Mats Leijon and Harry Frank are number 6 and 4 from the left. Kjell Isaksson from Vattenfall is the first from the left.

9.2 Exclusion of transformers

The value of the Powerformer concept should be reviewed from a system perspective more than just the machine as such. An electric power system consists basically of power plants, transmission lines, distribution nets and loads. Most of the power is generated in large power plants with outputs of several tens, hundreds and even thousands of MW. These power plants are often located far from the load centers and that means that the power has to be transported long distances via transmission lines. It is economical to use high transmission voltages, e.g. 220 or 400 kV, which both are typical levels in Sweden and also in many other countries. Power plant generators are usually built for voltages in the range of 10 – 25 kV which means that transformers are required to increase the generator voltage to the transmission voltage. A transformer is also needed in the receiving end to enable connection between a transmission line and a local distribution grid. Figure 9.2 gives an example of a simple power system.

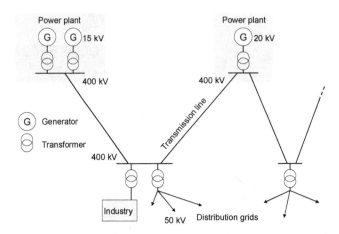

Figure 9.2 Diagram showing a simple power system.

It has since long been common to use so-called block transformers in power plants, i.e. there is one transformer for each generator which together constitute a block. The generator is connected to the transformer's "low-voltage" side via a generator busbar and necessary switchgear such as circuit breakers and disconnectors. The high-voltage side of the transformer is then connected to the transmission line through additional switchgear. Figure 9.3 shows an example. This kind of block will, of course, be much simplified if the generator can be built for the transmission voltage, thus eliminating the need for the transformer, the "low-voltage" busbar and switch gear. This was the basic motivation for Powerformer. The block thus becomes simpler when a Powerformer replaces the normal generator and transformer as illustrated in figure 9.4.

Figure 9.3 Diagram of generator/transformer block in power plant

Figure 9.4 Diagram of "power plant block" with Powerformer

Powerformer, as such, is bulkier and more expensive than a corresponding conventional generator, but the total investment can be reduced as the transformer, the generator busbar and the "low-voltage" switch gear are deleted. Furthermore, there is no longer any space required for these items, so the building volume can also be reduced. This is extra important in underground hydropower plants. It should also be possible to reduce operation costs because there is no maintenance needed for the eliminated equipment. The total efficiency can also be improved as the transformer losses disappear, assuming that a Powerformer has more or less the same efficiency as a conventional generator. ABB has also claimed that the availability will be improved because the number of system components is reduced. Figure 9.5 contains a diagram used by ABB for illustrating the economical advantages of a Powerformer [482].

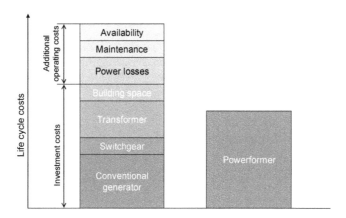

Figure 9.5 Cost comparison between Powerformer and conventional solution.

9.3   High voltage machines

Powerformer was not the first electrical machine in the world created for very high voltages, even though ABB claimed that it was when it was introduced [479]. Experimental generators with voltages up to 121 kV had been built in the USSR around 1970. This will be dealt with later in this section.

Higher voltages mean lower currents, which often is an advantage. Early Swedish examples of machines with unusually high voltages are a couple of 1830 kVA synchronous generators built by Asea for Brattfors hydropower plant in 1903 and 1906. The first one had a voltage of 18 kV and the second one 20 kV [483]. However, the voltage has not increased much since then. Before Powerformer was introduced, the highest generator voltage used by Asea/ABB in Sweden was 21.5 kV, used in the turbogenerators for Ringhals and Forsmark built in the 1970's.

International development was not much different. The Hungarian manufacturer Ganz built four 30 kV, 5.2 MVA hydropower generators for Dalmatia in 1904 and two for Rome in 1905 [484]. Parsons built a few 36 kV turbogenerators during the period between 1928 – 34. Both examples concern generators connected directly to the grid, i.e. without transformers [485]. The successive growth in generator size, up even in the GW range, has been achieved through a corresponding increase in current. The voltage has remained below 30 kV due to lack of practical insulation systems for much higher levels.

A serious attempt to develop generators for direct grid connection was made in the USSR as mentioned above. A couple of generators were built; one hydropower generator rated 20 MW, 121 kV and later another one rated 100 MW, 165 kV. The Russian engineers tried to apply transformer insulation technology, i.e. it was basically an oil/paper system [284]. The results are unknown, but were probably less successful, because the concept became never widely used.

9.4   Powerformer® concept

Powerformer is based on modern cable technology, because the entire 3-phase stator winding consists of high voltage cables. Cables based on different polymer materials, e.g. cross-linked polyethylene (XLPE), were introduced in the 1960's. XLPE has, since then, formed a basis for the continued development of high-voltage cable insulation. Modern XLPE cables have proved to be very reliable and have high dielectric strength [486]. The cables used for Powerformer are, in principle, commercially available standard products. Such a cable has a circular cross-section with a copper conductor in the centre made up of a number of twisted strands. Radially outside the conductor is a semi-conducting equipotential layer of XLPE mixed with black carbon and then the main XLPE insulation. There is also an outer equi-potential layer of the same type as the inner one. A cross-section of such a cable is shown in figure 9.6. The two semi-conducting layers guarantee a uni-form electric stress distribution around the cable. The radial electric stress is also shown in the figure. An important property of this type of cable is that it is mechanically flexible enough for the curved end-windings.

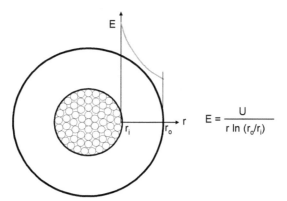

Electric field strenght as a function of the cable radius

Figure 9.6 Cross-section of XLPE cable including a diagram showing the electric stress

Stator windings in large generators usually have single-turn coils of diamond type with two coil sides per slot. If the voltage is to be radically increased, it is necessary to use multi-turn coils, which in the case of Powerformer means placing several turns of the cable in each slot. The slot section naturally be-comes quite different for a Powerformer compared to a conventional ma-chine. Figure 9.7 illustrates the difference. A normal large generator has two rectangular coil sides made from a large number of rectangular strands. The main insulation usually consists of glass backed mica tape impregnated with epoxy resin. Powerformer has a larger number of circular conductors per slot and the slot becomes much deeper and its width varies.

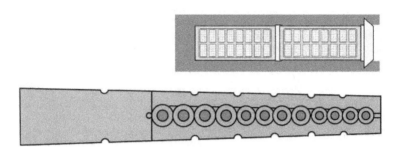

Figure 9.7 Cross-sections of conventional slot and Powerformer slot

This special slot configuration will have an impact on the properties of the machine in several respects. Theoretically it should make no difference for an electrical machine of a certain size if the voltage, for instance, is doubled, because the current will then be halved. There will be twice as many turns in each circuit but the conductor area can be halved, so in principle, the amount of copper and current in each slot will be unchanged. The losses and the temperature rise would also be unaffected. A provision is that the amount of insulation and the shape of the slot remain as they were. This is not at all the case in Powerformer. First of all, there is much more insulation in each slot, partly due to increased thickness and partly due to the fact that a larger number of conductors (turns) must be provided with the main insulation. Secondly, there is more copper in each slot. Theoretically, as mentioned above, it should be unchanged, but much thicker insulation and the fact that the XLPE insulation only allows a temperature of 90 °C (limited to 70 °C in this application) while normal class F insulation allows 155 °C implies that the current density must be reduced. The amount of copper must thus be increased. The shape of the slot will also add to its depth.

The deep slots may increase the mmf required for driving the flux through the teeth depending on the flux density. The back of the stator core will be moved radially outwards so Powerformer will end up with a larger outer diameter and significantly heavier stator compared to a conventional generator. Another difference is that a much deeper slot usually gives a larger leakage reactance especially affecting the transient and sub-transient reactances of the machine, but not necessarily in a negative way. It is difficult to make a general comparison of data for a Powerformer and a corresponding conventional generator. It has to be a case to case comparison. Table 9.1 below can therefore only be seen as an example of such a comparison.

Table 9.1   Comparison of Powerformer and conventional
hydropower generator for Porsi

Generator type	PFH	GS
Commissioning year	2002	1962
Rated power [MVA]	75	90
Power factor	1.0	0.9
Speed [rpm]	125	115.4
Rated voltage [kV]	155	15
Rated current [A]	279	3460
Excitation [V, A]	150, 1050	330, 880
Stator diameter, outer [mm]	10250	10500
Stator diameter, inner [mm]	9060	9730
Active length [mm]	2400	1500
Utilization factor [kVA/rpm m^3]	3.05	5.49
Weight, stator active parts [ton]	546	380

Powerformer concept is applicable both to hydropower generators and turbo-generators. All parts, except from the stator winding and the core remain, in principle unchanged if compared with a conventional generator. A critical factor is, as always, the cooling. It has to be very efficient to keep the stator winding at a low temperature. Powerformer that have been built so far, therefore have a system for axial water-cooling of the stator slot section consisting of parallel plastic cooling tubes placed axially through the teeth. Even the back of the stator core is provided with such axial cooling tubes. This water-cooling is additional to the more conventional, closed air-cooling that cools the rotor, the end windings and some other parts.

Presentations of Powerformer have often underlined the fact that there is uniform electric field strength in the circular conductor insulation in contrast to the non-uniform field around a rectangular conductor that is caused by stress concentrations at the corners as shown in figure 9.8. This is of course true, but its importance seems to have been exaggerated as it is not only the electric field strength which determines the thickness of conventional slot insulation; the risk for mechanical wear is, for instance, of great importance.

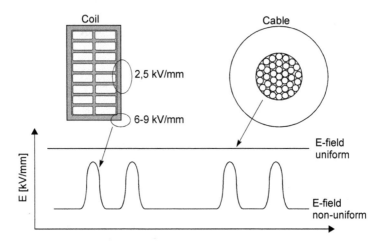

Figure 9.8 Diagram showing kV/mm in circular and rectangular coil insulation

The primary difference between Powerformer and conventional generators is the high voltage cable; all other differences are secondary. The intention was to use standard XLPE cables but, according to experts from ABB High Voltage Cables, this was not quite as easy as first thought. The electrical machine application required a conductor made up of thinner, insulated copper strands in order to reduce eddy current losses. The tolerance required, for the outer diameter, was also much more stringent than for a standard cable. Normally there is no problem if a large power cable is somewhat too thick but in this particular case it would be impossible to get the cable into the slots. The cable is pulled into semi-closed slots and it must therefore be divided into manageable sections. Figure 9.9 contains a photo of the winding process. The winding is also made with graded insulation, which means that the turns most adjacent to the neutral point have thinner insulation than those close to the terminals, where the voltage to earth is much higher. This may be efficient but it implies a large number of cable joints in the end windings. The technology and devices for making these joints turned out to be a complicated part of the winding. It must be mentioned that conventional machines also have a lot of joints in the stator windings as each coil has to be connected to the next. One problem, for the cable manufacturer, is the small production volumes needed for Powerformer and to make use of the full production line. Much higher volumes would be needed to economically motivate a separate production line. This also applies to the accessories needed for joints and terminations.

Figure 9.9 Winding process

Cables that could withstand higher temperatures and even higher electrical stresses would be beneficial for this particular application, especially the latter. As an example, silicon cables are made for temperatures up to several hundreds centigrade, but not for high voltages. It is not likely that such cables will be developed because the market does not require high temperature power cables. The resistive losses will be too large if the working temperature is raised. It seems, therefore, that XLPE cables will remain as the only realistic alternative for future Powerformer and Motorformer.

9.5   The development

In interviews, Mats Leijon has mentioned that he started to think about the idea of making a directly connected high voltage generator sometime in 1990 - 91. During his work within ABB Corporate Research, he had been exposed to problems with various high-voltage products and saw possibilities for improvements. In the case of Powerformer, Mats Leijon told the author, that the discovery of the possibility came first, not the ambition to get rid of the transformer. He described that he used an un-orthodox approach by applying field theory to analyze the power flow radially through the airgap and then axially through the windings. The analysis was based on the use of "Poynting's vector" [487], which is well-known from wave propagation and antenna theory [488]. Leijon has therefore often referred his invention back to an interpretation of Maxwell's equations and has claimed that this is a more direct way from physics to product than the traditional method applying circuit analysis. This can be questioned because the traditional analytic algorithms are also basically derived from Maxwell's equations. Both approaches would give similar results assuming the same prerequisites, but there is undoubtedly a pedagogic difference.

Mats Leijon was carrying his ideas for a long time until he consulted some colleagues. He filed the first internal report on his invention in June 1993. Harry Frank admitted, in an interview to the author, that Leijon's invention did not receive much attention at the beginning; if it had been a great idea, someone else would already have made it. The next steps, the oral presentations to Frank and Svanholm, have already been mentioned in section 9.1. The president of ABB Generation, Billy Johansson, was informed about the project in the autumn of 1994. He mentioned, in an interview, that he had then consulted some of his leading engineers, who first considered the concept as foolish, but soon admitted that it could be feasible.

Development started in the last quarter of 1994. The project was first staffed entirely by personnel from Corporate Research in Västerås until ABB Generation became engaged the following year. The project management preferred persons who were considered to have no old technology to defend. Several qualified researchers were employed directly from the universities. Lars Gertmar and later Claes Ivarson, both from Corporate Research, initially contributed with necessary electrical machine competence, but additional engineers with such experience joined the team in 1995, mainly from ABB Generation. The design group from Generation was located at Corporate Research. Eventually other Swedish ABB units needed to be involved, primarily ABB High Voltage Cables in Karlskrona. The project was, in spite of this, kept strictly confidential. Every person had to sign a special secrecy agreement and no information was released, not even to the Group management in Zürich. It was feared that the Swiss business area management would oppose the Powerformer development if it had been asked for comments. The CEO, Percy Barnevik was, however, informed in the summer of 1995 in his role as chairman of ABB Sweden. He approved maintaining the secrecy and he strongly urged the organization to work for a wide patent protection. It took almost another year until the executive vice president of ABB's power segment, Armin Mayer, was informed by Billy Johansson in June 1996. Mayer was positively surprised but wanted a review of the invention by Swiss generator specialists. The ABB Group's technical director, Craig Tedmon, received his first information regarding the project the following day [489]. According to Billy Johansson, approximately 250 persons signed secrecy agreements.

The project grew rapidly and engaged in the order of 20 persons late 1995 and 60 persons by 1999. The latter figure does not only include Powerformer but also Motorformer, Windformer and a transformer called Dryformer. ABB Sweden financed the entire development during the first years, but later the project received substantial corporate support. Project meetings were held regularly every Friday morning from 1995 until 2000. The main purpose of these meetings was for team members to inform each other about progress and discuss various technical issues. Several persons have testified that these meetings had a central function for the coordination of the work and to address potential problems long before they had grown serious. No really large problems have been reported from Powerformer development.

Various test rigs and mockups were built, but the project management realized that a full scale prototype would be of great importance as it focuses the development work, gives a lot of practical experience and demonstrates the technology in a clear and understandable way. As in several earlier occasions ABB turned to Vattenfall, its traditional Swedish development partner, when the time had come to discuss practical testing and demonstration of the new technology. Contact was established in May 1995 with Vattenfall's CEO Carl-Erik Nyqvist and Dr. Kjell Isaksson, manager for Vattenfall Hydropower, e.g. long before Powerformer was presented publicly. It is interesting to note that this took place before anyone in ABB's own executive management had been informed. The ambition to establish the concept as a Swedish technology was obviously very strong but also to protect the project at a sensitive stage. Half a year later the discussions between ABB and Vattenfall led to an agreement to install a prototype machine in Porjus hydropower plant in northern Sweden. The location was convenient because Vattenfall, together with ABB, Kvaerner Turbine and some authorities, had established an education and research unit there called Porjus Hydropower Centre, where the installation of two 11 MVA units, one for educational and one for development purposes, had already been planned. ABB suggested building the latter as a Powerformer with a rated voltage of 45 kV, a suggestion that was accepted and implemented. Kjell Isaksson mentioned, in an interview, that this new technology seemed to have potential to meet Vattenfall's goal to reduce investment costs by 10 percent, operation and maintenance costs by 10 percent and improve efficiency by one percent unit.

The design and manufacture of the "Porjus prototype" was handled by ABB Generation, which established a group, early in 1996, headed by Ivan Jonas, mechanical engineer with extensive experience with generator design. The prototype was ready for workshop tests in the fall of 1997.

As already mentioned the project had been carried out under strict confidentiality and the prototype was built in a restricted area of the workshop. On November 24th 1997 the Powerformer was presented for all personnel in ABB Corporate Research and ABB Generation in Västerås, e.g. several hundred of persons. It is remarkable that the confidentiality could be maintained for an additional three months until the press conference in Zürich, mentioned in section 9.1, took place; especially since one of the Swedish business papers published a notice, in the beginning of December, that ABB had applied for patents on a new "super motor without transformer" [490]. This notice was not very revealing but nevertheless, the local newspaper VLT missed a scoop.

It had been discussed to send the entire "Porjus prototype" to the launching press conference in Zürich, or more exactly to ABB's Corporate Research Centre in Dättwil, but it was replaced by a mockup. The event became anyway very spectacular and the invention aroused a lot of interest, not the least from competitors like Siemens and Alstom. The real prototype was transported to the power plant in Porjus, situated in the far north of Sweden, where it was installed and then started up and connected to the local grid on June 10th 1998.

The project received full support from ABB's executive management, at least as long as Göran Lindahl was CEO. However, new inventions can be hurt from the "not invented here" syndrome and Mats Leijon and others have claimed that this happened to Powerformer, especially from foreign parts of ABB. The influence this had on the transition from the development phase to commercial marketing and production can neither be proved, but nor can it be ignored. The development did not only comprise Powerformer, but a whole family of "Formers" of which Motorformer already has been mentioned. Other products were the oil-free transformer called Dryformer and the wind power generator Windformer.

9.6   Powerformer® deliveries

The prototype for Porjus has been dealt with above. It has, since its start, been in trouble-free operation for more than 27 000 hours and it is still in use [491]. A sketch of the machine is presented in figure 9.10.

Figure 9.10 X-ray sketch of the Porjus Powerformer

In June 1998 Vattenfall and ABB signed a letter of intent concerning delivery of a full-sized Powerformer. The final contract was awarded in March the following year. Vattenfall had plans for a total renovation of two old units in the Porsi power plant located at the Lule River downstream of Porjus. It was then decided, on initiative of ABB, to replace one of the old units with a Powerformer rated 155 kV, 75 MVA and 125 rpm. Vattenfall's motivation for a full-scale test of the new technology was increased plant efficiency, better availability and reduced maintenance, all of them contributing to lower operation costs. The higher cost for Powerformer was partly covered by a contribution from the Swedish National Energy Agency. Vattenfall payed a price corresponding to that of a conventional generator. The project was regarded as a very interesting demonstration applicable on many old hydropower units in Vattenfall's system. It was originally planned to have the Porsi unit ready for commissioning in 2001, but it was not put into service until May 2002 [492].

Powerformer for Porsi is a vertical shaft salient pole synchronous machine. It has a closed cooling system with air as the primary and water as the secondary coolant. This cools the rotor, stator end windings and some other parts. The heat created in the active part of stator winding and the stator core is, however, cooled by a separate water-cooling system as described in section 9.4. The heat exchangers are placed at the outside of the stator frame as for conventional hydropower generators. It has a combined guide and thrust bearing below the rotor and a guide bearing above the rotor. The excitation current is provided from a rotating exciter. Table 9.1 contains a comparison of some main data for Powerformer and the corresponding conventional generator at Porsi. The transformer removed from Porsi had a weight in the order of 100 tons [493].

Powerformer at Porsi has now been in service for more than four years and had accumulated 27 000 operation hours in the autumn of 2006. The generator has, after some initial problems, which had nothing to do with the new concept, performed very well. Kjell Isaksson mentioned, in an interview, that the measured performance had been in accordance with the predictions to an extent that was surprising for a radically new machine type. This can probably be attributed to very comprehensive computer simulations as a basis for the design.

The first, and up till now only, Powerformer of turbotype was supplied to a municipal bio-mass power plant in Eskilstuna, a city located 40 km south of Västerås. Anders Björklund, president of Eskilstuna Energi & Miljö, said, in an interview, that the Swedish government had decided, in June 1997, on a program to subsidize the building of new combined heat and power plants. Eskilstuna filed an application the same month and soon started to prepare requests for bids etc. ABB invited Björklund and his predecessor for an information about Powerformer in November, i.e. before it had been launched. Eskilstuna had had some negative experience from stator winding failures in large heat pump motors and therefore saw the new cable winding as very interesting. Calculations indicated that it would be profitable to change to the new concept which was also supported by the Swedish Energy Agency, which granted Eskilstuna 112 MSEK (14 – 15 MUSD) for the project. ABB received the order in October 1998. The Eskilstuna generator had a nominal voltage of 136 kV. The rated output was 42 MVA and the speed 3000 rpm. This unit was installed and put into operation in December of 2000.

Powerformer in Eskilstuna is a horizontal shaft 2-pole synchronous machine with cylindrical rotor. It has, like the hydropower generator at Porsi, a closed cooling system with air as the primary coolant. Heat exchangers, air/water, are placed in the corners at both ends of the stator. The rotor winding is directly air-cooled while the stator winding is indirectly water-cooled by means of cooling tubes placed axially through the teeth. There are also such axial cooling tubes through the back of the stator core. Figure 9.11 presents a sketch of the machine. Table 9.2 contains some main data for the Eskilstuna machine as well as a couple of other turbogenerators for reference purposes. It is doubtful if Powerformer technology is at all suitable for 2-pole machines as the stator becomes very large and heavy [494]. The 42 MVA generator for Eskilstuna had a stator weight of approximately 205 tons. This can be compared with 234 tons for the 12 times more powerful 540 MVA generator for Ringhals 2. Indeed, the transformer is eliminated but the weight of an ordinary 50 MVA transformer is only around 50 tons and the transformer efficiency is at least 99.7 percent.

Table 9.2   Main data for turbogenerators

Power plant	Eskilstuna	Helsingborg	Ringhals
Rated output [MVA]	42	75	540
Power factor	0.933	0.8	0.85
Voltage [kV]	136	11.5	19.5
Active length [mm]	3250	2700	4100
Rotor diameter [mm]	910	900	1150
Outer stator core diameter [mm]	3200	2200	2905
Weight, stator + rotor [ton]	231	101	284
Utilization factor [kVA/rpm, m^3]	4.31	9.56	23.95
Specific power [kVA/kg]	0.18	0.74	1.9
Efficiency, measured [%]	98.24	98.21	98.59
Delivery year	2001	1981	1974

Figure 9.11 X-ray sketch of the Eskilstuna Powerformer

Swedish newspapers reported in December of 2005 that Eskilstuna had decided to replace Powerformer with a conventional generator. The reason for
this decision was that the stator winding had failed a second time. The first
failure had already occurred in November of 2001 due to overheating of the
stator winding cable, a problem that was caused by a malfunction in the
cooling system. The stator was re-wound and Powerformer was then put back
into service. The second failure, an earth fault close to the slot end in one of
the phases, came after 2½ years of operation Sparking had occurred between
semi-conducting fixations and adjacent cables, which damaged the outer
semi-conducting coating of the cable which then broke down due to dielectric
stresses. It is more difficult to avoid this kind of problem in a turbogenerator
that has a higher induced emf per unit length [V/m] than a low speed hydropower generator. The decision was to replace the cable wound stator with a
standard one, but not because the customer was dissatisfied. Anders
Björklund regrets that it became necessary to abandon Powerformer, but they
were forced to do so as the insurance companies refused to offer insurance
for interrupted operation, primarily due to the expected long repair time that
this kind of failures requires. Eskilstuna Energi och Miljö needs such an insurance because the company is simply too small to carry the risk by itself
[495].

Three more Powerformer have been built and delivered to power companies,
two of them outside Sweden, all for hydropower plants. Table 9.3 below includes a list of all Powerformer delivered up until now. All hydropower units
had performed without any trouble until a failure occurred in Höljebro in
September of 2006. It was a stator winding failure, but the reason has not yet
been reported.

Table 9.3   List of Powerformer

Location	Type	Power [MVA]	Voltage [kV]	Commissioned [year]
Porjus, SE	Hydro	11	45	1998
Eskilstuna, SE	Turbo	42	136	2001
Porsi, SE	Hydro	75	155	2002
Höljebro, SE	Hydro	25	78	2001
Miller Creek, CA	Hydro	33	25	2002
Katzurazawa, JP	Hydro	9	66	2003

9.7   Other applications

As already mentioned, Powerformer has a couple of followers; Motorformer™ and Windformer™. A Motorformer is, in principle, not different from a Powerformer, rather the difference lies in its application. In reality there are practical differences, but these have very little to do with the high-voltage winding. Motorformer built so far have been salient 4-pole synchronous machines with horizontal shafts and built-on air/water heat exchangers. ABB markets such motors under the type designation AMT in the output range 5 – 45 MW and for voltages 20 – 70 kV [496]. The first Motorformer rated 6.5 MW, 42 kV was supplied to a Swedish air-compressor plant in 2001.

The next delivery was particularly interesting because it was for an offshore application, which is especially suitable for this type of high voltage machines. In 2004 two Motorformer rated 40 MW, 56 kV were delivered to Statoil, Norway for installation in Troll A, a gas platform in the North Sea. Each of these motors is fed from the grid on the mainland through a so-called HVDC Light™ transmission link as shown in figure 9.12. The inverters open the possibility to operate the motors with variable speed, something which improves the system efficiency of such compressor units. The speed can, in this case, be varied between 1290 - 1890 rpm. The main advantages of using Motorformer for this type of application is that it saves very expensive space at the oilrig and it can be directly connected to an HVDC link, which is cost effective for long offshore power transmission [497]. A photo of the stator is shown in figure 9.13. The two Troll machines have been in successful commercial operation since October 2005.

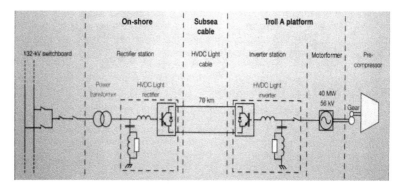

Figure 9.12 Simplified diagram for Motorformer in combination with HVDC Light transmission.

Figure 9.13 Motorformer stator with cable winding

Robert Larsson, who is president of ABB's Automation Product Segment in Sweden, mentioned, in an interview, that ABB has a contract from Statoil to deliver two more Motorformer. Other oil companies have expressed interest, but are waiting for the technology to be further proven. Larsson also said that Motorformer cannot compete with conventional motors in general as it is too expensive, but there are several applications where the use of Motorformer becomes advantageous in comparison with a combination of a conventional motor and transformer. The weight penalty seems to be less for a Motorformer than for Powerformer, at least in the studied cases. The machines built for Troll weigh around 110 tons each while a corresponding conventional motor would have a weight of approximately 80 tons. One reason is probably that the voltage is much lower, 56 kV compared to 136 and 155 kV for the generators. The latter require much thicker insulation, which affects the entire machine size. Transformers for conventional motors of this size would have had a weight of approximately 40 tons each.

ABB is also marketing the AMT machine type as generators under the name of Powerformer Light but has so far not received any orders [482]. According to Robert Larsson 4-pole generators of this size are mainly sold for gas turbine power units which are extremely standardized. The gas turbine, gearbox and generator along with auxiliaries are mounted on a skid. The transformer is used for the adaptation of such standard units to various grids. The market for Powerformer Light is hence very limited.

Windformer is a wind power generator for high voltage. The intention was to build large offshore wind farms and transmit the power onshore via HVDC Light links. The Windformer was to be a low speed, permanent magnet generator directly driven by the wind turbine and provided with a cable winding for such high voltages that it could be connected to the electric transmission without any transformer. The concept would thus eliminate both the gear-box and the transformer. ABB presented Windformer at a press conference in London on June 8th, 2000. Like Powerformer it was also in this case announced that a prototype plant was planned together with Vattenfall. A 3.5 MW unit should be installed at Gotland, the large island in the Baltic, during the following year [498]. To build large, directly driven, wind power generators is not easy as the speed is extremely low, in the order of 20 – 25 rpm, and the machine diameter therefore becomes large, several meters. It is, at the same time, necessary to design such a generator with a very small airgap and this is difficult to maintain due to differential thermal expansion in combination with other existing forces. In addition, the cable wound stator would be heavy, in the order of 100 tons for a 3.5 MW unit, and to put that on top of an 80 m high tower would create severe problems [499]. The prototype project, which involved several actors, was cancelled before the unit was built. Whether this decision depended on feared technical difficulties or a re-evaluation of the profitability is not clear and is neither of importance for this thesis. An interesting question is nevertheless if Windformer was introduced as an optimal solution for large offshore wind power units, or were these units identified as another area for application of cable wound machines? The concept seems, however, to have been too optimistic and ABB abandoned it completely in March 2002 [500, 501].

9.8   Patent strategy

It has been mentioned in section 9.5 that Percy Barnevik requested the project management to systematically strive for a comprehensive patent protection. This was a new approach for ABB who had usually applied for patents for suitable inventions, when such were reported by the employees. In this case a patent engineer was allocated to the project team which had regular review meetings in order to identify possible inventions for which patent applications should be filed. This activity was structured in a number of different subjects such as winding technology, cooling system etc. The process resulted in a large number of patent applications. Harry Frank mentioned, in the interview, that the first round, filed in 1994, contained as many as 35 applications.

ABB aimed at obtaining world wide patent protection for Powerformer and the other products in the "Former family". Some press information claim that ABB has more than 200 patents for this technology, but this is probably somewhat exaggerated. More detailed information can be retrieved from various patent registers, e.g. from the European patent office [502].

9.9 New company structures

The situation was fairly simple as long as ABB had its own power generation segment. Indeed, ABB in Switzerland had the responsibility for generator development, but Powerformer was allocated to the Västerås factory. It remained like that even after ABB and Alstom had merged their power generation activities and formed ABB Alstom Power in 1999. The major change took place in May of 2000 when Alstom acquired ABB's part of the jointly owned company. ABB's CEO Göran Lindahl wanted to divest heavy parts, like large steam and gas turbines, and focus more on IT related business areas. It has been mentioned, in some interviews, that Lindahl had wanted to keep generator activities in ABB but was forced to sell them with the turbines, including the exclusive rights to design, sell and manufacture Powerformer. An illustration of the importance that was attributed to the latter is the following quotation from a book about the crisis in ABB, which received a lot of attention. *"The price tag .... rises from the fact that ABB contributes [to the joint venture with Alstom] with one of the most advanced large gas turbines in the market and a new, revolutionary generator technology – the in Sweden developed Powerformer, which can reduce investment costs for a power plant by perhaps 30 percent."* (My translation) [503]. ABB was, according to the agreement with Alstom, for a limited time prevented from re-entering the market as a manufacturer of large power generation equipment, but the Powerformer license that ABB gave to Alstom is no longer exclusive. ABB maintained the rights to the build Motorformer.

Since five years back Powerformer has been an Alstom product and the business area management, located in Switzerland, has the formal responsibility. No machine has been sold during this period. All activities, up till now, have been carried out by the Swedish unit in Västerås, which at present has no right to sell any generators for new plants anywhere in the world, whether or not the generators are Powerformer or conventional units. Alstom's current strategy is obviously not to pursue marketing of Powerformer.

9.10 Results, status and prognoses

It is an understatement to claim that Powerformer has not lived up to the commercial expectations. Several sources mentioned an expected billion dollar business in less than 5 years. Sofar, six units have been delivered, including the prototype to Porjus. At an estimated price of 1000 kSEK/MVA, the total sales would be in the order of 200 MSEK or close to 30 MUSD. The reasons for the huge discrepancy can be discussed. The original prognoses were certainly very optimistic, that could be one reason, but no sales have in fact been made since 2000. As mentioned in the previous section, Alstom does not try to market Powerformer for two main reasons. The first reason is a cautious and conservative manufacturer attitude; why take the efforts and risks in introducing new concepts, which could potentially cannibalize the existing products? The second is that there is no demand from the market. Powerformer presumes a quite different system solution and that is not specified, by customers or consultants, when they request proposals from different suppliers. The threshold for new technologies can often be very high, especially as long as one manufacturer is the only available source. It could very well have been an advantage if ABB had licensed Powerformer to more than one company or even had made the development in partnership with a competitor. Several of the key persons interviewed have claimed that Powerformer technology was too different and ABB did not posess the right structure for handling this. It could have been better to establish a separate company for development and market introduction of Powerformer. It is difficult to compete against a dominant technology. It is also difficult for a large organization to accept radically new concepts [504].

The technical results have been good; at least all hydropower generators have fine operation records. Tests have shown good agreement with calculated performance data. Powerformer has higher iron losses, but lower copper losses than conventional generators which mean that it has its best efficiency at high loads while it is somewhat inferior at low loads. Reduced airflow, due to the water-cooling of the stator, also decreases the ventilation losses. It has been claimed that Powerformer has better efficiency than corresponding conventional generators, but it must then be kept in mind that it also contains much more material, which reduces current and flux densities. It would not be difficult to increase the efficiency in conventional generators at the sacrifice of more material, i.e. weight and cost.

Simple comparisons based on the machines delivered to Porsi, Eskilstuna, Höljebro and Troll indicates, as could be expected, that the weight penalty for Powerformer and Motorformer compared to conventional machines increases with the voltage. A higher voltage requires thicker cable insulation, which increases the size and weight of the entire stator. The pole number seems, however, also to have a big impact and the output have probably some influence too, but that would require a study of more examples to be verified. The diagram in figure 9.14 gives therefore only a brief indication of the influence from the voltage and the pole number on the weight relation with and without transformer.

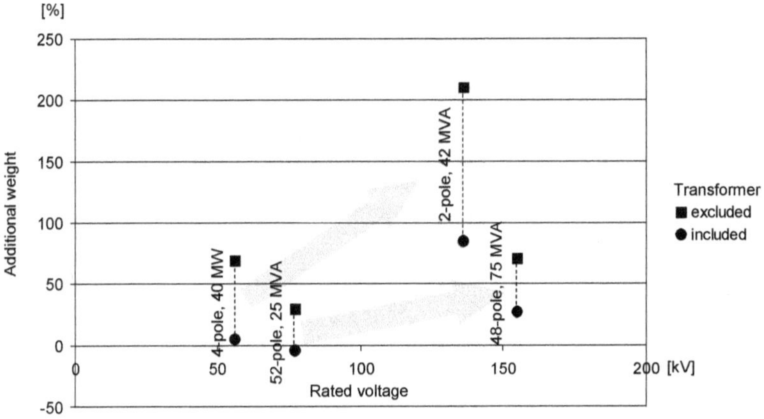

Figure 9.14 Weight relation (Powerformer/conv. machine) vs. voltage

The manufacturing costs for a Powerformer become much higher than for a conventional machine. Published documentation indicates that it will be roughly twice as expensive [494]. Powerformer can therefore hardly compete on a machine to machine basis; it must instead always be a comparison of system costs as figure 9.5 illustrates. Whether ABB and Alstom have reached their targets for production costs is not publicly known, but as long as so few units have been built, it is probable that they still are rather high on the learning curve. Neither ABB nor Alstom have revealed the development costs, it seems to be a sensitive matter, but there are reasons to believe that they have been substantial. It is, in this respect, not relevant to try to separate Powerformer from the other products in the "Former family". Neither is there any motive to differ between formal development and order design. It has been a process, for almost 10 years, that has engaged 20 – 60 persons, i.e. probably in the order of 500 man years. In addition, there are costs for experiments, test rigs, prototype, deficit on orders, and the extensive patent portfolio. A total cost of well above 500 MSEK (70 MUSD) can therefore be assumed.

The most crucial question for Powerformer is about the potential markets. The domestic market is limited to retrofit units. There is hardly any new hydropower and thermal power plants built in Sweden anymore. Kjell Isaksson, from Vattenfall, who is positive to the technology, said that there are unfortunately very few renovation projects where both generator and transformer need to be replaced; it is thus difficult to obtain an acceptable payback of the extra cost for a Powerformer. It is most suitable for new plants, for which it is possible to take full advantage of reduced system costs. Jan Boivie, design manager for hydropower generators at Alstom in Sweden, claimed in an interview, that there may have been too much stress on getting rid of the transformer. Instead of using a Powerformer for 150 – 400 kV, it could in several cases be better to take one for 50 – 100 kV and arrange a voltage step-up at a suitable location. It would still be possible to replace the expensive busbar, which a conventional generator would require, by cost-effective HV-cables. The possible wide range of generator voltages introduces new possibilities in terms of electrical plant and system layout. Conventional high current applications require short distances between generators and transformers due to cost and efficiency reasons. This is not the case in a Powerformer application where the voltage can be chosen more freely; the local transmission losses can be reduced and the step-up transformer can be conveniently placed.

The large market for new powerplants can be found mainly in China, India and other non European countries to which it is difficult to export Swedish built generators. The best chance for the survival of Powerformer technology is probably to license it to manufacturers in Asia. From a strictly Swedish point of view, Motorformer seems to have a more promising future.

What are the intentions of the manufacturers concerned? This is a question of fundamental importance for the future of these new machines. Where are the competence and the resources today? ABB continues to develop and market Motorformer; Robert Larsson confirmed that it is included in the long-term strategy, but the company has chosen to move cautiously in pace with the increased experience gained from the units in operation. ABB has, for a long time, been the world's leading manufacturer of 4- and 6-pole machines in the output range 5 – 70 MW. The company has thus such strength, technically and commercially, that it ought to constitute the best possible platform for Motorformer. No official comments are available from Alstom regarding their strategy for Powerformer.

9.11 Conclusions

Powerformer history covers not much more than 10 years, but it contains, nevertheless, many interesting results. Some observations from the study of this period are listed below:

- Powerformer, as well as Motorformer, represent a radical step in electrical machine development and the machines, which have been built and put into operation, prove that the concept is technically sound.

- The concept requires a system approach taking into account advantages thanks to the absence of transformer and other equipment. Powerformer and Motorformer can never compete with conventional machines on a "machine to machine" basis.

- The extra weight, and thus also extra costs, required for Powerformer and Motorformer, increase with the voltage level. Their competitiveness seems therefore to be most pronounced at moderate voltages, probably below 100 kV for machines with low pole numbers, but perhaps up to 200 kV for machines with high pole numbers.

- The development of Powerformer was well organized, and it was carried out very systematically. It is a good example of what can be achieved, when a development project receives ample resources and the development team gets possibilities to focus on a single target, in this case a full scale prototype.

- Characteristic for Powerformer project was the way it was carried out under strict confidentiality, while a comprehensive global patent protection was obtained. When the time was right, a well directed product launching succeeded in presenting Powerformer as the most sensational electrical machine for almost a century.

- As many times before, ABB turned to Vattenfall as a partner and first customer, when the new technology needed to be demonstrated and tested in commercial operation.

- The predicted commercial success has not occurred, partly depending on an overestimation of the available market and partly due to lack of efforts to pursue the marketing of Powerformer. To which extent the latter depends on the transfer from ABB to Alstom is not obvious; the business control was in both cases allocated outside of Sweden.

The future for Powerformer technology does not look bright. The current level of activity is minimal and there are no signs that it will change within the next few years. The following comments summarize the author's view:

- From a technical point of view, Powerformer would be more competitive if cables, sustaining much higher electrical field strength, were available. This seems, however, not to be the case within the foreseeable future.

- The market for Powerformer is much more limited than originally anticipated. It is difficult to take into account Powerformer advantages except for completely new power plants.

- The domestic market is, in the case of Powerformer, very important but unfortunately extremely small. A political decision to exploit one of the protected rivers for hydropower production could change the situation and at least give Powerformer a good chance.

- The marketing of Powerformer has suffered from ABB's and Alstom's internal structures and strategies, which have prevented the Swedish units from pursuing international sales efforts. This situation is not likely to change unless, for instance, a company like VG Power acquires a right to use the technology.

- Motorformer has, at present, a more promising future, but it is also in this case a matter of a niche market. It is probably Motorformer that will help the technology to survive.

The development of Powerformer, and the other "-former products", has received much attention. It can serve as a good example of the technical accomplishment of the development of an advanced new electrical machine. It can, however, also be concluded that the business plan must have been too weak. Why it was like that is not known, but it is no surprise that it is difficult to introduce a new technology competing with established solutions, especially in a large multi-national organization.

10   Synthesis and conclusions

The focus for this thesis has been on the development of four electrical machine types. All four of them represent substantial efforts by Swedish industry even though the span, in this respect, is very wide. If the measure of success is to have reached and maintained a leading position as a global supplier of these different machines, only one of the them is qualified. All four machine types have, however, performed well from technical point of view. The intention of the research has primarily been to study and present the development history of the selected machine types. It is difficult to condense these into meaningful summaries. This final chapter will instead concentrate on comparisons of important results and aggregated conclusions that can be drawn from the project.

10.1 Important similarities and differences

A few research questions were formulated in section 1.5. The questions are relevant but not very exact. The same can be said of the answers. The lack of preciseness can depend on difficulties to find the truth, but it more often depends on the fact that the development processes are partly irrational even for a rational product such as the electrical machine. The first four questions were much related to each individual machine type, and the answers have therefore been presented in the respective chapters. The three remaining questions were more general and of comparative nature, and will therefore be addressed in this chapter.

The first of these general questions was: which factors have had the most influence on the development of the different machine types? Is the development of smaller machines more market driven and the development of larger machines more technology driven? A comparison of how a number of key factors have influenced the development of the machines is presented in section 10.3; therefore only some general comments, referring to the second part of the first question, will be made here. Different interpretations of "development" were discussed in chapter 5 and it is important to keep in mind which ones are relevant. The main objective has been to study the technical development of the actual machines in a wide context, and whether the development has been influenced primarily by technical or commercial factors. It is necessary to make a distinction between initiating and influencing factors. Three of the machine types have been initiated by the customer needs while one, Powerformer, was developed due to the invention of a technical concept. In spite of this, it can not be generally concluded that machine size has any relationship with the initiating factor. It is, however, easier to conclude that technical factors are more influential on the development of large machines while commercial factors dominate in case of smaller machines. The standard induction motor is a mature product with a huge market and many competing suppliers. The product differentiation is small and the dominating production development is chiefly market driven. The matrix in figure 10.1 is a somewhat subjective illustration of the position of the four machine types with respect to driving factor, machine size and technical content. The way all the studied machines have developed in a commercial sense is primarily related to market factors but also influenced by each manufacturer's strategic decisions.

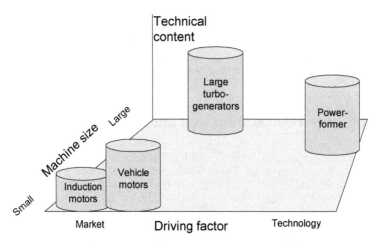

Figure 10.1 Positioning of the development of the machine types

The next question was related to the acquirement and maintenance of knowledge and competence. Had the transfer of know-how between development teams for large and small machines been negligible? The base for the studies has been Asea/ABB's development of the chosen machine types, though brief information has also been obtained from a few other Swedish manufacturers. Asea/ABB has been the dominant electrical machine manufacturer in the country for approximately 120 years and its competence and resources have been created step by step. It is obvious, from all four cases, that the company rarely has acquired technology through licenses or other transfer from external sources. Inputs from consultants and universities have also been very limited. Asea/ABB has been enough unto itself and has looked upon itself as the domestic centre of excellence in this particular field. Therefore, the traditional way of building and maintaining sufficient competence has been to recruit recently graduated engineers and teach them through their practical work. Most other Swedish manufacturers have been more dependent on input of know-how from external sources, e.g. through cooperation with universities, employment of engineers from Asea/ABB or simply being followers of the leading companies.

The question about transfer of technology between departments for smaller and larger machines is only relevant for Asea/ABB. After a common start, separate design offices were organized for various machine types in the early 20th century. They were located in the same building until after World War II, when a geographical dislocation to various outskirts of Västerås began. Asea was organized in divisions during the 1960's with the standard motors and the large generators in separate divisions. These units were later transformed into subsidiaries and have belonged, from the beginning of this century, to quite different corporations. The base for formal cooperation has thus decreased more and more over the years. Through the years a few persons have been transferred from the design departments for large machines to the departments for smaller machines and vice versa, but not as part of any planned technology transfer. The transfers have instead more often been managers rather than specialists. It is obvious that the direct flow of technology, between large and small machine development, has been very limited. However, the access to common technical bases summarized, in section 10.2, has been beneficial for Asea/ABB.

The last question was: how has the development of the selected machine types changed as a function of time? Has the development of smaller machines become more innovative in later years compared to that of large machines? It is difficult to give a consistent answer to the question regarding the change of development over time, except for trivial replies such as improvements related to the introduction of computers. A review of the four studied cases points in different directions. The standard induction motors have not undergone much development during the last 50 years as the focus instead has been on improved production methods. The "golden age" for development of large turbogenerators was the 1970's. Since then development has mainly been a matter of refinement of the designs. Electrical machines for automotive drivelines represent a very expansive application and the development of these machines has been intensive during recent years. Powerformer, finally, speaks for itself. It is a radical development applicable to large electrical machines. The result from this study is that there is no big difference between small and large machine with respect to the innovative level of development. This is illustrated by the matrix in figure 10.2. This conclusion can not be generalized. It is relevant for the situation in Sweden during the last 10 – 15 years, but an international study would most probably have shown a more intensive development of smaller machines.

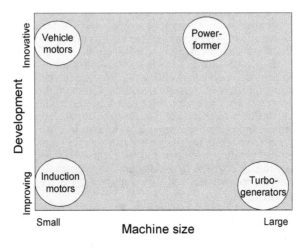

Figure 10.2 Character of recent development

10.2 Common technical base

In a large electrical industry such as Asea/ABB, common technical bases exist that have been of great importance for the development of motors and generators. The Corporate Research Organization, earlier known as the Central Laboratories, has been the most important. It has, in all the studied cases, been mentioned as the primary source of special know-how and services. Typical areas for its contributions have been material technology, insulation systems, flow dynamics, measurement technology, and acoustics but even advanced calculations and simulations. A central department for production development has played a major role in the creation of production lines and the design of special tools. General computer programs, e.g. for FEM analysis, constitute another common base for the development engineers. It has been testified too, in several interviews, that the informal network, which consisted of various Asea/ABB specialists who could be found through the internal telephone directory, was often of great help. The need for external assistance was very limited.

Documentation of theoretical studies and investigations as well as practical experiments and tests has been an important source of knowledge. Abstracts of all technical reports and memoranda, during the Asea period, were therefore distributed to all relevant design departments and laboratories. The full documents have been archived in the central archive since the late 19th century.

Contacts with the technical universities have neither been regular nor systematic. They can not be considered as a base for Asea/ABB's electrical machine development in any other sense than as a recruiting ground for graduated engineers and researchers.

10.3 Comparison of certain key factors

A major question for the entire study has been; which factors have been most influential on the machine development? Table 1.2 lists a large number of such factors and chapters 6 – 9 contain some explicit, but also several implicit, answers to this question. The interpretation becomes easily subjective, but the author has tried to compensate for this by consulting a few of the interviewed specialists. Strategic, commercial and technical factors are presented separately subsequent to an aggregated appraisement. The results are presented without further comments in graphical form in figures 10.3 – 10.7.

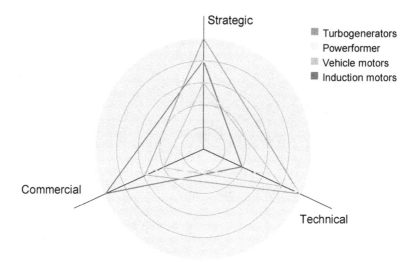

Figure 10.3 Impact of general factors on product and process development

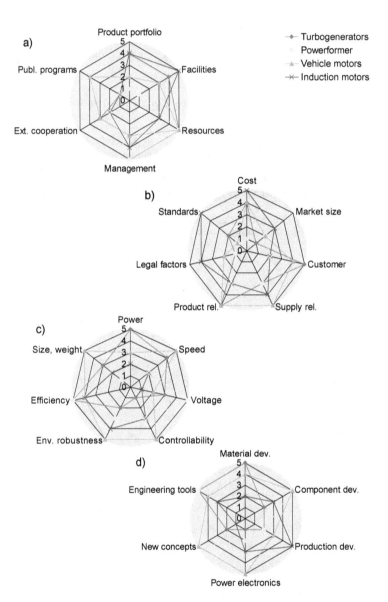

Figure 10.4 a) Impact of strategic factors b) Importance of commercial factors c) Importance of performance factors d) Influence from other technical factors

10.4 Fulfillment of objectives, technical and commercial

All product development has an aim, usually to meet a number of both commercial and technical targets. The detailed reviews, presented in the previous chapters, have shown to which extent this has been fulfilled in each of the cases studied. Only some brief comments shall be made here.

Asea/ABB has been a successful manufacturer of standard induction motors. These motors have belonged to the company's base products for more than a hundred years and ABB today maintains a position as one of the 2 – 3 largest induction motor manufacturers in the world. It is a profitable business in spite of competition from all kinds of companies, including those from low-cost countries. ABB's motors meet the performance requirements and the customers' expectations on reliability. It can be concluded that the development of these induction motors have fulfilled both the commercial and technical objectives.

The development of the large, directly water-cooled, turbogenerators was difficult and the initial problems were severe, both technically and commercially. Asea succeeded to overcome the first technical difficulties and as a result the generators have performed well since the early 1980's. The original commercial expectation has, on the other hand, not been met. The global market for large generators decreased drastically 20 – 25 years ago and the formation of ABB, and later its transfer to Alstom, has limited the Swedish subsidiary to a supplier of service and retrofit units for existing plants.

ABB has had a vague and hesitant attitude towards electrical machines for automotive drivelines. The company has, however, made attempts to approach this expansive market and ABB in Sweden has developed some induction and PM motors for such projects. The technical results have been good taking into account that the attempts have been limited to prototypes. None of the projects managed to reach commercial status, which was the intention for two of the projects. The decisions to terminate the projects were, in both cases, made by the car manufacturers.

The Powerformer development is another example of a technically successful development that was a commercial disappointment. Six machines have been built and put into operation and at least four of the hydropower units have performed well and met the performance expectations. In spite of this, the huge projected sales volumes have simply not materialized. The reasons for this could be an overestimation of the available market, customers' reluctance to accept unique solutions and/or lack of efforts to market Powerformer.

The studied cases show that the development has, in general, been successful from a technical perspective. The reasons for partial or complete failures have instead been commercial and more specifically related to the market and the marketing. An important observation is therefore that the market analyses have most likely been insufficient in comparison with the technical feasibility studies carried out before the development projects have been started. The induction motor was the exception and that is easily explained by the fact that it is a well proven product in a very well established market.

10.5 The development process

The development of induction motors, described in chapter 6, covers a wide time span. A review of the decision and development process during the last decades is the most interesting for this study. The results have proven that both the management and development teams at large have done the correct things. An example of a shift of motor generation is given in section 6.4.5. It contained three major steps, a pre-study, a pre-project and the main-project, all of them managed by a project manager reporting to a steering committee, which had regular meetings throughout the entire process. It was also mentioned, in section 6.4.3, that Asea's motor division had transferred the responsibility for the entire development projects until the start of commercial production from the line organization to dedicated development teams. Market analyses are explicitly included in the development work. Production development, prototype manufacturing and testing are other important activities included.

The development of large turbogenerators was very different. Most of it was carried out as design work on specific customer orders. Only a few experiments and test rigs were handled in separate development projects. The line organization held the main responsibility but it was, in reality, a shared responsibility between the turbogenerator office and the development office, at least during the first years. No formal steering committee had been appointed. For long periods of the time, the work became too much of a burden when focused on trouble shooting activities. This can be attributed to extrapolation problems, lack of a full scale prototype, and too many customer orders at the same time. It took around 10 years to overcome all the technical problems at a cost that was much higher than a development project with a large prototype would have required. It is more difficult to criticize the lack of a realistic market analysis since all prognoses for future generating capacity, made during the 1960's and early 70's, predicted the sky to be the limit.

ABB's projects for the development of machines for electric and hybrid vehicles are briefly described in chapter 8. Most of the projects were carried out during the second half of the 1990's with the small development company ABB Hybrid Systems as main supplier and ABB Motors and Corporate Research as important sub-suppliers. None of the projects left the prototype stage, however, primarily due to the automotive OEMs. The technical results were not free from problems but were, in general, acceptable. The concept of establishing a special development company was unusual for ABB, at least in the case of electrical machines and drive systems which traditionally belonged to the core business. The advantage of this arrangement was a focused unit with its own board and the results were very visible. The disadvantages were limited resources and the fact that the automotive business never became a fully accepted part of ABB's activities as mentioned in section 8.4.4. The main problems, that led to the termination of the company's efforts, seem to have been lack of commitment from the executive management combined with insufficient knowledge of the automotive market.

The Powerformer project is in many respects a fine example of technical development. The target was well defined and a focused team was appointed. Thorough theoretical analyses were combined with several experiments and test rigs. Even a large prototype was built, tested and put into operation. The project team had weekly meetings to follow the progress and discussing possible problems. Systematical work resulted in a very wide patent protection. Furthermore, the Powerformer development had full support from the executive management, even though there was certain reluctance from foreign ABB units. Nevertheless, the market analysis and strategy seems to have been wrong. It can be argued, but not proven, that the situation would have been better if the power generation business had remained within ABB instead of being sold to Alstom.

The four cases studied have yielded a couple of general results. A well organized and implemented project with full scale prototypes is important for technical success; it is difficult to gain anything through short cuts. Perhaps a less obvious result is the insight that a large enterprise, such as ABB, fails in market strategy and analysis even if the technical development is successful. Can the industry learn from such mistakes without becoming overcautious?

10.6 Future electrical machine development

Electrical machines can, in general, be considered as mature products which would imply that their future development can be expected to be an extrapolation of current trends. However, an analysis of the development history of the four studied machine types will modify such a simplified statement. The development intensity has varied over the years, differently for different machine types. Some of the study cases indicate clearly that it could be difficult to make forecasts even within a time frame as short as 10 years. Inventions of new concepts and requirements from new applications have been driving factors. It is impossible to predict new inventions while new applications and new requirements from existing applications ought to be found through systematic investigations.

Material technology has always been a cornerstone for electrical machine development and it will remain so also in the future. It is a matter of electric conductors, magnetic materials, insulators and structural materials. All four categories are subject to development, but in a short or medium term perspective, new or improved magnetic and structural materials can be expected to have the largest impact. Cost-effective high-temperature superconductors may also become important for certain types of larger machines. Inverter technology, power electronics and control principles have had a huge impact, primarily on electric motors, during the last 40 – 50 years. Various solutions have become well established and it is therefore not likely that the ongoing development will lead to any radically new situation for rotating machines. However, a reservation must be made for the influence that SiC components may have on integrated designs for high temperature applications.

Concerning the machine types that have received most attention in this thesis, there are no reasons to foresee any increased development intensity for standard induction motors and large turbogenerators. The Powerformer activities are, at present, very limited and there are no signs that the situation will change quickly. A renaissance is perhaps possible when sufficient long term durability has been proven by the existing units. Machines for automotive drivelines have been in focus for development during the last two decades and will continue to receive much attention for quite some time as this continues to be an expanding application. Its potential impact can be illustrated by the simple fact that the annual global production of power generators amounts to approximately 100 GW, while the total production of alternators for cars and trucks corresponds to 50 GW (50 millions 1 kW alternators), i.e. the same order of magnitude. A wide introduction of HEVs or FCVs will change the situation drastically as the installed electrical machine capacity will be in the order of 10 – 100 kW per vehicle.

It has been a trend for several years to improve the efficiency of electrical machines. This efficiency is already high compared with most other types of machinery, but the development can nevertheless be expected to continue. The main reason for this development is, of course, that the energy costs have risen rapidly during the last years and are predicted to increase further in the future. A normal method to improve the efficiency is to design machines for lower current and flux densities, which leads to larger and heavier machines. The extra costs are compensated for by the reduced capitalized costs for losses. Even material, both copper and steel, has recently become more expensive and a question is if it will remain so. A requirement to reduce both energy and material will probably favour more complicated and labor intensive designs, independent of whether it is a matter of manual or automatic work.

10.7 Swedish manufacturers in an international perspective

A list of companies engaged, at least to some extent, in the development and manufacture of electrical machines in Sweden was presented in table. 2.1. The companies located in Västerås represent more than 2/3 of the total production volume and number of employees in Sweden. Furthermore, these companies have electrical machines as their core business while the largest ones among the other companies are manufacturing motors as components integrated in their main products, chiefly pumps and tools. The discussion in this section is therefore limited to the companies in Västerås, which also is consistent with the earlier parts of the thesis.

ABB has two factories for electrical machines in Västerås, usually referred to as ABB Motors and ABB Machines. Alstom has a unit for the sales and engineering of service and retrofit of turbo- and hydropower generators. VG Power has a corresponding activity limited to hydropower units. GenerPro is a contract manufacturer of large machines. Bombardier, finally, develops and designs their own traction motors, which are manufactured by ABB Machines. Until the mid 1980's all of the above mentioned companies were part of Asea, which had its roots in a century long development and had gained a position among the internationally leading electrical machine manufacturers. The accumulated competence and resources were very large. The present actors have built on this base and added to it through recent development and some transfer of technology from their present foreign sister companies. It can be concluded that, viewed as a group, these companies maintain a strong position, even internationally.

The situation has changed quite a bit from the Asea period also in other respects than the split-up into several independent units. The production of both small and large machines was previously very integrated; all steps were conducted in-house, casting, welding, machining, punching, stacking, winding, assembly, testing etc. etc. Today, on the other hand, substantial parts of the production are outsourced to domestic or foreign suppliers. Another important difference, in the present situation, is that the executive power over all but the smallest company is now abroad.

Strengths	Opportunities
Competence and facilities	Large Swedish OEMs
Industrial tradition	Increased need for electrical
Wide market presence	machine applications due to CO_2
Service and support	Strong Swedish process industry
Simple, efficient work structure	Multinational ownership
Weaknesses	**Threats**
Small domestic market	Low cost competition
High labor costs	Lack of specialists
	Foreign business control

Figure 10.5 SWOT-matrix, Swedish electrical machine industry

The Swedish electrical machine industry stands and falls with the units in Västerås, but not necessarily within existing frames. This was proven when Alstom intended to close down generator activities resulting in the establishment of GenerPro and VG Power. The current situation is healthy for all of the companies but what are the prospects for the future? Is there a risk of a transfer of both engineering and manufacture to low-cost countries? There are certainly such risks, but it has also been witnessed that outsourcing to such suppliers can create problems, which ultimately eliminate the profit. It is in this respect interesting to note that several large foreign companies have contracted the manufacture of complete generators to GenerPro. Swedish industry is obviously competitive in this field. Staying in business requires, of course, the maintainence and development of existing competence and resources. It is impossible to predict the future for Swedish electrical machine industry, but it is possible to point at some factors of influence. These factors can be understood from the so-called SWOT matrix presented in figure 10.5 (SWOT = Strengths, Weaknesses, Opportunities, Threats).

10.8 Conclusions

History includes all that lies behind us. It can therefore be re-written again and again. This is particularly true when the intention is to describe something as dynamic as technical development. This specific thesis could not have been written 10 years ago and its content would be different 10 years from now as well. This study has focused on the development of a few machine types, the selection of which was of course dependent on when the selection was made. The four cases, reported in chapters 6 – 9, contain a lot of technical and organizational facts but also a number of assumptions and opinions. The latter two are, to a large extent, based on interviews with initiated key persons but in the end these assumptions and opinions are results of the conclusions made by the author. The chapters mentioned should give correct and comprehensive pictures of the development of the selected machine types, but the reports are far from complete. The aim has been to briefly describe the machines and their properties, essential phases in the development and important decisions made, technical and commercial problems, development of competence and resources, and the industrial context domestically and internationally. Most of this thesis, but not all, is based on Asea's, and later ABB's, activities in Sweden. It has been the intention to document and explain selected parts of the Swedish electrical machine development, but this thesis will hopefully also give the readers a good understanding of the different sides of electrical machine development in general. It is possible that a few readers will disagree with some of the authors conclusions. This is perfectly acceptable provided all available facts have been considered, because electrical machine development is not, in every respect, an exact science.

It could have been of interest to include other types of electrical machines in the study, but in such a case, it would have become too cumbersome. This is instead left for possible future studies. Even if details would be different, the following general conclusions are valid. Development of electrical machines, in Sweden and elsewhere, is a multidisciplinary industrial activity that requires not only several technical competences, but also the ability to manage large projects and deep insight into the markets and the customers' needs.

References

[1]   Thelin, Peter *Design and evaluation of a compact 15 kW PM integral motor,* PhD thesis, TRITA-ETS-2002-02, ISSN 1650-674X, Royal Institute of Technology, Stockholm 2004

[2]   Bäckström, Thomas *Integrated energy transducer drive for hybrid electric vehicles,* PhD thesis, TRITA-EME-0004, ISSN 1404-8248, Royal Institute of Technology, Stockholm 2000

[3]   Harnefors, Lennart *On analysis, control and estimation of variable-speed drives,* PhD thesis, TRITA-EMD-9702, ISSN 1100-1631, Royal Institute of Technology, Stockholm 1997

[4]   Nee, Hans-Peter *On rotor slot design and harmonic phenomena of inverter-fed induction motors,* PhD thesis, TRITA-EMD-9602, ISSN 1100-1631, Royal Institute of Technology, Stockholm 21996

[5]   Dittmann, Frank *Geschichte der Elekrischen Antriebstechnik in Deutschland* pp. 7-126. Geschichte der Elektrotechnik Band 16, ISBN 3-8007-2287-9, VDE-Verlag, Berlin 1998

[6]   Fridlund, Mats *Den gemensamma utvecklingen* Stockholm Papers in the History and Philosophy of Technology, TRITA-HOT 2036, ISBN 91-7139-463-X, Symposion, Stockholm, 1999

[7]   Glete, Jan *Asea under hundra år, 1883 - 1983, En studie i ett storföretags organisatoriska, tekniska och ekonomiska utveckling* Asea AB, ISBN 91-7260-764-5, Västerås, 1983

[8]   Helén, Martin *Asea:s historia, 1883 - 1948 Vol I - III* Asea, Västerås, 1955 - 1957

[9]   Lindner, Helmut, *Strom, Erzeugung, Verteilung und Anwendung der Elektrizität,* Deutsches Museum, 1680-ISBN 3 499 17723 4, Munich 1985

[10] Israel, Paul, *Edison - A Life of Invention,* John Wiley & Sons, ISBN 0-471-36270-0, New York 1998

[11] Neidhöfer, Gerhard, *Michael von Dolivo-Dobrowolsky und der Drehstrom,* VDE-Verlag, ISBN 978-3-8007-2779-7, Berlin 2004

[12] Jäger, K., *Alles bewegt sich, Geschichte der Elektrotechnik Band 16,* ISBN 3-8007-2287-9, VDE-Verlag, Berlin 1998

[13] Vickers, V. J. *Recent trends in turbogenerators* pp. 1273-1306, Proc. IEE, Vol. 121, No. 11R Nov. 1974, IEE Reviews

[14]   Ekelöf, Stig *Catalogue of Books and Papers in Electricity and Magnetism*, 638 pages, Chalmers University of Technology, Gothenburg, 1991

[15]   *Electricity Market 2002* Report from the Swedish National Energy Agency, Eskilstuna, 2002

[16]   *Elmarknaderna runt Östersjön 1997* pp. 11, Report R1997:81 from the Swedish National Board for Industrial and Technical Development, Stockholm, 1997

[17]   *Energy Statistics of non-OECD Countries 2000- 2001* pp. 514-515, IEA Statistics, International Energy Agency, OECD, 2003

[18]   Helén, Martin *ASEA:s Historia 1883-1948 Vol. I* pp. 116-132, Västerås 1955

[19]   Jäger, Kurt *Alles bewegt sich* pp. 31-33, 128-132, Geschichte der Elektrotechnik Band 16, ISBN 3-8007-2287-9, VDE-Verlag, Berlin 1998

[20]   Hansson, Staffan *Den skapande människan* pp. 430-438, Studentlitteratur, ISBN 91-44-02148-8, Lund 2002

[21]   Hoffmann, Lars *Elmotorer i bilar* E-mail från Saab, 2006-07-08

[22]   Meneroud, P. et al *Micromotor based on film permanentmagnets* pp. 491-494, Actuator 2004, 9th International Conference on New Actuators, Bremen, Germany, June 2004

[23]   Joho, Reinhard *Advances in synchronous machines* pp. 398-400, Vol. 1, Proceedings from IEEE Power Engineering Society Winter Meeting, Jan. 2002

[24]   Baldwin, Samuel F. *Energy-efficient electric motor drive systems* pp. 21-58, "Electricity" edited by Thomas B. Johansson et al, Lund University Press, 1989

[25]   Tsakiridou, Evdoxia *Energy misers* pp. 66, Pictures of the Future, Siemens Magazine for Research and Innovation, München, 2006

[26]   *Energiläget 2005 - ppt. presentation* pp. 23, Report from the Swedish Energy Agency, Eskilstuna, 2006

[27]   *High efficiency electrical motors*, Report from the Swedish Energy Agency, Eskilstuna, 2003

[28]   *Energy Policy and Conservation Act of 1992, Part 431, Public law 94-163*, U.S. Department of Energy, Washington

[29]   Jansson, Lars G. *Att köra traditionellt var det aldrig tal om* pp. 28-29, Electrolux 50 år i Västervik, Electrolux, Västervik, 1998

[30]   Ekelöf, Stig, *Hans Christian Örsted och elektromagnetismens födelse – ett*

150 årsminne pp. 13-38, Dædalus 1970, Tekniska Museets Årsbok, Stockholm 1970

[31] Lindner, Helmut, *Strom, Erzeugung, Verteilung und Anwendung der Elektrizität* pp. 52-66, Deutsches Museum, Munich 1985

[32] Ekelöf, Stig and Rosell, Göte, *Elhistoria* pp. 3:5-3:11, Chalmers Tekniska Högskola, Göteborg 2000

[33] Maxwell, James C. *A treatise on electricity and magnetism*, Clarendon Press, Oxford, 1873

[34] Lindner, Helmut, *Strom, Erzeugung, Verteilung und Anwendung der Elektrizität* pp. 76-81, 91-99 Deutsches Museum, Munich 1985

[35] Wicks, Frank, *The blacksmith's motor* Mechanical Engineering, The American Society of Mechanical Engineers, July 1999

[36] Mahr, Otto, *Die Entstehung der Dynamomaschine* pp. 88, GEE Bd. 5, Berlin, 1941

[37] Siemens, Georg *Der Weg der Elektrotechnik, Geschichte des Hauses Siemens, Bd 1*, pp. 93-94, Verlag Karl Alber, Freiburg/München, 1961

[38] Lindner, Helmut, *Strom, Erzeugung, Verteilung und Anwendung der Elektrizität* pp. 114-131, Deutsches Museum, Munich 1985

[39] *Zénobe Theophile Gramme* pp. 1-6 http://chemch.huji.ac.il/eugeniik/history/gramme, Feb. 2005

[40] *Sigmund Schuckert (1846-1895)* pp. 1, www.med-archiv.de/infos/sigmundschuckert, Feb. 2005

[41] *Edward Weston* pp. 1-11, http://chem.ch.huji.ac.il/~eugeniik/history/weston.htm, 2003-03-25

[42] La Favre, Jeffrey, *The Brush Dynamo* pp. 1-7, www.lafavre.us/brush/dynamo.htm, Feb. 2005

[43] *Elihu Thomson* pp. 1-12, www.geocities.com/bioelectrochemistry/thomson, Feb. 2005

[44] Israel, Paul, *Edison – A Life of Invention* pp. 167-190, 321-337, John Wiley & Sons, New York 1998

[45] *About George Westinghouse* pp. 1-2, http://memory.loc.gov/ammem/papr/west/westgorg, Feb. 2005

[46] Lindner, Helmut, *Strom, Erzeugung, Verteilung und Anwendung der Elektrizität* pp. 152-156, Deutsches Museum, Munich 1985

[47] Lindner, Helmut, *Strom, Erzeugung, Verteilung und Anwendung der Ele-

ktrizität pp. 197-211, Deutsches Museum, Munich 1985

[48] Parsons, R.H. *The steam turbine and other inventions of Sir Charles Parsons* Published for the British Council by Longmans, Green & Co, London, 1942

[49] Asztalos, Peter *Product Development of the Century-Old Ganz Electric Works* pp. 3-20, Ganz Electric Works, Budapest 1978

[50] Hopkinson, John *Dynamo-Electric Machinery* Philosophical Transactions of the Royal Society, London, May 6, 1886

[51] Thompson, Silvanus P. *Dynamo-electric Machinery: a Manual for Students of Electrotechnics,* London, 1884

[52] Reiman, Dick *Charles Proteus Steinmetz: The wizard of General Electric* pp. 1-4, http://ieee.cincinnati.fuse.net/reiman/05_1991, Feb. 2005

[53] Arnold, E. and la Cour J. L. *Die Wechselstromtechnik, Vierter Band, Die Synchronen Wechelstrommaschinen,* Verlag von Julius Springer, Berlin, 1904

[54] Arnold, E. and la Cour J. L. *Die Gleichstrommaschine, Zweiter Band, Konstruktion, Berechnung und Arbeitsweise,* Verlag von Julius Springer, Berlin, 1907

[55] Helén, Martin *ASEA:s Historia 1883-1948 Vol. I* pp. 13-22, 68-77, Västerås, 1955

[56] *"Sköldpaddan" 60 år* pp. 110, Aseas Tidning, Västerås ,1942

[57] Karlsson, Petter and Erseus, Johan *Svenska uppfinnare* pp. 45, 115, 161, 237-261, ISBN 91-89204-36-0, Bokförlaget Max Ström, Stockholm, 2003

[58] Nyberg, Anders *Från Wenström till Amtrak. Den anspråkslösa början i Arboga.* pp. 11-13, Asea, Västerås, 1983

[59] Glete, Jan *Asea under hundra år* pp. 24-26, Asea, Västerås ,1983

[60] Vrethem, Åke *Jonas Wenström and the three-phase system* Stockholm Papers in History and Philosophy of Technology, Report TRITA-HOT-2004, Stockholm, 1980

[61] Stål, Gustaf *Aseas första trefasmotor* pp. 36-38, Aseas Tidning, Västerås, 1933

[62] Ferraris, Paolo *Von Galileo Ferraris Drehfeld zu Ossannas Kreisdiagramm: Die Theorie der Asynchronen Maschinen und die Mathematischen Methoden. ...* pp. 1-10, www.regione.taa.it/giunta/enel/ferraris

[63] Passer, Harold C. *The electrical manufacturers 1875 - 1900* pp. 277-278, Harvard University Press, Cambridge, MA, USA, 1953

[64]   Martin, Thomas C. *The Inventions, Researches and Writings of Nikola Tesla* pp. 3-25, Barnes & Noble Books, ISBN 1-56619-812-7, New York 1995 (The original written in 1893)

[65]   Hosemann, Gerhard *Michael von Dolivo-Dobrowolsky – Leben und Bedeutung* pp. 1-5, Elektrotechnische Zeitschrift, ETZ-A Bd. 91 (1970) H.1

[66]   Neidhöfer, Gerhard *Michael von Dolivo-Dobrowolsky und der Drehstrom* pp. 55-94, VDE Verlag, Berlin 2004

[67]   Kline, Ronald *Science and Engineering Theory in the Invention of the Induction Motor, 1880 - 1900* pp. 283-313 Technology and Culture, Vol. 28 No. 2, The University of Chicago Press, Chicago, April, 1987

[68]   Hedberg, Petrus, *Uppfinningarnas bok* pp. 479, Wilh. Siléns Förlag, Stockholm, 1909

[69]   Sadarangani, Chandur *Electrical Machin, Design and Analysis of Induction and Permanent Magnet Motors* pp. 1-667, Royal Institute of Technology, Stockholm, 2000

[70]   Gustavson, Fredrik *Elektriska maskiner,* pp. 1-544, Kungliga Tekniska Högskolan, Stockholm, 1996

[71]   Fitzgerald, A.E. et al *Electric Machinery* pp. 1-599, Mc-Graw-Hill College, 1990

[72]   Bödefeld, Theodor and Sequenz, Heinrich *Elektrische Maschinen*, Achte Auflage pp. 1-813, Springer-Verlag, Wien, 1971

[73]   Anpalahan, Peethamparam *Design of transverse flux machines using analytical calculations & finite element analysis* Licentiate thesis, KTH TRITA EME-0101, ISBN 91-7283-048-4, Royal Institute of Technology, Stockholm, 2001

[74]   Schröder, Tim *Cruising on cold power - Superconducting generators* pp. 60-61, Pictures of the Future, Siemens, München, Spring 2006

[75]   *Katalog över elektroplåt och annat material för magnetiska ändamål* Surahammars Bruks Aktiebolag, Surahammar, 1950

[76]   *Non-oriented Electrical Steels* pp. 1-54, Catalogue from Surahammars Bruk, Surahammar Sweden, 1992

[77]   Lindenmo, Magnus *Lean non-oriented electrical steel grades* pp. 178-182, Journal of Magnetism and Magnetic Materials 304, 2006

[78]   Hopkinson, John *Dynamo Machinery and allied subjects* pp. 89, Whittaker & Co. London 1893

[79]   Wuppermann, Carl-Dieter *Steel in the 21. century - status and perspec-*

tives pp. 323-331, Proceedings 2nd International Workshop, Magnetism and Metallurgy, WMM 06, Freiberg, Germany, 21-23 June 2006

[80] *Soft magnetic materials and semi-finished products* pp. 7-8, 15 Brochure from Vacuumschmelze GmbH, Hanau, Germany, 2006

[81] Magnussen, Freddy *On Design and Analysis of Synchronous Permanent Magnet Machines for Field-weakening Operation in Hybrid Electric Vehicles* pp. 59-68, 99-110, 121-131, 241-248, Doctoral thesis, TRITA-ETS-2004-11, Royal Institute of Technology, Stockholm, 2004

[82] Martinez, David *Design, Modelling and Control of Electrical Machines* pp. 1-167, Doctoral Thesis, ISBN 91-88934-35-7, Lund University, Lund, 2004

[83] Persson, Mats et al *Development of Somaloy Components for a BLDC motor in a scroll compressor application* Paper presented at PM2006, Busan, Korea, September 2006

[84] *Rare earth permanent magnets Vacodym Vacomax* Brochure PD-002, Vacuumschmelze GmbH & Co, Hanau, Germany, 2003

[85] Jönsson, Kenneth *MICAPACT II - coils for rotating high-voltage machines* pp. 97-102, Asea Review, 1980

[86] *Recommendations for the classification of materials for the insulation of electrical machinery and apparatus in relation to their thermal stability in service* IEC Publication 85, Geneva, 1957

[87] Helén, Martin *ASEA:s Historia 1883-1948 Vol. I* pp. 278-282, 357-361, Västerås 1955

[88] *SKF General Catalogue* GC 500, SKF, Gothenburg, 2003

[89] Wiedemann, Eugen and Kellenberger, Walter *Konstruktion elektrischer Maschinen* pp. 61-71, Springer-Verlag, Berlin, 1967

[90] Saari, Juha *Thermal analysis of high-speed induction machines* PhD thesis, Acta Polytechnica Scandinavica, ISBN 952-5148-43-2, Helsinki University of Technology, 1998

[91] Lindner, Helmut *Strom, Erzeugung, Verteilung und Anwendung der Elektrizität* pp. 124, Deutsches Museum, Munich, 1985

[92] *Webster's New World Dictionary of the American Language* pp. 386, Simon and Schuster, New York, 1984

[93] *Nationalencyklopedins ordbok, tredje bandet* pp. 491, Bokförlaget Bra Böcker, Höganäs, 1996

[94] Fridlund, Mats *Den gemensamma utvecklingen* pp. 15-23, Stockholm

Papers in the History and Philosophy of Technology, TRITA-HOT 2036, Symposion, Stockholm, 1999

[95] *Nationalencyklopedien, sjätte bandet* pp. 522, Bokförlaget Bra Böcker, Höganäs, 1991

[96] Clough, Ray W. and Wilson, Edward L. *Early finite element research at Berkley* pp. 1-35, Presentation at the Fifth U.S. National Conference on Computational Mechanics, Aug. 1999

[97] Häggström, G *Underlag för program R30093 Hållfasthetsberäkning av GTP rotor* pp. 25-29, Laboratorieprotokoll KYBA LP 4110-1018, Asea. Västerås, 31.3.1971

[98] Chari, M.V.K. and Silvester, P.P. *Finite Elements in Electrical and Magnetic Field Problems* J. Wiley & Sons, ISBN 0-471-27578-6, New York, 1980

[99] Touma Holmberg, Marguerite *Three-dimensional finite element computation of eddy currents in synchronous machines* Doctoral thesis at Chalmers University of Technology, ISBN 91-7197-702-3, Gothenburg, 1998

[100] Elfving, Gunnar et al *Elektriska drivsystem, Stora växelströmsmaskiner* pp. 184-185, ABB Handbok Industri, ISBN 91-970956-5-6, ABB AB, Västerås, 1993

[101] Toader, Stefan *Optimum design of electrical machines using nonlinear programming* Doctoral thesis at Chalmers University of Technology, ISBN 99-0176736-9, Gothenburg, 1979

[102] *Motors, Drives and Power electronics*, ABB's web-site: www.abb.com, Nov. 2004

[103] Stål, Gustaf *Aseas första trefasmotor* pp. 36-38, Aseas Tidning, 1933

[104] Glete, Jan *Asea under hundra år* pp. 30-49, Asea,Västerås, 1983

[105] Helén, Martin *ASEA:s Historia 1883-1948 Vol. I* pp. 120-121 Asea, Västerås, 1955

[106] Helén, Martin *ASEA:s Historia 1883-1948 Vol. III* pp. 58-70 Asea, Västerås, 1957

[107] Hedström, Hans *Asynkronmotorns utveckling* pp. 155-158, Aseas Tidning 1961

[108] *Trefas-Motorer, Typ RMC-RXC*, Allmänna Svenska Elektriska Aktiebolaget, Annonsblad No. 55, Västerås, 7 Jan. 1905

[109] *Vattentätt inneslutna trefasmotorer*, ASEAs Cirkulär No. 238, Västerås, 18 July 1908

[110] Helén, Martin *ASEA:s Historia 1883-1948 Vol. II* pp. 144-145, 153-159, 338, Asea, Västerås, 1956

[111] Glete, Jan *Asea under hundra år* pp. 56-63, 82-89, 102-104, 113-119 Asea, Västerås, 1983

[112] Thunholm, Lars-Erik *Ivar Kreuger* pp. 165-169, 244-262, Fischer & Co. Stockholm, 1995

[113] la Cour, Jens Lassen *Gleichstrommaschinen, Erster Band, Theorie und Untersuchungen* Verlag von Julius Springer, Berlin, 1919

[114] la Cour, Jens Lassen *Gleichstrommaschinen, Zweiter Band, Konstruktion, Berechnung und Arbeitsweise* Verlag von Julius Springer, Berlin, 1927

[115] Wallenberg, Marcus J. *Sigfrid Edström* pp. 9-10, Asea, Västerås, 1970

[116] Helén, Martin *ASEA:s Historia 1883-1948 Vol. II* pp. 302-304, Västerås, 1956

[117] Glete, Jan *Asea under hundra år* pp. 123-128, 247-248, 293 Asea, Västerås, 1983

[118] Olsson, Hjalmar *Historik om företaget, anförande hållet av Dir. Olsson vid sammanträde med Elektromekanos styrelse d. 14.6.1949* pp. 1-14, Typed presentation, Elektromekano, Helsingborg, 11.6.1949

[119] Helén, Martin *ASEA:s Historia 1883-1948 Vol. II* pp. 163-164, 168-169, 174-177, Asea, Västerås, 1956

[120] Glete, Jan *Asea under hundra år* pp. 74-75, 143-147, Asea, Västerås,1983

[121] Helén, Martin *ASEA:s Historia 1883-1948 Vol. I* pp. 34, 81-82, 307-309, Asea, Västerås, 1955

[122] *Allmänna Svenska Elektriska Aktiebolagets Förvaltningsberättelse för år 1929. 47:de räkenskapsåret* pp. 1, Asea, Västerås, 1930

[123] Strunk, Peter *Die AEG* pp. 57, 62, 102, 214, Nicolaische Verlagsbuchhandlung Bauerman, Berlin, 1999

[124] Siemens, Georg *Der Weg der Elektrotechnik, Geschichte des Hauses Siemens Band II* pp. 163-164, Verlag Karl Alber, Freiburg/München, 1961

[125] Weissheimer, Herbert *Oberflächenbelüftete Drehstrommotoren* pp. 291, Die Entwicklung der Starkstromtechnik bei den Siemens-Schuckertwerken, Siemens AG, Berlin/Erlangen, 1953

[126] Weiher, Sigfrid v. and Goetzeler, Herbert *Weg und Wirken der Siemens-Werke in Fortschritt der Elektrotechnik 1847 – 1980* pp. 60, 119, 123, Siemens AG, Berlin/München, 1981

[127] *Our products in industry, trade and agriculture. A. Drives by electric motors* pp. 39-40, The Brown Boveri Review, Jan./Feb. 1936

[128] Alder, Ken *Världens mått* pp. 11-21, 379-406, ISBN 91-1-301008-5, Nordstedts Förlag, Stockholm, 2003

[129] *Rating of Electrical Machinery* International Electrotechnical Commision, Published by Waterlow & Sons Ltd. London, Aug. 1911

[130] *IEC Recommendation 72-1, 72-2, Dimensions and output ratings of rotating electrical machines,* International Electrical Commision, Geneva, 1959

[131] *Voluntary agreement of CEMEP,* EU - CEMEP, Brussels, 1999

[132] *IEC Recommendation 60072-1, Dimensions and output series for rotating electrical machines, Part I Frame number 56 – 400* Published as IEC 72-1 Sixth Edition, International Electrical Commission, Geneva, 1991

[133] *Low Voltage General Purpose Motors,* ABB Catalogue BU/General purpose motors EN 12-2005, Waasa, Finland, 2005

[134] *Energy-saving Motors,* Siemens' website: www.automation.siemens.com/sd/motoren/html_76/energiespar.htm, Jan. 2006

[135] Hedström, Hans *ASEA:s nya motorserier* pp.151-157, Aseas Tidning, 1955

[136] Hedström, Hans and Broström, Elon *Nya asynkronmotorer för 150 – 1500 hk* pp. 155-158, Aseas Tidning, 1959

[137] Rösel, Arvid *Aseas nya småmotorserie* pp. 159-164, Aseas Tidning, 1961

[138] Hedström, Hans *Årsrapport för Konstruktionsavdelningen MK 1964* pp. 5, Västerås, January 1965

[139] Krecker, Wolfgang *Stora M-motorn* pp. 103-106, Aseas Tidning, 1967

[140] Marup Jensen, Gunnar *MT-motorn, en ny standardmotor* pp. 123-127, Aseas Tidning, 1968

[141] Becker, Lage *Årsrapport för Konstruktionsavdelningen MK 1968* pp. 13, Västerås, January 1969

[142] Becker, Lage *Reduktion av produktprogrammet* Meddelande, Asea M, 1970-08-31

[143] Möller, Finn *Årsrapporter för Konstruktionsavdelningen MK 1970, 1971, 1972, 1973, 1974, 1975,* Västerås in January, following years

[144] Krecker, Wolfgang *Helkapslade motorer storlekar 250 – 355* Meddelande, Asea MKX, 1971-09-29

[145] Carlsson, Sune *Årsrapport från M-sektorn 1976* pp. 2, Asea M, 1977-01-31

[146] Ancker, Johan *Nya koncept till standardmotorer för M-sektorn* Sammanträdes-protokoll, Asea KYYU SP 442-7021, 1977-02-09

[147] Möller, Finn *Årsrapport från MK för 1977* Asea MK, 1978-01-18

[148] Spira, Lennart *"Öststaterna säljer motorer för billigt!"* pp. 24-25, Elteknik 1979:2, Stockholm

[149] Möller, Finn *Förstudie gällande ny småmotorserie* pp. 1-13, Asea MK, 1979-05-10

[150] Wilhelmsson, Bengt *Förprojekt gällande ny småmotorserie* pp. 1-16, Asea MKA, 1980-04-16

[151] Becker, Lage *Protokoll, Styremöte med sektor M 1980-06-02, Paragraph 5.1,* Asea Västerås, 1980-06-03

[152] Becker, Lage and Möller, Finn *Projektgrupp* Meddelande Asea M/MK, 1980-08-20

[153] *User's Manual – PC-OSKAR* Technical Report SECRC/KEC/TR-92/092, Rev. 2.3 ABB Corporate Research, Västerås, 1999-12-01

[154] *Asea MBT Helkapslade kortslutna trefas asynkronmotorer* Asea Motors Katalog MK 20-108, Västerås, 1985

[155] Porteus, Tom *Kravspecifikation MBT 180-250* Kravspec. Asea MKA, 1984-10-31

[156] Johansson, Lennart *Protokoll vid extra styrgruppsmöte, Projekt MBT ....* Protokoll Asea MKX, 1985-06-11

[157] Gertmar, Lars *Magnetljud MBT, lägesrapport mätningar och beräkningar* RM KYEC 80-128, Asea KYEC, 1980-12-17

[158] Olofsson, Ingemar *Rotorledarkonfigurationens inverkan på magnetljudet hos burlindade asynkronmotorer* Asea Rapport KM 87-003, 1987-08-13

[159] *Växelströmsmotorer Axelhöjd 400 – 630 mm 200 – 5000 kW* Asea Motors Katalog B10-3000, Västerås, 1984

[160] Sadarangani, Chandur *Beskrivning av program EDDY som beräknar momentkurva i kortslutna asynkronmotorer med beaktande av tvärströmmar i rotorplåt* Asea Teknisk Rapport TR 4123 003 CS, Asea MKC, 1984-09-25

[161] Sadarangani, Chandur *Contributions to the analysis of magnetic field problems in electrical machines* PhD thesis, Chalmers University of Technology, Technical Report No. 89, Göteborg, 1979

[162] Sadarangani, Chandur *Tomgångstillsatsförluster i rotorledare för korts-lutna asynkronmotorer* Asea Teknisk Rapport TR 4123 001 CS, Asea MMKC, 1981-05-14

[163] *Utökad forskning* Information published in "vi aseater" no. 19, magazine for Asea employees, Asea, Västerås, 5 Dec. 1986

[164] *Asea Annual Report 1980* pp. 3 and *1987* pp. 4, Asea, Västerås

[165] *Årsredovisning för 1986* ABB Motors AB, Västerås, 1987-02-18

[166] Olofsson, Ingemar *M2000 steg 1, storlek 112, 132*, Anslagsbegäran MOT/MK, 1991-02-22

[167] Edquist, Harald *Do hedonic price index change history? The case of elec-trification* SSE/EFI Working Paper Series in Economics and Finance No. 586, Stockhom School of Economics, Feb. 2005

[168] Folkhammar, Jan *Re: Fråga om motorpriser* e-mails from BEVI, 2006-12-11 and 2007-02-22

[169] Glete, Jan *Asea under hundra år* pp. 24, Asea, Västerås, 1983

[170] *Hägglunds 1899 – 1989* pp. 7-50, Glimten, House magazine of the Hägglund Group, Örnsköldsvik, 1989

[171] Anderson, Helén *En produkthistoria!* pp. 36-39, 51-56, 114-117 PhD the-sis, ISBN 91-7258-376-2 EFI, Handelshögskolan, Stockholm, 1994

[172] Almroth, Wilhelm *Besök vid AB Elmo den 17/6 1965* PM, Asea M, 1965-06-18

[173] *Elmo Handboken* Folder with company information and product cata-logues, Elmo AB, Flen, 1993 and 2000

[174] *Excellence in electric drives and power*, Company presentation, BEVI AB, Blomstermåla, Sweden, 2005

[175] Carlsson, Curt *Elmotorverkstaden – kort historik* pp. 1-3, ITT Flygt Prod-ucts, 2004

[176] *Motors built for submersible pumps*, Brochure 894325, ITT Flygt AB, Sundbyberg, Sweden, 2003

[177] Helén, Martin *ASEA:s Historia 1883-1948 Vol. I* pp. 339-341 Västerås, 1955

[178] Aschenbrenner, Norbert et al *Mighty Motors* pp. 48-52, 57, 66-67, Pictures of the Future, Siemens Magazine for Research and Innovation, München, 2006

[179] Jones B. L. and Brown J. E. *Electric variable-speed drives* pp. 516-558,

IEE Proceedings, Vol. 131, Pt. A, No. 7, Sep 1984

[180] Elfving, Gunnar et al *Varvtalsreglering* pp. 191-197, ABB Handbok Industri, ISBN 91-970956-5-6, ABB, Västerås, 1993

[181] Helén, Martin *ASEA:s Historia 1883-1948 Vol. III* pp. 178-180 Västerås, 1957

[182] Jahns, Thomas M. and Owen, Edward L. *AC Adjustable-Speed Drives at the Millenium: How Did We Get Here?* pp. 17-25, IEEE Transactions on Power Electronics, Vol. 16, No. 1, January 2001

[183] Lamm, Uno and Lindahl, Gunnar *En valsverksutrustning med gallerstyrda likriktare* pp. 35-, Aseas Tidning, 1937

[184] Alm, Emil *Asynkronmaskinen. Elektroteknisk Handbok* pp. 280-282, 327-330 Bokförlaget Natur och Kultur, Stockholm 1956

[185] Borg, Lennart *Presentation av Aseas tyristorprogram* pp. 71, Aseas Tidning, 1962

[186] Mellgren, Gunnar *Asea och kraftelektronik* pp. 113-137, Industriminnen av Asea-Präntarna, ISBN 91-630-9817-2, Västerås, 2000

[187] Bylynd, Per-Åke and Mellgren, Gunnar *Rectifier assembly comprising semi-conductor rectifiers with two separate heat sinks.* Patent No. 3364987, U.S. Patent Office, Jan 23, 1968

[188] Mohan, Ned, Undeland, Tore and Robbins William *Power Electronics, Converters, Application and Design* pp. 16-32, 505-666, John Wiley & Sons Inc, New York, 1995

[189] Ållebrand, Björn On SiC JFET converters: components, gate-drives and main-circuit considerations PhD thesis, Trita-ETS-2005-18, KTH, Stockholm, 2005

[190] *Kansai Electric and Cree demonstrate a 100 kVA silicon carbide three phase inverter* Press release from Cree Inc. Durham NC, 2006-01-25 www.cree.com/press/39, Jan. 2006

[191] Ström, Nils-Erik *Varvtalsreglering av mindre likströmsmaskiner med transduktorstyrd likriktare* pp. 114-117, Aseas Tidning, 1957

[192] Nordin, Tore *Världsledande lokteknik i samverkan mellan SJ och Asea* pp. 30-34, Industriminnen av Asea-Präntarna, ISBN 91-630-9817-2, Västerås, 2000

[193] Brisby, Kurt and Friman, Carl-Johan *Tyristorströmriktare för magnetisering av likströmsmaskiner* pp.83-86, Aseas Tidning, 1962

[194] Mårtensson, Heine and Karlsson, Arvid *Tyristorströmriktare för drift av*

likströmsmaskiner pp. 75-82, Aseas Tidning, 1962

[195] Nordin, Tore *Aseas tyristorlok erövrar världen* pp. 95-112, Industriminnen av Asea-Präntarna, ISBN 91-630-9817-2, Västerås, 2000

[196] Reibring, Göran *Tyristorströmriktare i valsverk* pp. 158-161, Aseas Tidning, 1964

[197] Eriksson, Sture *Stora likströmsmaskiner typ LAA* pp. 119-124, Aseas Tidning ,1970

[198] Söderberg, Christer *DMI – a milestone in the development of DC motors* pp. 9-13, ABB Review, Special Report on Motors and Drives, ABB Ltd, Zurich, 2004

[199] *ABB Catalogue DMI 180 - 400*, 05/09/ rev. 2.0, ABB Automation Technologies AB, Västerås, 2005

[200] ABB LV Motors/ Cat. BU/Low-Voltage General Purpose Motors EN 12-2005, ABB Automation Technologies, Wasaa, Finland, 2005

[201] Krabbe, Ulrik *Eftersläpningsreglering av asynkronmotorer medelst tyristorer* pp. 81-82, Aseas Tidning 1962

[202] Becker, Lage *Årsrapport för MK år 1967* pp. 10, Asea's M division, Västerås, Jan. 1968

[203] Schönung, A. and Stemmler, H *Geregelter Drehstrom-Umrichtertechnik mit gesteuertem Unterschwingungsverfahren* , BBC-Nachrichten 1964/12

[204] Mohan, Ned, Undeland, Tore and Robbins William *Power Electronics, Converters, Application and Design* pp. 200-248, John Wiley & Sons Inc, New York, 1995

[205] Pettersson, Tore and Frank, Kjell *Starting of large synchronous motors using static frequency converter* pp. 172-179, IEEE Transactions, Power Apparatus and Systems, 91, 1972

[206] Högberg, Kjell-Erik *Frekvensomriktare för start av aggregaten I Foyers pumpkraftverk* pp. 3-8, Aseas Tidning, 1976

[207] Becker, Lage *Protokoll, Styremöte med sektor M 1980-06-02, Paragraph 5.1,* Asea Västerås, 1980-06-03

[208] Barnevik, Percy *PM VD, Kort kommentar till ökad motorsatsning på de säljande dotterbolagen,* Asea Västerås, 1980-09-20

[209] Harmoinen, Martti *SAMIntarina* pp. 46-47, 91-226, 248-250, ISBN: 952915405-1 Helsinki, 2002

[210] Blaschke, Felix *The principle of field orientation as applied to the new "transvector" closed-loop control system for field machines* pp. 217-220,

Siemens Review vol. 34, May 1972

[211] Depenbrock, Manfred *Direckte Selbst-regelung (DSR) für hochdynamische Drehfeldantriebe mit Stromrichterspeisung* pp. 211-218, ETZ Archiv BD7, 1985

[212] *Direct Torque Control – the world's most advanced AC drive technology* pp. 1-32, ABB Technical Guide No. 1, Helsinki, 2002

[213] Tiitinen, Pekka et al *The next generation motor control method – Direct Torque Control, DTC* Presentation at EPE Symposium in Lausanne, Switzerland, Oct. 19, 1994

[214] Tiitinen, Pekka *A winning formula* pp. 4-8, ABB Review, Special Report on Motors and Drives, ABB Ltd, Zurich, 2004

[215] Brännback, Malin et al *Adjusting local and regional to national and global – the Turku innovation environment* pp. 1-15, Åbo Academy University School of Business, ISBN 952-12-1711-1, Turku, Finland, 2006

[216] *VäxelströmsMotorn* pp. 38, Product information, A10-2004, ABB Motors, Västerås, 1988

[217] Gertmar, Lars *The influence of the harmonics in inverter-fed asynchronous machines* PhD thesis, Chalmers University of Technology, Gothenburg, 1976

[218] *Bearing currents in modern AC drive systems* Technical Guide No. 5, EN 01.12.99, ABB Industry Oy, Helsinki, 1999

[219] Ottersten, Rolf *On control of back-to-back converters and sensorless induction machine drives* PhD thesis, Chalmers University of Technology, ISBN 91-7291-296-0, Gothenburg, 2003

[220] Gertmar, Lars and Sadarangani, Chandur *Rotor design for inverter-fed high speed induction motors* pp. 51-56, Proceedings EPE, Aachen, Germany, 1989

[221] Sadarangani, Chandur *Kortsluten asynkronmotor för frekvensomriktardrift* Patentansökan 8803665-2, Stockholm, 1990-05-28

[222] Pyrhönen, Juha *The high-speed induction motor: calculation effects of solid-rotor material on machine characteristics* PhD thesis, Lappeenranta University of Technology, 1991

[223] Lähteenmäki, Jussi *Design and voltage supply of high-speed induction machines* PhD thesis, Acta Polytechnica Scandinavica, ISBN 951-666-607-8, Helsinki University of Technology, 2002

[224] Huppunen, Jussi *High-speed solid rotor induction machine – electromagnetic calculation and design* PhD thesis, ISBN 951-764-981-9,

Lappeenranta University of Technology, 2002

[225] *Rotatek Finland Oy* Homepage: www.rotatek.vacon.com, 2005

[226] Henze, Michael *The Integral Motor – a new variable-speed motor drive* pp. 4-8, ABB Review no. 4, 1996

[227] *ABB General Purpose Motors GB 12-2004, Integral Motor Section* Doc. No. B5 0274, Section 9, pdf, www.abb.com/ Product Guide, 2004

[228] IEC Recommendations 34-2 *Losses and efficiency, §9.1.3* pp. 31, 3rd edition International Electrotechnical Commission, Geneva, 1972

[229] Williamson, Steve and Smith, Sandy *The effect of inter-bar current on induction motor losses* pp. 48-57, Conference proceedings, vol. 1, ISBN 3-8167-6904-7, eemods 05, Fraunhofer IRB Verlag, Heidelberg, Germany, Sep. 2005

[230] Lefevre, Louis *Design of line-start permanent magnet synchronous motors using analytical and finite element analysis* Licentiate thesis, TRITA- EME-0003, ISBN 91-7170-574-0, Royal Institute of Technology, Stockholm, 2000

[231] Wallmark, Oskar *Control of permanent-magnet synchronous machines in automotive applications* PhD thesis, No. 2528, Chalmers University of Technology, Gothenburg, 2006

[232] Ikäheimo, Jouni *New roles for permanent magnet technology* pp. 37-40, ABB Review, Special Report on Motors and Drives, ABB Ltd, Zurich, 2004

[233] Haring, Tapio et al *Direct drive – opening a new era in many applications* IEEE, IAP Pulp and Paper Technical Conference, Charleston, June 2003

[234] Zellman, Per *SR-motorer* E-mail from Emotoron, 2006-03-21 and 2006-03-24

[235] Helén, Martin *ASEA:s Historia 1883-1948 Vol. III* pp. 32, 51-53, 58, Asea, Västerås, 1957

[236] *ASEAs Tidning 50 år, Jubileumsnummer 1958* pp. 42, 101, 107-108, Asea, Västerås, 1958

[237] *ASEA GENERATORER* Beskrivning utgiven av Allmänna Svenska Elektriska Aktiebolaget, Asea, Västerås, 1919

[238] Glete, Jan *ASEA under hundra år* pp. 39-44, 90-92, 167, Västerås, 1983

[239] Strömberg, Tage *Aseas generatorleveranser till Snowy Mountains* pp. 23-30, Aseas Tidning 1965

[240] Helén, Martin *ASEA:s Historia 1883-1948 Vol. I* pp. 103-111, Asea,

Västerås, 1955

[241] Hard, Franz *75 Jahre Brown-Boveri-Dampfturbinen* Neue Züricher Zeitung, 1976-10-12

[242] Sommerscales, Euan *The First 500 Kilowatt Curtis Vertical Steam Turbine* pp 1-7,The American Society of Mechanical Engineers, New York 1990

[243] Siemens, Georg *Der Weg der Elektrotechnik, Geschichte des Hauses Siemens Band I* pp. 267-268, Verlag Karl Alber, Freiburg/München, 1961

[244] Ljungström, Olle *Fredrik Ljungström 1875-1964 Uppfinnare och inspiratör* pp 92-102, Stockholm, 1999

[245] Helén, Martin *ASEA:s Historia 1883-1948 Vol. II* pp. 107-113, Asea, Västerås, 1956

[246] Andersson, Bengt *Product Development, Papers and Discussions, Fourth Annual Industrial Power Seminar* pp. IX:1-IX:22, Stal-Laval, Finspong, 1971

[247] *Turbogenerators driven by steam turbines* Reference list OK 10-102 E, Asea Generation, Västerås ,1985

[248] Hargett, Y. S. *Large steam turbine-driven generators* pp. 4, GET-3459A, General Electric Company, Schenectady, NY, 1982

[249] Liljeblad, Jan *Annual Report 1966 för the Design Departement OK*, Asea, Västerås, 1967

[250] Fisher, Jim and Eriksson, Sture *Asea Turbogenerators, Papers and Discussions, Fourth Annual Industrial Power Seminar* pp VI:1-VI:28, Stal-Laval, Finspong, 1971

[251] Asztalos, Péter *Product Development of the Century-Old Ganz Electric Works 1878 – 1978,* pp. 12, Ganz, Budapest, 1978

[252] Andersson, Anders R. and Helén, Camillo *Micapact insulation* pp. 3-9, Asea Journal No. 1, 1965

[253] Tengstrand, Claes *Micapact in large synchronous machines* pp. 10-14, Asea Journal No. 1, 1965

[254] *MICAREX® Insulation System for High-Voltage Machines* Pamphlet OG 01-0020E, ABB Generation AB, Västerås, 1991

[255] Östmar, Eric *Gasturbinutvecklingen vid Stal-Laval* pp. 70-80, Stal-Laval Turbin AB, Finspång, 1971

[256] Fechheimer, C. J. *Liquid cooling of turboalternators* pp. 969-974, Electrical Engineering Volume 66, AIEE, New York, 1947

[257] Asztalos, P. A. *Direct cooling systems for turboalternator rotors in view of the maximum rating of hydrogen cooling* pp. 1935-1945, IEEE Transactions on PAS, Vol. PAS-80, No. 8 Nov./Dec. 1970

[258] Holley, C. H. and Willyoung, D. M. *Conductor-cooled rotors for large turbine-generators, Experience and Prospects* pp. 1-26, Special paper from General Electric presented at the Cigrè conference in Paris, 1970

[259] Noser, R. and Kranz, R-D. *Turbo alternator with liquid cooled rotor* Cigrè, Report 111, Paris, 1966

[260] *Stromrichtererregte Turbogeneratoren* Reference list, No. KWU 121, Kraftwerk Union AG, Mühlheim, Germany, July, 1971

[261] Laurent P. and Ruelle G. *Progress report of study committee No. 17 GENERATORS* Cigrè, Paris, 1968

[262] Holley C. H. *Comments on turbogenerator development* pp. 36, Proceedings of discussions, Cigrè, Paris, 1968

[263] Concordia, Charles *Future development of large electric generators* pp. 39-49 Phil. Transactions of the Royal Society in London, Series A Mathematical and Physical Sciences, Vol. 275 No. 1248, London, Aug. 30, 1973

[264] Angelin, Stig et al *Hydro Power in Sweden* pp. 10, The Swedish Power Association and The Swedish State Power Board, ISBN 91-7186-064-9, Stockholm, 1981

[265] Leijonhufvud, Sigfrid *(parantes?* pp. 20-23, 46-47, 59-70, 71-77, ISBN 91-630-2976-6, ABB Atom, Västerås, 1994

[266] Sundqvist, Cnut *Kärnkraft – från ingenting till hälften av Sveriges Elförsörjning* pp. 223, Industriminnen, ISBN 91-630-9817-2, Västerås, 2000

[267] Leijonhufvud, Sigfrid *(parantes?* pp. 81-87, 104-107, 113, 115-116, 137–139, ISBN 91-630-2976-6, ABB Atom, Västerås, 1994

[268] Wivstad, Ingvar *Letter to Asea's technical director Torsten Lindström* Vattenfall, Stockholm, 1976-01-23

[269] Glete, Jan *ASEA under hundra år* pp. 244 –246, Västerås 1983

[270] Fridlund, Mats *Den gemensamma utvecklingen* pp. 221-224, Stockholm Papers in the History and Philosphy of Technology, TRITA-HOT 2036, Symposion, Stockholm, 1999

[271] *Protokoll nr 461 fört vid sammanträde med styrelsen för Allmänna Svenska Elektriska Aktiebolaget på bolagets kontor I Västerås den 22-23 februari 1963. § 6 Produktion och marknadsläge   Årsberättelse för*

konstruktionsavdelningen, bilaga 3 sid 49, Atomkraftavdelningen Ka Asea, Västerås, 1963

[272] Dahl Madsen, Kristian *Preliminär utredning rörande stora synkrongeneratorer med supraledare* Tekniskt meddelande TM 10567, Asea, Västerås, 22.8.1962

[273] *Högeffektsynkronmaskin med hög medelinduktion i luftgapet* Svenskt Patent nr 315 654

[274] Liander, Halvard *VVDKs årsrapport 1964* pp. 24a, Asea, Västerås, January 1965

[275] Liander, Halvard *VVDKs årsrapport 1965* pp. 14, Asea, Västerås, January 1966

[276] Tengstrand, Claes and Rönnevig, Carl *Direct cooling of water-wheel generators, influence on dimensions and generator parameters* pp. 1-8, Cigrè, Report 11-03, Paris, 1968

[277] Liljeblad, Jan *Årsrapport 1967 för konstruktionsavdelningen OK*, Asea, Västerås, 24.1.1968

[278] Liljeblad, Jan *Årsrapport 1968 för konstruktionsavdelningen OK*, Asea, Västerås, 21.1.1969

[279] Tjernström, Ove, Sivertsen, Richard and Jonsson, Birger *Vattenkyld provrotor, Projekt 862.210-26*, PM OKUP, Asea, Västerås, 22.7.1966

[280] Liljeblad, Jan *Vattenkyld turbogenerator – ett realistiskt alternativ för Oskarshamn* Meddelande OK Ref.nr. 5860.2188, Asea, Västerås, 9.8.1966

[281] Sjökvist, Eric *Sammanträde 6/66 med styret för sektor O hållet 20.12.66 och 22.12.66, §3. Verksamhetsbudget,* Sammanträdesprotokoll, O, Asea, Västerås, 6.3.1967

[282] Oliver, J.A., Ware, B.J. and Carruth, R.C. *345 MVA fully water-cooled synchronous condenser for Dumont station, Part I. Application Considerations* pp.2758-2764, IEEE Transactions on Power Apparatus and Systems, Vol. PAS-90 No 6, Nov/Dec 1971

[283] Landhult, Hans and Nordberg, Birger *345 MVA fully water-cooled synchronous condenser for Dumont station, Part II. Design, Construction and Testing* pp.2765-2772, IEEE Transactions on Power Apparatus and Systems, Vol. PAS-90 No 6, Nov/Dec 1971

[284] Larsson, Bertil *Besök vid All Union Research Institute of Electrical Machinery samt på Electrosila i Leningrad* Reserapport, OK1, Asea, Västerås, 1975-02-20

[285] Noser, R. *Comments on turbogenerator development* pp. 2-4, Session

Proceedings Group 11, Cigrè, Paris, Aug. 29, 1970

[286] Nicolin, Curt *Besök vid BBC's fabriker* Meddelande, Asea VD, 13.1.1965

[287] *Elkraftförsörjningen i Sverige,* Prognos från Central Driftledningen CDL, Stockholm, Oct. 1972

[288] Abegg, K. *The growth of turbogenerators* pp. 51-67, Phil. Transactions of the Royal Society in London, Series A Mathematical and Physical Sciences, Vol. 275 No. 1248, London, Aug. 30, 1973

[289] Liljeblad, Jan *Vattenkyld turbogenerator – ett realistiskt alternativ för Oskarshamn* Meddelande OK Ref.nr 5860.2188, Asea, Västerås, 9.8.1966

[290] Liljeblad, Jan *Årsrapport 1966 för konstruktionsavdelningen OK,* Asea, Västerås, 24.1.1967

[291] Helén, Martin *ASEA:s Historia 1883-1948 Vol. I* pp. 269-301, Asea, Västerås 1955

[292] Strömberg, Tage *Bruchloch Wellenwicklungen: eine systematische Untersuchung ihres praktischen Entwurfes* PhD thesis, Chalmers University of Technology, Gothenburg, 1942

[293] Britsman, C. et al *Handbok i FMEA* pp. 12-14 Industri Litteratur, ISBN 91-7548-317-3, Stockholm, 1993

[294] Richardson P. and Hawley R. *Stator core and end winding vibrations* IEEE Winter Power Meeting, C 72 241-3, 1972

[295] Glebov, I. A. et al *Investigations of the losses and temperature rises in the end parts of direct-cooled turbo-generators* Cigrè, Report 11-05, Paris, 1970

[296] Nicolin, Curt *Letter to Professor Herbert H. Woodson at MIT,* Asea, Västerås Oct. 1, 1968

[297] Morath, Erik *Partresistanser och reaktanser för beräkning av utjämningsströmmar mellan spårledares parter* Asea TM 11073, Västerås, 22.4.1968

[298] Klein, Hans and Nordberg, Birger *Kraftfördelning, rörelser och formförändringar i statorlindningsspår i stora växelströmsmaskiner. Teoretisk genomgång och praktisk utprovning i statorhärvattrapp* Asea TM 11283, Västerås, 8.11.1971

[299] *Alloyed rotor steel,* Material specification 1102 2598, Asea, Västerås, week 49, 1969

[300] Chapman, Stephen J. *Electric Machinery and Power System Fundamentals* pp. 591-615, ISBN 0-07-229135-4, McGrawHill, New York, 2002

[301] Barret, Ph. et al *Stresses on turbo alternators under unbalanced conditions* Cigrè, Report 11-11, Paris, 1970

[302] IEC Standard *Rotating electrical machines – Part 1: Rating and performance* IEC 60034-1, pp. 63, Geneva, 2004

[303] *Retaining rings*, Material specification No. 1102 2189E, Asea, Västerås, week 18, 1970

[304] Eriksson, Sture *Direct water-cooling of two-pole turbine generator rotors* Presentation at AIM Conference Centrales Electriques Modernes, Liege, Belgium, 1981

[305] Kredell, Bengt and Holmdahl, Gunnar *Från hålkort till superdator – från rutiner till integrerade system* pp.175-193, Minnenas Mosaik, ISBN 91-631-5947-3, Västerås, 2004

[306] Morath, Erik *Av belastningsströmmarnas minusföljd inducerade tandströmmar i en lång turborotor utan dämplindning* Asea TM 11180, Västerås, 21.3.1970

[307] Morath, Erik *Contraction Factors for Shallow Machine Slots* pp. 69-84, Wiss. Z. Elektrotechn., Leipzig 17 2/3, 1971

[308] Morath, Erik *Der elektrische Widerstand rechteeckiger Leiter mit geschwächtem Querschnitt* pp. 98-104 E und M (Elektrotechnik und Maschinenbau), Springer-Verlag Wien, Heft 3, 1971

[309] Wehrlin, H-P. *Temperaturfördelning i rotor och stator i GTD 125/490 för Barsebäck* Asea TM 11272, Västerås, 24.9.1971

[310] Carlsson, P-O, Spännings- och kraftberäkning för en turbostator vid tvåfasig kortslutning Asea TM 11424, Västerås, 1976-11-29

[311] *Summary of design hours L8860.227 Ringhals and L9860.2252 Aroskraft,* OER Asea, Västerås, W48, 1973

[312] *Asea Telephone Directory* pp. 110-111, Asea, Västerås, 1971-09-03

[313] *Verkstadsprov och prov på platsen för generator GTD 125/490 för Barsebäcksverket* Technical provisions 4104 001-10, OKTB, Asea, Västerås, Dec. 1972

[314] Jonsson, Uno *Letter to Sture Eriksson*, Stockholm, 2004-09-09

[315] Sjökvist, Eric *Sammanträde 2/68 med styret för sektor O den 30.5.68, §3. Eventuellt samarbete med English Electric,* Sammanträdesprotokoll, O, Asea, Västerås, 21.8.1968

[316] Sjökvist, Eric and Wådell, Tor *Aroskraft- och Ringhalsgeneratorerna* Meddelande, O/OP, Asea, Västerås, 2.5.1972

[317] Sjökvist, Eric and Eriksson, Sture *Turbogenerator för Aroskraft. Åtgärder med anledning av jordfel i rotor* Message, O/OKT, Aseaq, Västerås, 15.11.1972

[318] Wivstad, Ingvar *Beträffande generatorerna för Ringhals 2, Letter to Asea*, Vattenfall, Stockholm, 1.12.1972

[319] Eriksson, Sture *Rotorspårisoleringar för stora turbogeneratorer* Meddelande OKT, Asea, Västerås, 1973-08-06

[320] Larsson, Bertil I. *Årsrapport 1973 för konstruktionsavdelningen OK*, Asea, Västerås, 1974-01-23

[321] Larsson, Bertil I. *Ringhalsgenerator R21* Meddelande, OK, Asea, Västerås, 1974-08-19

[322] Eriksson, Sture *Sammanfattande historik för rotor 21 Ringhals*, Meddelande, OK 1, Asea, Västerås, 1979-06-26

[323] *Operation records for turbogenerator Ringhals 21* Printed statistics from Alstom Power's data base DEPRO, Västerås, 2006-09-21

[324] Sjökvist, Eric *Troliga kostnadsöverskridanden för stora turbogeneratorer*, Meddelande, O, Asea, Västerås, 1970.09.29

[325] Eriksson, Sture, Klein, Hans and Piensoho, Lauri *Rapport rörande kostnadsläget för Ringhals- Aros- och Barsebäcksgeneratorerna*, Meddelande, OKT, OKU, OER, Asea, Västerås, 1971.10.18

[326] Piensoho, Lauri *Turbogeneratorer typ GTD*, Diagram, OER, Asea, Västerås, 1974-09-05

[327] Hansson, Lennart *Generatoroffert – TVO Finland*, Meddelande, OFO, Asea, Västerås, 10.10.72

[328] *L-order 3860.3312* OFO, Asea, Västerås, 1973-09-01

[329] Sjökvist, Eric *Turbogenerator för Aroskraft. Åtgärder med anledning av jordfel i rotor*, Protokoll, O, Asea, Västerås, 1972-11-01

[330] Larsson, Bertil I. *Rapport om personalsituationen på OK*, Meddelande, OK, Asea, Västerås, 1973-03-20

[331] *GTD 1725KT Assembly drawing no 4422 168*, OKTK, Asea, 1976

[332] Lambrecht, Dieter et al *Design and performance of large four-pole turbogenerators with semi-conductor excitation for nuclear power stations*, Cigrè, Report 11-04, Paris, 1972

[333] *Mataraggregat komplett Drawing no 4894 818*, OKTK, Asea, 1976

[334] Krick, Norbert et al *Advanced design of large 4-pole turbogenerators*

Cigrè, Report 11-06, Paris, 1966

[335] Eriksson, Sture *Utveckling av 2-poliga turbogeneratorer i effektområdet 1000 – 1350 MW* Meddelande, OKT, Asea, Västerås, 1973-10-31

[336] Larsson, Bertil I. *Årsrapport för konstruktionsavdelningen OK för 1974*, Asea, Västerås, 1975-01-20

[337] *Vertrag ... zwischen Allmänna Svenska Elektriska Aktiebolaget ... und Ganz Elektrotechnische Werke*, License agreement signed 15.4.1971

[338] Heinrichs, Friedhelm *Turbogeneratoren*, ETG-Fachberichte 3, Fachtagung Kraftwerks-Generatoren, Nov. 8 and 9, 1977, VDE-Verlag, Berlin

[339] *Completely water-cooled turbine generator*, Brochure, Hitachi (no date stated)

[340] Toader, Stefan *Superconducting generators*, Technical Report TR KYE 80-018, KYE, Asea, Västerås, 1980-10-21

[341] McCown, W. R. and Edmonds, J. S *300 MVA superconducting generator – plan for design, testing and long term operation* Cigrè, Report 11-08, Paris, 1980

[342] Gray, R. F. et al *Designing the cooling systems for the world's most powerful turbogenerator - Olkiluoto unit 3* Proceedings IEEE Power Engineering Society General Meeting, June, 2006

[343] Larsson, Bertil I. *Årsrapport för konstruktionsavdelningen OK år 1974*, pp. 2-4, Asea, Västerås, 1975-01-20

[344] Eriksson, Sture *Generatorrotor nr 1 för TVO* pp. 1-6, Meddelande OK 1, Asea, Västerås, 1979-03-23

[345] Eriksson, Sture and Wetterfall, Sven-Erik *Retaining ring failure in a 710 MVA, 3000 r/min turbogenerator in Barsebäck* Information OK 140-585 E, Asea, Västerås, 1980-05-28

[346] Jolly C. B. et al. *The failure of an end ring of a generator rotor operating in a hydrogen atmosphere* Technical paper presented to CEA Thermal & Nuclear Power Section, Halifax, Oct. 7th 1975, C. A. Parsons & Co. Ltd. Newcastle upon Tyne, England, 1975

[347] Unnerberg, Lars *Anteckningar från EPRI-workshop om kapselringar för turbogeneratorer i Paulo Alto 82-10-11—12* Report RR KYD 82-017, Asea, Västerås, Oct. 1982

[348] Viswanathan R. *Retaining ring failures* Paper presented at EPRI Workshop on Generator retaining Rings, Paulo Alto, Oct.1982

[349] Unnerberg, Lars *Diskussion om kapselfrågor med KWU* Report RR KYD

79-9207, Asea, Västerås, 1979-12-18

[350] Eriksson, Sture and Sundstrand, Arne *Cracks in rotor teeth in TVO rotor 1 and 2* Report R OK 80-01, Asea, Västerås, 1980-10-03

[351] Murphy M. C. *The development of fretting fatigue cracks in the Drax generator rotor* pp. 13-19, Report published at NEI Parsons/CEGB Symposium on Fretting Fatigue in Generator Rotors, Newcastle upon Tyne, England, Jan. 1979

[352] Eriksson, Sture *Report regarding cracks in TVO generator rotors*, OK, Asea, Västerås, 1980-07-04

[353] Jagger, M. et al *Development of Major cracks in the Drax generator rotor* pp. 29-35, Report published at NEI Parsons/CEGB Symposium on Fretting Fatigue in Generator Rotors, Newcastle upon Tyne, England, Jan. 1979

[354] Eriksson, Sture *Spricka i generatorrotor i Frankrike* Meddelande, OK1, Asea, Västerås, 1980-03-24

[355] *Catastrophic rotor fracture in Porchville* Further information through EdF, Paris

[356] Eriksson, Sture *Förebyggande åtgärder mot rotorhaveri* Meddelande, OK 1, Asea, Västerås, 1979-10-25

[357] Carlsson, Janne, Eriksson, Sture and Sundstrand, Arne *Fatigue cracks in electric generator rotors – a case study* pp. 989-1004, Proceedings of International Conference on Fatigue Thresholds, Stockholm, June 1981

[358] Carlsson, Janne *Growth rate for large cracks in TVO-rotors*, Report, KTH, Stockholm, 1980-08-15

[359] Eriksson, Sture *Tekniska aspekter på fortsatt drift med TVOs generatorrotor 3*, Report R OK 81-01, Asea, Västerås, 1981-02-25

[360] Sundstrand, Arne *Utmattningsspricka i axeltapp på rotor 1 till TVO* Technical Report TR KY 83-001, Asea, Västerås, 1983-01-10

[361] Eriksson, Sture *Meeting with EdF, Alsthom-Atlantique and BBC regarding rotor shaft end cracks* Handwritten notes, 1982-02-24

[362] Johnsson, Arne *Vibrationsbild vid stopp av TVO II 1981-05-20*, Tb 10/81, Stal-Laval, Finspång, Sweden, 1981-05-22

[363] Alenfelt, Bengt *TVO-rotor, Sprickor i rotorbalkar* Report R OKT 1981-49, Asea, Västerås, 1981-09-16

[364] Nilsson, Leif *On the vibration behavior of a cracked rotor IFT o MM International conference on rotor dynamic problems in power plants*, Rome,

1982

[365] Eriksson, Sture *Beskrivning av de nya generatorrotorerna till TVO och jämförelse med tidigare levererade rotorer* Report R OK 81-02, Asea, Västerås, 1981-11-20

[366] *Asea Annual Report 1980*, pp 10, 13, Västerås, 1981

[367] Glete, Jan *ASEA under hundra år* pp. 305, Västerås 1983

[368] *Generatorutredningen, Slutrapport*, Vattenfall – Asea – Stal-Laval, Stockholm/Västerås, June, 1979

[369] *Licence agreement between BBC Brown Boveri & Company Limited and Allmänna Svenska Elektriska Aktiebolaget concerning turbogenerators with gas-cooled rotors*, Baden 19th December, 1975

[370] Lundblad, Erik *Letter to Mr. Combeau Alsthom-Atlantique* Asea, Västerås, 1978-10-19

[371] Lundblad, Erik *Memorandum of meeting between specialists from USSR Ministry for Electrical Engineering Industries and representatives of the Asea Company.* Asea, Västerås, 1978-09-28

[372] Glebov I. A. and Eriksson, Sture *Memorandum of meeting between the specialists of the Asea Company and the USSR Ministry for Electrical Engineering Industries held in Leningrad from October 27 to October 31, 1980* Leningrad, USSR, Oct. 31, 1980

[373] Carlsson, Bengt and Nachemson-Ekwall, Sophie *Livsfarlig ledning, Historien om kraschen i ABB* pp.80-101, Ekerlids Förlag, Stockholm, 2003

[374] Svensson, Bo *Generator manufacturer specializes in outsourced work* pp. 79, Diesel and Gas Turbine Worldwide, June, 2005

[375] *AMS Synchronous Generators, 4- and 6-pole, 3 - 70 MVA, 3 - 15 kV* Brochure 3BSM 006 540 R002, ABB Automation Technologies AB, Västerås, Apr. 2005

[376] *TVO placed an order on a new generator for the Olkiluoto 2 unit* Press Release, TVO, Helsinki, 2006-10-05

[377] *50 years of Nuclear Energy* pp. 3-5, Special report, International Atomic Energy Agency, Vienna, 2006

[378] Lindner, Helmut, *Strom, Erzeugung, Verteilung und Anwendung der Elektrizität* pp. 196, 235-236 Deutsches Museum, Munich 1985

[379] *Electric Vehicles: Technology, Performance and Potential* pp. 19-20, OECD/IEA, Paris 1993

[380] *History of hybrid vehicles* www.hybridcars.com/history, 2005

[381] Wilson, Kevin *Electric Car* Microsoft ®Encarta ®Online Encyclopedia 2004

[382] von Frankenberg, Richard and Neubauer, Hans-Otto *Geschichte des Automobils*, pp. 16-21, 159-162, Sigloch Edition, Künzelsau, Germany, 1999

[383] Stein, Ralph, *The Automobile Book* pp. 98-102, Paul Hamlyn Ltd, London, 1967

[384] *Från Wenström till Amtrak* pp. 49, Asea, Västerås 1983

[385] *The California Low-Emission Vehicle Regulations*, §1960.1, California Air Resource Board, Sacramento, CA, Sep.1990

[386] *Electric Vehicles: Technology, Performance and Potential* pp. 23-40, 61-79, OECD/IEA, Paris 1993

[387] *Think Nordic AS* Home page, www.think.no, Aurskog, Norway, 2006

[388] Sutula, R. A. et al *U.S. Department of Energy Advanced Battery Systenms Program: Meeting the Critical Challenge of Development* pp. 297-306, Proceedings Vol. 2, International Electric Vehicle Symposium (EVS-12), Anaheim, CA, USA, Dec. 1994

[389] Eriksson, Sture and Birnbreier, Hermann *Batteries and Drive Systems for Electric Vehicles, Experience and Future Prospects* pp. 285-291, The Urban Electric Vehicle, Conference Proceedings, OECD/IEA, Paris 1992

[390] Linden David *Handbook of batteries* Second Edition, ISBN 0-07-037921-1, McGraw-Hill, New York, 1994

[391] Wouk, Victor *Hybrids: then and now* pp. 16-21, IEEE Spectrum, July 1995

[392] Jaura, Arun and Levin, Michael *Starter alternator evolution and interface in hybrid vehicles* CD-ROM Proceedings from EVS 15, Brussels, 1998

[393] Lehna, Marius and Heindl. P *Erfarungen aus dem Flottenversuch mit der Konzeptstudie Audi Duo* VDI-GET Tagung, Rationelle Energieausnutzung bei Hybridfahrzeugen, München, Oct. 1995

[394] Mizuno, Yota *Development of a new hybrid transmission for FWD Sports Utility Vehicles* Presentation at 4. International CTI-Symposium, Berlin, Dec. 2005

[395] *The Civic Hybrid – concept and Outline* pp. 20-21, AutoTechnology, Special 2007, Wiesbaden, Germany, 2006

[396] Pessis, José *PSA Peugeot Citroen Stop & Start* CD-ROM Proceedings from EVS 21, April Monaco 2005

[397] SEP *Tracked and wheeled modular armoured vehicles* Brochure 04.06

SEP 005.ID.3103, BAE Systems Hägglunds, Örnsköldsvik, Sweden, 2006

[398] Chudi, Peter and Malmquist, Anders *A hybrid drive for the car of the future* pp. 3-12, ABB Review, No. 9, 1993

[399] Berg, Niclas Environmentally friendly hybrid vehicles in commercial service pp. 40-47, ABB review, No. 3, 1998

[400] *Prius* Brochure 0511/PRU/SE Toyota Sweden AB, Sundbyberg, Sweden, 2005

[401] Hofman, Theo *A fundamental case study on the Prius and IMA drivetrain concepts* CD-ROM Proceedings from EVS 21, April Monaco 2005

[402] Eriksson, Sture and Sadarangani, Chandur *A four quadrant hybrid electric drive system* Published in Proceedings of IEEE Vehicular Technology Conference, Vancouver, Canada, Sept. 2002

[403] Nordlund, Erik *The Four-Quadrant Transducer System* KTH, TRITA-ETS-2005-4 Doctoral Thesis in Electrical Machines and Drives, Stockholm, 2005

[404] Magnussen, Freddy *On design and analysis of synchronous permanent magnet machines for field-weakening operation in hybrid electric vehicles* KTH, TRITA-ETS-2004-11 Doctoral Thesis, Stockholm, 2005

[405] Lindbergh, Göran et al *Elektrokemi för bränsleceller- och batterier* pp. 89-101, Compendium, Dep. for applied electrochemistry, Royal Institute of Technology, Stockholm, 2001

[406] *Climate Change 2007: The Physical Science Basis* pp. 1-21, Intergovernmental panel on climate change, UNEP, Paris, Feb. 2007

[407] Schück, Johan *Sinande tillgångar driver upp oljepriset* Dagens Nyheter, Stockholm, 2005-08-19

[408] Fulton, Lew *Reducing oil consumption in transport: combining three approaches* pp. 4-6, IEA/EET Working Paper, International Energy Agency, April 2004

[409] *Production purchasing global terms and conditions, PPGTC,* Ford, Dearborn, MI, Jan. 2004

[410] *Orgalime S2000: General conditions for the supply of mechanical, electrical and electronic products,* Orgalime, Brussels, Aug. 2000

[411] *Technical specification of EDS (Electrical Drive System)* BMW, München, 1997-08-04

[412] Hoffmann, Lars *Re: Kort fråga* E-mail från Saab, 2006-06-26

[413] Reckhorn, Thomas *Stromeinprägendes Antriebssystem mit fremderregter*

Synchronmaschine Dissertation RWTH Aachen, ISBN 3-86073-061-4, 1992

[414] Assouline M. et al *Synchronous drive with electric excitation* pp. 487-496, Proceedings Vol. 1, International Electric Vehicle Symposium (EVS-12), Anaheim, CA, USA, Dec. 1994

[415] Jezernik, Karel and Rodic, Miran *Torque sensorless IM control for EVs* CD-ROM Proceedings from EVS 21, April Monaco 2005

[416] Harman, R. T. C. *Experience with electric vehicle concepts and operation* pp.301-310, The Urban Electric Vehicle, Conference Proceedings, OECD/IEA, Paris 1992

[417] Skudelny, Hans-Christoph *Untersuchungen an Drehstromantrieben für Elektrospeicherfahrzeuge, Teil 1 und 2*, Wissenschaftliche Mitteilungen vom Institut für Stromrichtertechnik und elektrische Antriebe, RWTH, Aachen, 1982

[418] King, R. D. and Konrad, C. E. *Advanced on-road electric vehicle AC drives – concepts to reality* Paper 12.11, Proceedings Vol. 2, International Electric Vehicle Symposium (EVS 11), Florence, Italy, Sep. 1992

[419] *Finnish Electric Car on the Horizon* Pamphlet from Neste, Imatran Voima, Strömbergs and Saab-Valmet, Helsinki 1986

[420] Young, R. W. et al Fourth generation propulsion subsystem for commercial electric vehicles, pp. 279-286, Proceedings Vol. 1, International Electric Vehicle Symposium (EVS 13), Osaka, Japan, Oct. 1996

[421] Sims, R. I., Kelm, B. R. and Konrad, C. E. *The development of the Ecostar powertrain* Paper 12.08 Proceedings Vol. 2, International Electric Vehicle Symposium (EVS 11), Florence, Italy, Sep. 1992

[422] Willis R. L. and Brandes, Jürgen *Ford next generation electric vehicle powertrain* pp. 449-458, Proceedings Vol. 1, International Electric Vehicle Symposium (EVS-12), Anaheim, CA, USA, Dec. 1994

[423] Ishikawa, T. and Furutani, M. *AC motor propulsion system for electric vehicle* pp. 228-234, Proceedings, International Electric Vehicle Symposium (EVS-10), Hongkong, Dec. 1990

[424] Junge, G. and Schäfer, U. *An advanced drive system for electrical vehicles* Paper 12.01, Proceedings Vol. 2, International Electric Vehicle Symposium (EVS 11), Florence, Italy, Sep. 1992

[425] Walker, Frank and Smith, Brian *Power trains for EVs, Design for flexibility* pp. 20-25, Proceedings Vol. 2, International Electric Vehicle Symposium (EVS-12), Anaheim, CA, USA, Dec. 1994

[426] Yamamura, H et al *Development of powertrain system for Nissan FEV*, Proceedings Vol. 2, International Electric Vehicle Symposium (EVS 11), Florence, Italy, Sep. 1992

[427] Rajashekara, Kaushik et al *Propulsion control system for a 22 foot electric/hybrid shuttle bus* pp. 169-178 Proceedings Vol. 2, International Electric Vehicle Symposium (EVS-12), Anaheim, CA, USA, Dec. 1994

[428] King, R.D. et al *Development of a 225 kW hybrid drive system for a low-floor, low-emission transit bus* pp. 179-188 Proceedings Vol. 2, International Electric Vehicle Symposium (EVS-12), Anaheim, CA, USA, Dec. 1994

[429] Mayrhofer, J. et al *A hybrid drive based on a structure variable arrangement* pp. 189-200, Proceedings Vol. 2, International Electric Vehicle Symposium (EVS-12), Anaheim, CA, USA, Dec. 1994

[430] Bitsche, Otmar *Fully integrated electric vehicle control unit* pp. 650-659, Proceedings Vol. 2, International Electric Vehicle Symposium (EVS-12), Anaheim, CA, USA, Dec. 1994

[431] Reckhorn, Thomas et al *Drive Systems for Electric, Hybrid and Fuel Cell Vehicles* CD-ROM Proceedings from EVS 18, Berlin 2001

[432] Lowry, Michael et al *Electric Drive System for Electric and Hybrid Vehicles* CD-ROM Proceedings from EVS 18, Berlin 2001

[433] Paris, Christophe et al *Influence of a die cast copper rotor in the behavior of induction machines such as starters and generators* Paper FP3-2, CD-ROM Proceedings IEEE VTS – VPP 2004, Paris, October 2004

[434] Ådanes, Alf Kåre *High efficiency, high performance permanent magnet synchronoous motor drives* pp. 79-95, Doctoral dissertation, The University of Trondheim, The Norwegian Institute of Technology, Trondheim, 1991

[435] Chin, Yung-Kang Robert *A permanent magnet traction motor for electric forklifts – Design and iron loss analysis with experimental verification* pp. 14-20, KTH, TRITA-EE-2006-59 Doctoral Thesis in Electrical Machines and Drives, Stockholm, 2006

[436] Gertmar, Lars *Unique Mobility Electrical Car Drives – PM Motor Technology Assessment* pp. 1-57, ABB Technical Report, SECRC/KEC/TR-93-011, Västerås, 1993-02-08

[437] Eriksson, Sture *Antriebssysteme mit permanenterregten Synchronmotoren* pp. 733-753, Tagungsband 4. Aachener Kolloquium Fahrzeug- und Motorentechnik, Aachen, Germany, Oct. 1993

[438] Lutz, Jon and Cambier, Craig *Phase advanced operation of permanent*

magnet motor drive system pp. 279-287, Proceedings Vol. 2, International Electric Vehicle Symposium (EVS-12), Anaheim, CA, USA, Dec. 1994

[439] Eriksson, Sture *Samtal med BMW om PM motorer* ABB/ISY/XEL Internal message 1995-01-31

[440] Zelinka R. and Erhart P. *Stadtomnibus mit dieselelektrischem Antrieb und Schwungradspeicher* pp. 851-879, Tagungsband 4. Aachener Kolloquium Fahrzeug- und Motorentechnik, Aachen, Germany, Oct. 1993

[441] Lagerström, Gunnar and Malmquist, Anders *Advanced hybrid propulsion system for Volvo ECT* pp. 51-61, Technology Report, Volvo, Gothenburg, No. 2, 1995

[442] Chudi, Peter and Malmquist, Anders *Development of a small gas turbine-driven high-speed permanent magnet generator* KTH, TRITA-EMK-89-03 Licentiate Thesis in Electrical Machines and Power Electronics, Stockholm, 1989

[443] Chudi, Peter *Calculation of a 38 kW, 90 000 rpm permanent magnet generator* Report R HSG 90-07, HSG Project, Stockholm, 1990-12-14

[444] Magnussen, Freddy *Design of a 50 kW interior permanent magnet motor for an electric vehicle* Technical report SECRC/G/TR-98/039E, ABB, Västerås, 1998-03-05

[445] Staunton, R. H. et al *PM motor parametric design analyses for a hybrid electric vehicle traction drive application* Report ORNL/TM-2004/217, Oak Ridge National Laboratory for U.S. Dep. of Energy, Oak Ridge, TN, Sep. 2004

[446] Takizawa, Keiji *New hybrid transmission* CD-ROM Proceedings from EVS 21, Monaco, April 2005

[447] *TM4* Home page, www.tm4.com, Boucherville, Canada, 2006

[448] Zhou, Rao S. et al *Integrated electric drives for use in automotive driveline applications* CD-ROM Proceedings from EVS 21, April Monaco, 2005

[449] Nagashima, James M. *Wheel hub motors for automotive applications* CD-ROM Proceedings from EVS 21, Monaco, April 2005

[450] Cirani, Maddalena *Analysis of an innovative design of an axial flux Torus-machine* KTH, TRITA-ETS-2002-05 Licentiate Thesis in Electrical Machines and Power Electronics, Stockholm, 2002

[451] Masmoudi Ahmed and Elantably Ahmed *A claw pole TFPM drive for hybrid bus propulsion systems* CD-ROM Proceedings from EVS 21, Monaco, April 2005

[452] EL-Refaie, Ayman M. and Jahns, Thomas M. *Comparison of synchronous PM machine types for wide constant-power speed range operation* pp. 1015-1022, Proceedings IEEE, IAS 2005

[453] Lipman, Thimothy and Hwang, Roland *Hybrid electric and fuel cell vehicle technological innovation: Hybrid and zero-emission vehicle technology links* CD-ROM Proceedings from EVS 20, Long Beach, Nov. 2003

[454] Bauer, Stefan E. et al *Design of an integrated switched reluctance traction drive for autonomous freight wagon* CD-ROM Proceedings from EVS 21, Monaco, April 2005

[455] Inderka, Robert and Keppler, Stefan *Extended power by boosting with switched reluctance propulsion* CD-ROM Proceedings from EVS 21, Monaco, April 2005

[456] McClelland, Mike et al *The design of cost effective, switched reluctance drive system for mild hybrid-electric vehicles* CD-ROM Proceedings from EVS 21, Monaco, April 2005

[457] Altendorf, Jens-Peter et al *Assessment criteria for electric drives in electric-, hybrid- and fuel cell vehicles within the OKOFEH project* CD-ROM Proceedings from EVS 20, Long Beach, Nov. 2003

[458] Kristiansson, Urban *Re. En fråga utanför programmet* e-mail from Volvo Cars, 2006-08-17

[459] Glete, Jan *Varför har svensk starkströmsindustri blivit högteknologisk?* pp. 41, Dædalus/Tekniska Museets Årsbok 1984, Stockholm, 1984

[460] Hira, Ron *R&D 100* pp. 30-35, IEEE Spectrum, December 2005

[461] *A leader in hybrid electric technology* www.wavecrestlabs.com/ wc-presskit-sep06, Wavecrest laboratories, Rochester Hills, MI, 2006

[462] Hideshiitazaki *The Prius that shook the world* pp. 3-6, 18-22, 49-55, PDF-book, 1999

[463] Mauro, Kanehira *Re: Diverse frågor* e-mail to Sture Eriksson, 2006-02-27--28

[464] Wouk, Victor *Hybrid Electric Vehicles* pp. 70-74, Scientific American, Oct. 1997

[465] *Visionary vehicles* pp. 16-18, ABB Drives World, Company magazine for BA ABB drives, No. 3, 1992

[466] Centerman, Jörgen *Automotive Drive Systems* Memo, ABB Industrial and Building Systems Management, Zurich, 1994-07-05

[467] *ABB Hybrid Systems - a forward-looking company* Brochure, ABB,

Västerås, 1996

[468] Persson, J C and Lund, P D *Energy systems in road bound vehicles research programme* Evaluation report, Swedish Energy Agency, Eskilstuna, Sweden, Dec. 2002

[469] *Programbeskrivning för forskningsprogrammet Energisystem i vägfordon, Period 2 (2004 - 2006)* Document, Swedish Energy Agency, Eskilstuna, Sweden, 2004-04-19

[470] Eriksson, Sture and Johansson, Göran *R&D activities in Sweden on hybrid and fuel cell vehicles* PowerPoint presentation, Green Car program, 2004-05-18

[471] Helén, Martin *ASEA:s Historia 1883-1948 Vol. III* pp. 68-70, 214-236) Asea, Västerås, 1957

[472] Nordin, Tore *Aseas tyristorlok erövrar världen* pp. 95-112, Industriminnen ISBN 91-630-9817-2 Västerås, 2000

[473] Östlund, Stefan *Elektrisk Traktion* pp. 54-82, Department of Electrical Machines and Power Electronics, Royal Institute of Technology, Stockholm, 2001

[474] Nordin, Tore et al *Svenska Ellok* pp. 296, ISBN 91-85098-84-1, Svenska järnvägsklubben, Stockholm, 1998

[475] Andersson, Per-Gunnar and Johansson, Thomas *Trådbuss Landskrona* pp.36-47, ISBN 91-631-5367-X, Trivector, Lund, 2005

[476] *Traction Motors* pp. 7-8, Broschure from Ganz Transelectro Electric Co. Ltd. Budapest, Hungary, www.ganztrans.hu, 2005

[477] Gothnier, Ulf *Världsnyhet från Västerås, En transformator och generator i ett, Mannen bakom uppfinningen* Newspaper articles from Vestmanlands läns tidning, Västerås, 1998-02-26

[478] Wallerius, Anders *ABB har hittat sitt Losec, Ny generator producerar högspänning direkt till nätet* Newspaper article from Ny Teknik, Stockholm, 1998-02-26

[479] *Stort tekniskt genombrott – ABB lanserar världens första högspänningsgenerator* Press Release, ABB 1998-02-25

[480] Leijon, Mats *Powerformer™ - a radically new rotating machine* pp. 21-26, ABB Review,2/1998

[481] *Annual Report 1998, Financial Review* pp. 35, 51 ABB Asea Brown Boveri Ltd, Zürich, 1999

[482] *AMT Synchronous Generator, 4- and 6-pole, 5 – 55 MVA, 20 – 150 kV* pp.

4, ABB Motors AB Machines Division, Brochure 3BSM006542R001, Västerås, Jan. 2001

[483]   Helén, Martin *ASEA:s Historia 1883-1948 Vol. III* pp. 39, Västerås ,1957

[484]   Asztalos, Peter *Role of the Ganz factory in popularizing the use of alternating current* pp. 9, Product Development of the Century-Old Ganz Electric Works 1878-1978, Budapest, 1978

[485]   Leijon, Mats et al *Powerformer™ is based on established products and experiences from T&D* pp. 3-4, IEEE T&D Committee, Summer Meeting, Edmonton, Canada, July 1999

[486]   Dellby, Björn et al *High voltage XLPErformance cable technology* pp. 35-44, ABB Review 4/2000

[487]   Bolund, Björn Electric *Power Generation and Storage Using a High Voltage Approach* pp. 41-50, PhD thesis, Uppsala Universitet,ISBN 91-554-6552-8, Uppsala, Sweden, 2006

[488]   Skilling, Hugh H. *Fundamentals of Electric Waves* pp. 131-136, John Wiley & Sons, New York, 1960

[489]   Frank, Harry *Powerformer (HÖG) – Dryformer (TORR) historien*, List of important events achieved from ABB on 2006-04-21

[490]   Dietl, Thomas *ABB tar patent på vassare elmotor utan transformator* Dagens Industri, Stockholm, 1997-12-06

[491]   Boivie, Jan *Statistik Porjus U9* e-mail Alstom Power, 2006-

[492]   Magnell, Lars *Powerformer – generatorkonceptet som gick mot strömmen* pp. 26-28, Perspectives, Forskning och utveckling inom Vattenfalls-koncernen, Vattenfall AB, Stockholm, 2004

[493]   Hallberg, Hans A. *Re: Transformatorvikter* e-mail från ABB Transformers, 2007-03-19

[494]   Andersson, Lars et al *Powerformer™ chosen for Swedish combined heat and power plant* pp. 19-23, ABB Review 3/1999

[495]   Wallerius, Anders *Eskilstuna skrotar ABBs framtidsgenerator*, Newspaper notice, Ny Teknik, Stockholm, 2005-12-21

[496]   *AMT Synchronous Motor, 4- and 6-pole, 5 – 45 MW, 20 – 150 kV* ABB Motors AB Machines Division, Brochure 3BSM006541 R001, Västerås, Jan. 2001

[497]   Ahlinder, Johannes and Johansson, Thomas L. *Record-breaking electric motors give heavy industry more drive* pp. 45, ABB Review Special Report, ABB Zurich, 2004

[498] *ABB lanserar nydanande teknik för vindkraft*, ABB Press Release 2000-06-08

[499] Hellström, Rolf *Beskrivning av teknikutvecklingen i projektet Näsudden III*, (Report to the Swedish Energy Agency), ABB, Västerås, 2002-02-04

[500] Bunne, Tobias *Headwind – Background to the interaction that affected a Swedish wind-power venture* pp. 39-56, Diploma thesis for a Bachelor of Business Administration, University College of Gävle, Sweden, 2002

[501] Köhler, Niclas *ABB skrotar vindkraftsatsning, nyutvecklad teknik för jättesnurror blev för dyr* Ny Teknik, Stockholm, 2002-03-06

[502] *esp@cenet* patent register, http://www.espacenet.com

[503] Carlsson, Bengt and Nachemson-Ekwall, Sophie *Livsfarlig Ledning, Historien om kraschen i ABB* pp.89, Ekerlids Förlag, Stockholm, 2003

[504] Utterback, James M. *Mastering the Dynamics of Innovation* pp. 23-55, 79-102, 160-163, 215-232, Harvard Business School Press, ISBN 0-87584-740-4. Boston, 1994

Figure sources

All figures and tables, not specified below, have been made by the author. Permission to use photos and other pictures has been obtained from the sources listed below. Photos from the 19th century and a few others, available from numerous public sources, are referred to as "public domain". Copies of the photos for chapter 7 were in many cases found in both ABB's and Alstom's archieves.

Chapter 1
Fig. 1.3   Photo from Alstom
Fig. 1.4   Photo from ABB

Chapter 2
Fig. 2.1   Photo from Alstom
Fig. 2.2   Photo from ABB
Fig. 2.5   Photos from Bosch and Electrolux
Fig. 2.8   Photos Sture Eriksson, Märklin and public domain

Chapter 3
Fig. 3.1, 3.4, 3.5, 3.8, 3.9, 3.11, 3.16, 3.17 from public domain
Fig. 3.2, 3.3, 3.7, 3.10 Photos Sture Eriksson at Deutsches Museum
Fig. 3.12, 3.13, 3.15 from ABB (Asea)

Chapter 4
Fig. 4.7   Photos from ABB (Asea)
Fig. 4.12   Photos from ABB and Honda
Fig. 4.16   Diagram from Magnus Lindenmo, Surahammars Bruk
Fig. 4.17, 4.18 Diagrams from Vacuumschmelze
Fig. 4.30   Photo from ABB

Chapter 5
Fig. 5.7, 5.8, 5.9, 5.11 from ABB
Fig. 5.10 from KTH

Chapter 6
Fig 6.1, 6.2, 6.3, 6.4, 6.5, 6.6, 6.7, 6.8, 6.10, 6.11, 6.12, 6.15, 6.16, 6.17, 6.18, 6.19, 6.21, 6.22, 6.23, 6.24, 6.25, 6.35, 6.36, 6.39, 6.40, 6.56 from ABB (Asea) in Västerås
Fig. 6.27   Photo from Danaher Motion
Fig. 6.28   Picture from ITT Flygt
Fig. 6.34   Diagram from Elektroteknisk Handbok, Natur och Kultur
Fig. 6.38   Diagram from Undeland et al, see reference [188]
Fig. 6.45, 6.48, 6.49, 6.51, 6.58 from ABB, Finland

Chapter 7
Fig. 7.1, 7.3, 7.8, 7.24, 7.26, 7.27, 7.28, 7.46, 7.47 from ABB (Asea) in Västerås
Fig. 7.2, 7.7, 7.11, 7.12, 7.13, 7.20, 7.31, 7.32, 7.33, 7.34, 7.35, 7.36, 7.37, 7.38, 7.40, 7.43, 7.44, 7.45, 7.48, 7.49, 7.52, 7.53, 7.58, 7.59, 7.60, 7.61, 7.62, 7.63, 7.64, 7.65, 7.66, 7.67, 7.69, 7.71 , 7.72 from Alstom in Västerås
Fig. 7.5   Photo Sture Eriksson at Deutsches Museum
Fig. 7.6, 7.18 Sketches from reference [244]
Fig. 7.17   Sketches from a) GE, b) Alstom

Fig. 7.22   Photo from Forsmark Kraftgrupp
Fig. 7.23   Photo from K. Dahl-Madsen
Fig. 7.54, 7.55   Sketches from reference [338]

Chapter 8
Fig. 8.1, 8.6, 8.7 Pictures from public domain
Fig. 8.2, 8.30, 8.42, 8.43 from ABB (Asea)
Fig. 8.4   Photos Sture Eriksson and Think Nordic AS
Fig. 8.5   Photos ABB and public domain
Fig 8.11   Photo and diagram from BAE Systems Hägglunds
Fig. 8.12   Photos from Volvo Cars
Fig. 8.13   Photos from Volvo
Fig. 8.14   Photo from Toyota
Fig. 8.19   Photo from Honda
Fig. 8.22   Photo from GM
Fig. 8.23, 8.33, 8.34 Photos Sture Eriksson
Fig. 8.25   Photo from reference [438]
Fig. 8.27   Sketch from Magnet Motors
Fig. 8.28   Photo from reference [443]
Fig. 8.29   Photos by Anders Malmquist and Birgitta Eriksson
Fig. 8.31   Photo ABB and sketch BMW
Fig. 8.36   Photo Sture Eriksson and sketch from reference [449]
Fig 8.38   Photo from SR Drives Ltd
Fig. 8.45   Photo from Bombardier

Chapter 9
Fig. 9.1   Photo from Harry Frank
Fig. 9.3, 9.4, 9.5, 9.7, 9.8, 9.12, 9.13 Pictures from ABB
Fig. 9.9, 9.10, 9.11 Photos from Alstom

List of persons interviewed

The list includes persons interviewed in direct meetings or via telephone
(marked with d or t) as well as persons who have contributed with specific
information (marked with i). The affiliation refers, in general, to the periods
dealt with in the interviews.

Name	Affiliation	Chapter	Interview
Erik Agerman	Asea O-division	7	d
Ola Aglén	ABB Machines	6	d
Bengt Alenfelt	Asea/ABB Generation, Alstom	7	d
Jerry Atterklint	Lidköping Machine Tools	2	i
Lage Becker	Asea Motors	6	d
Anders Björklund	Eskilstuna Energi & Miljö	9	t
Jan Boivie	ABB Generation/ Alstom	9	d
Urban Bokén	ABB Motors	6	i
Mats Bölja	Bombardier	8	d
Curt Carlsson	ITT Flygt	6	d
Kristian Dahl Madsen	Asea, Asea O-division	7	d
Kent Engvall	Asea/ABB Generation, Alstom	7	i
Jan Folkhammar	BEVI AB	2, 6	i
Harry Frank	ABB Corporate Research	9	d
Werner Fritz	Ankarsrum Industries	2	i
Claes Fröling	Bombardier	8	i
Lars Gertmar	ABB Corporate Research	6, 9	d
Andreas Goubeau	BMW, Germany	8	d
Anders Gustafsson	ABB High Voltage Cables	9	i
Fredrik Gustavson	KTH Electrical Machines	6	d
Esko Haapala	TVO, Finland	7	d
Hans Hallberg	ABB Transformers	9	i
Tapio Haring	ABB Motors, Finland	6	t
Martti Harmoinen	ABB Drives, Finland	6	d
Sven Helmersson†	Asea (electrical consultations)	7	i
Gösta Hesslow	GenerPro AB	2, 7	d
Stig Hjärne	Asea/ABB Generation, Alstom	7	d
Lars Hoffmann	Saab Automobile	8	t
Christer Holgersson	ABB Motors	6	d
Kjell Isaksson	Vattenfall Hydro Power	9	d
Billy Johansson	ABB Generation	9	d
Göran Johansson	Volvo Technology	8	d
Tomas Johansson	Victor Hasselblad AB	2	i
Niclas Jonsson	ABB Hybrid Systems / Motors	6, 8	i
Uno Jonsson	Vattenfall	7	t
Lars Karlsson	Spintec AB	2	i
Hans Klein	Asea/ABB Generation	7	d

Urban Kristiansson	Volvo Cars	8	t
Anders Kroon	Volvo Power Train	8	t
Per Lamell	Forsmark Kraftgrupp	7	t
Bertil Larsson	Asea Power Plant Department	7	t
Bertil I Larsson	Asea O-division	7	d
Gunnar Larsson	Saab, Volvo, Audi, Volkswagen	8	d
Robert Larsson	ABB Machines	9	t
Mats Leijon	ABB Corporate Reasearch	9	d
Jan Liljeblad	Asea O-division	7	d
Mats Lindgren	Kompositprodukter AB	2	i
Gösta Lindholm	Asea O-division	8	i
Rolf Linderborg	ITT Flygt	6	d
Torsten Lindström†	Asea	7	t
Freddy Magnussen	ABB Corporate Research	8	i
Bo Malmros	ABB Motors	6	d
Anders Malmquist	ABB Hybrid Systems / Motors	6, 8	i
Kanehira Mauro	ETC Battery & Fuel Cells	8	i
Gunnar Mellgren	Asea Electronics, ABB Corporate Research	6	d
Finn Möller	Asea Motors	6	d
Sven Nilsson	Asea/ABB Generation	7	d
Tore Nordin	Asea/ABB Traction	8	i
Carsten Olesen	Stal-Laval	7	d
Ingemar Olofsson	Asea/ABB Motors	6	t
Lars Overgaard	Scania	8	t
Mauno Paavola	TVO, Finland	7	d
Christer Parkegren	ABB/Generation, VG Power	2, 9	d
Folke Pettersson	Asea O-division	7	t
Lauri Piensoho	Asea O-division	7	d
Carl Rönnevig	Asea O-division	7	d
Chandur Sadarangani	Asea Motors, KTH Electrical Machines	6	d
Richard Sivertsen	Asea O-division	7	d
Sven Sjöberg	Asea/ABB Motors	6	d
Karl-Erik Sjöström	Asea/ABB Generation, Alstom	7, 9	d
Juliette Soulard	KTH Electrical Machines	6	i
Lennart Stridsberg	HDD Servodrives AB	2	i
Göran Ståhl	Danaher Motion i Flen	6	d
Arne Sundstrand	Asea Generation	7	d
Åke von Sydow	Stal-Laval	7	t
Gunnar Tedeståhl	Sydkraft	7	d
Bo Thunwall	ABB High Voltage Cables	9	d
Ove Tjernström	Asea O-division	7	d
Ingmar Waltzer	ABB Machines, Finland	6	d
Rune Vestlin	Hägglunds	6	d
Göran Westman	BAE Systems Hägglunds	8	t
Jan-Christer Zanders	Asea Motors	6	t
Per Zellman	Emotron	2, 6, 8	i

List of persons

List of companies and organizations

List of acronyms

AC	Alternating Current
Al	Aluminum
AlNiCo	Aluminum-Nickel-Cobalt
BEV	Battery Electric Vehicle
BWR	Boiling Water Reactor
CAD	Computer Aided Design
CAM	Computer Aided Manufacturing
CARB	Californian Air Resource Board
CEO	Chief Executive Officer
CO	Carbon-Oxide
CO_2	Carbon-Dioxide
Cu	Cupper
DC	Direct Current
DSP	Digital Signal Processor
DTC	Direct Torque Control
ECC	Environmental Concept Car
EMC	Electro-Magnetic Compatibility
EMF	Electro-Motoric Force
EV	Electric Vehicle
EVS	Electric Vehicle Symposium
FCV	Fuel-Cell Vehicle
FEM	Finite Element Method
FMEA	Failure Mode Effect Analysis
GTD	Generator Turbo Direct cooling (Asea)
GTO	Gate Turn-Off thyristor
HEV	Hybrid Electric Vehicle
HSG	High Speed Generation
HV	High-Voltage
HVDC	High-Voltage Direct Current
ICE	Internal Combustion Engine
IEC	International Electrotechnical Commission
IEE	Institution of Electrical Engineers
IEEE	Institute of Electrical and Electronics Engineers
IGBT	Insulated Gate Bipolar Transistor
ISAD	Integrated Starter Alternator Damper
ISG	Integrated Starter and Generator
JFET	Junction Field Effect Transistor
KTH	Kungliga Tekniska Högskolan (Royal Institute of Technology)
LTH	Lunds Tekniska Högskola (Lund University, Faculty of Technology)
LV	Low-Voltage
MIT	Massachusetts Institute of Technology
MMF	Magneto-Motoric Force
MOSFET	Metal-Oxide Semiconductor Field Effect Transistor

NaS	Sodium-Sulphur
NaNiCl	Sodium-Nickel-Chloride
NC	Numerically Controlled
NdFeB	Neodymium-Iron-Boron
NEMA	National Electrical Manufacturers Association
NiCd	Nickel-Cadmium
NiMH	Nickel-Metal-Hydride
NOx	Nitrogen-Oxides
OECD	Organization for Economic Cooperation and Development
OEM	Original Equipment Manufacturer
PEM	Proton Exchange Membrane
PM	Permanent Magnet
PWM	Pulse Width Modulation
PWR	Pressurized Water Reactor
R&D	Research and Development
RMS	Root Mean Square
SEK	Swedish Crowns
SiC	Silicon-Carbide
SmCo	Samarium-Cobalt
SR	Switched Reluctance
TIG	Tungsten Inert Gas
USD	U.S. Dollar
VDE	Verband der Elektrotechnik, Elektronik und Informationstechnik
XLPE	Cross-Linked Poly-Ethylene
ZEV	Zero Emission Vehicle

Map of Sweden

Kiruna

Seitevare **x**

x Porjus
x Harsprånget

x Porsi

☐ **Luleå**

Stornorrfors **x**
☐ **Umeå**

○ Örnsköldsvik

○ Härnösand

x Arbrå

Hofors
Nyhammar **x** ○ Vikmanshyttan ○ Forsmark
Ludvika ○ **x**
Grängesberg ○ **x** ○ Fagersta
Hallsjön ○ ☐ **Uppsala**
Surahammar ○ **x** Hamra
Arboga ○ **Västerås** ☐ **STOCKHOLM**
Örebro ☐
Hallsberg ○ **Eskilstuna**
Flen ○
Finspång ○ ☐ **Norrköping**
x Marviken
Vänersborg ○ Lidköping
Stenungsund ○
Partille ○ **Borås**
Göteborg ☐ ☐
Ankarsrum **x** ○ Västervik
x
Ringhals ○ Oskarshamn
x Blomstermåla

Emmaboda ○

Höganäs ○
Helsingborg ☐ ○ ☐ **Karlskrona**
Landskrona ○ Karlshamn
Barsebäck **x** ☐ **Lund**
Malmö ☐

Price level in Sweden

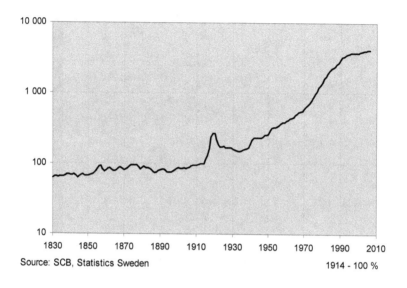

Source: SCB, Statistics Sweden

1914 - 100 %

Wissenschaftlicher Buchverlag bietet

kostenfreie

Publikation

von

wissenschaftlichen Arbeiten

Diplomarbeiten, Magisterarbeiten, Master und Bachelor Theses
sowie Dissertationen, Habilitationen und wissenschaftliche Monographien

Sie verfügen über eine wissenschaftliche Abschlußarbeit zu aktuellen oder zeitlosen
Fragestellungen, die hohen inhaltlichen und formalen Ansprüchen genügt,
und haben **Interesse an einer honorarvergüteten Publikation**?

Dann senden Sie bitte erste Informationen über Ihre Arbeit per Email
an info@vdm-verlag.de. Unser Außenlektorat meldet sich umgehend bei Ihnen.

VDM Verlag Dr. Müller Aktiengesellschaft & Co. KG
Dudweiler Landstraße 125a
D - 66123 Saarbrücken

www.vdm-verlag.de

www.ingramcontent.com/pod-product-compliance
Lightning Source LLC
LaVergne TN
LVHW022258060326
832902LV00020B/3152